THE CHEMOKINES
Biology of the Inflammatory
Peptide Supergene Family II

ADVANCES IN EXPERIMENTAL MEDICINE AND BIOLOGY

A Continuation Order Plan is available for this series. A continuation order will bring delivery of each new volume immediately upon publication. Volumes are billed only upon actual shipment. For further information please contact the publisher.

THE CHEMOKINES
Biology of the Inflammatory Peptide Supergene Family II

Edited by

I. J. D. Lindley
Sandoz Research Institute
Vienna, Austria

J. Westwick
University of Bath
Bath, United Kingdom

and

S. Kunkel
University of Michigan
Ann Arbor, Michigan

SPRINGER SCIENCE+BUSINESS MEDIA, LLC

Library of Congress Cataloging-in-Publication Data

The Chemokines : biology of the inflammatory peptide supergene family
  II / edited by I.J.D. Lindley, J. Westwick, and S. Kunkel.
      p.   cm. -- (Advances in experimental medicine and biology ; v.
  351)
    "Proceedings of the Third International Symposium on Chemotactic
  Cytokines, held August 30-September 1, 1992, in Baden bei Wien,
  Austria"--T.p. verso.
    Includes bibliographical references and index.
    ISBN 978-1-4613-6283-8   ISBN 978-1-4615-2952-1 (eBook)
    DOI 10.1007/978-1-4615-2952-1
    1. Chemokines--Congresses.   I. Lindley, I. J. D. (Ivan J. D.)
  II. Westwick, J. III. Kunkel, S. L. (Steven L.)   IV. International
  Symposium on Chemotactic Cytokines (3rd : 1992 : Baden, Austria)
  V. Series.
    [DNLM: 1. Cytokines--congresses.   2. Chemotactic Factors-
  -congresses. 3. Chemotaxis, Leukocyte--congresses.
  4. Inflammation--congresses.   5. Interleukin-8--congresses.   W1
  AD559 v. 351 1993 / QW 690 C5168 1993]
  QR185.8.C45C48   1993
  616.07'95--dc20
  DNLM/DLC
  for Library of Congress   94-9684
                                                                     CIP

Proceedings of the Third International Symposium on Chemotactic Cytokines, held
August 30–September 1, 1992, in Baden bei Wien, Austria

ISBN 978-1-4613-6283-8

© 1993 Springer Science+Business Media New York
Originally published by Plenum Press, New York in 1993
Softcover reprint of the hardcover 1st edition 1993

PREFACE

The first symposium in this series was held at the Royal College of Surgeons of England in December 1988 and was entitled "Novel Neutrophil Stimulating Peptides". That symposium successfully brought together the majority of laboratories working in the area of interleukin-8 and related peptides; see *Immunology Today* 10: 146-147 (1989). The Second International Symposium on Chemotactic Cytokines was held at the same venue in June 1990, and a much-increased attendance reflected the accelerating pace of work in the area of these chemotactic cytokines. The proceedings of that meeting were published in *Advances in Experimental Medicine and Biology, vol.* 305 (1991).

The rapid advances made in the field of chemotactic cytokines over the last 18 months necessitated a third Symposium in this series to collate and place in perspective an explosion of new data. The Third International Symposium on Chemotactic Cytokines was held between August 31 and September 1, 1992 in Baden-bei-Wien, Austria.

However, the lack of a clear nomenclature system was creating some confusion in the area, especially as new factors continue to be discovered and classified as family members. In the past, these inflammatory mediators had been placed arbitrarily under the broad heading of "intercrines" or "chemotactic cytokines" with no clear classification guidelines to follow. This nomenclature issue was addressed at the Symposium, where investigators in the field were invited to reach a consensus regarding a collective name for these mediators. The resulting decision was to identify the major family as **chemokines**, to replace all previous terms.

The chemokines are involved in recruitment and activation of specific immune cells and regulation of adhesion molecules on the cell surface, and have been implicated in many disease states, including atherosclerosis, asthma, renal disease, rheumatoid arthritis, psoriasis and allergy. Exciting new findings in the area involve cloning of new members of the family, new activities of known members, and cloning of specific receptors, in addition to new clinical and experimental data.

We are extremely grateful to the companies listed overleaf, for without their support the symposium and this book would not have been possible.

I.J.D. Lindley
S.L. Kunkel
J. Westwick

We are indebted to the following companies for their support:

Sandoz Forschungsinstitut

Bayer

Mallinckrodt

Austrian Airlines

Ciba-Geigy

Du Pont-Merck

Procter & Gamble

Monsanto

Glaxo

ICI

Genzyme

Genentech

Merck & Co. Research Laboratories

Synergen

Wellcome

Parke-Davis

Pfizer

Warner-Lambert

Lilly

CONTENTS

SYMPOSIUM PRESENTATIONS

ABSTRACTS

CHEMOTACTIC AND INFLAMMATORY CYTOKINES -- CXC AND CC PROTEINS

Marco Baggiolini

Theodor-Kocher Institute
University of Bern
P.O. Box
CH-3000- BERN-9, Switzerland

INTRODUCTION

This is the third meeting in about three years on a family of cytokines that became fancy with the discovery of interleukin-8 (IL-8). The IL-8-related cytokines are small, and relatively easy to work with. They are characterized by four conserved cysteine residues. Alignment of these residues differentiates two subfamilies: one with the first two cysteines separated by one amino acid (CXC cytokines), and the other with adjacent cysteines (CC cytokines). The genes for the two clans inhabit different chromosomes, number 4 for the CXC and number 17 for the CC, and the cytokines that they produce have different target cell preferences.

Platelet factor 4 (PF4) was the first protein of this class to be characterized. Its sequence was reported in 1977 (1-4), ten years before the discovery of IL-8 (5-7). PF4 is stored in the α—granules of blood platelets together with two other CXC proteins, platelet basic protein (PBP) and its N-terminal truncation derivative, connective tissue-activating peptide III (CTAP-III) (8,9). ß—Thromboglobulin, a truncation derivative of PBP and CTAP-III, was characterized early on (10). A gamma-interferon-inducible protein (IP10) was then identified. Unlike the CXC proteins stored in the platelet, IP10 was found to be *induced* in cells upon stimulation, and gamma-interferon is particularly potent (11,12).

This area of research was interesting, but rather quiet until several related proteins were found that are strongly chemotactic for leukocytes (13). When IL-8 came along, many research groups were apparently ready to enter the field, and within a few years our knowledge about CXC proteins increased enormously. The full sequence of IL-8 was quickly established in several laboratories, and today we know its three-dimensional structure (14), the structure and chromosomal location of its gene (15,16), the sequence and binding properties of its receptors (17-20), and we have gained detailed information on its biological activities, and the mechanism of signal transduction in neutrophils (21).

The structure of the first CC protein was deduced from the cDNA of the *LD78* gene which was cloned by subtractive hybridization from human tonsillar lymphocytes (22). Information on biological activities was subsequently obtained upon isolation of two proteins from murine macrophages, macrophage inflammatory protein (MIP) 1α and 1ß, which were found to induce inflammation and to activate leukocytes (23-25). Monocyte chemotactic protein-1 (MCP-1, MCAF) was isolated from several human sources, and shown to be chemotactic for monocytes, but not for neutrophils (26-28). Other related proteins which, like MCP-1, act on mononuclear cells, I-309 (29) and RANTES (30), were then isolated and characterized (31,32).

PROPERTIES OF INTERLEUKIN-8

Structure

The gene of IL-8 encodes a 99-amino acid protein that is secreted after cleavage of a leader sequence of 20 residues. The mature protein is processed at the N-terminus yielding biologically active variants consisting of 77, 72 70 and 69 amino acids (33,34). N-terminal truncation enhances the biological activity. As assessed by the ability to induce the release of elastase from cytochalasin B-treated neutrophils, IL-8(77) is about two- and seven-fold less potent than IL-8(72) and IL-8(69), respectively (35). The native, biologically active form of IL-8 has two disulfide bridges linking the first to the third and the second to the fourth cysteine. Nuclear magnetic resonance spectroscopy indicates that IL-8 forms a dimer in (concentrated) solution (14). The monomer has a short, flexible N-terminal domain that is anchored by the two disulfide bonds to a core structure made of three antiparallel ß—strands followed by a prominent C-terminal α—helix. Native IL-8 is resistant to inactivation by plasma peptidases, heat, pH extremes and other denaturing treatments. Reduction (and alkylation) of the disulfide bonds, by contrast, rapidly leads to the complete loss of activity (36,37).

Actions on Neutrophils

IL-8 is a neutrophil chemoattractant and, as such, induces several responses that are essential for antimicrobial host defence, namely shape change and directional migration, exocytosis, and the respiratory burst (13,21). *The shape change* reflects the activation of the cytoskeleton and the transient assembly of filamentous actin, a prerequisite for adherence to endothelial cells and migration. *Exocytosis* is a complex response involving the release of enzymes and other soluble proteins from several subcellular storage compartments, and the remodeling of the plasma membrane by fusion with subcellular membranes. IL-8 triggers the exocytosis of specific granules and secretory vesicles (21,36). Exocytosis of azurophil granules is also observed when the cells are pretreated with cytochalasin B (36,38,39). In addition IL-8 enhances the surface expression of integrins, CD11b/CD18 (complement receptor type 3), CD11c/CD18 (p150,95), and complement receptor type 1 (40-42). *The respiratory burst* is the most characteristic response of stimulated phagocytes. Like other chemoattractants, IL-8 elicits the rapid and transient activation of the NADPH-oxidase, leading to the formation of superoxide and H_2O_2 (21,36).

Signal transduction in IL-8 stimulated neutrophils depends on *Bordetella pertussis* toxin-sensitive GTP-binding proteins and the activation of a phosphatidylinositol-specific phospholipase C which delivers two second messengers, IP_3 and diacylglycerol. IP_3 induces a rise in cytosolic free calcium, and diacylglycerol activates protein kinase C (21). Treatment with *pertussis* toxin, depletion of mobilizable calcium, and exposure to inhibitors like staurosporine (39) and wortmannin (39) were used to modulate signal transduction, and to show that the responses elicited by IL-8, fMet-Leu-Phe and C5a are controlled by a similar mechanism (21).

Actions on Other Blood Cells

As compared to fMet-Leu-Phe, C5a, PAF and LTB$_4$, which act on several types of leukocytes, IL-8 is unusually selective for neutrophils. It has some chemotactic activity *in vitro* for basophils (43), but not for monocytes (6,38) or eosinophils (38,43). It has been reported, however, that monocytes saturably bind IL-8, albeit to a lesser extent than neutrophils (18,43), and respond to IL-8 with a moderate, transient rise in intracellular free calcium and the respiratory burst (44). IL-8 was also shown to induce cytosolic free calcium changes, a shape change and the release of peroxidase in eosinophils from patients with hypereosinophilia (45). Basophil leukocytes have IL-8 receptors (46) and release histamine and sulfido-leukotrienes in response to IL-8 when they are primed with IL-3 or GM-CSF (46,47). It was also reported that IL-8 is chemotactic for human lymphocytes in vitro (48,49). Another study, however, indicates that blood lymphocytes respond to IL-8 only after pretreatment (e.g. with anti-CD3) (50). No changes of cytosolic free calcium were observed in blood lymphocytes exposed to IL-8 (39), and only low numbers of IL-8 receptors (18) or none at all (51) were found in binding studies.

Effects in vivo

Human IL-8 acts on neutrophils of several species. Time course studies upon intradermal injection in rabbits revealed massive and long-lasting neutrophil accumulation and plasma exudation. The neutrophil immigration begins within minutes after injection and remains detectable for several hours (52,53). The unusually long duration of action presumably depends on two properties of IL-8, stability and slow clearance from the tissues due to interaction with acidic glycosaminoglycans. Studies in humans (54,55) show a similar inflammatory reaction with exclusive infiltration of neutrophils, which is particularly prominent around venules. Lymphocyte numbers are not increased, and no basophils, eosinophils or monocytes are found. In contrast to C5a, IL-8 causes no wheal and flare, itching or pain, suggesting that it does not induce histamine release from cutaneous mast cells.

OTHER CXC CYTOKINES

A surprising number of proteins share structural and biological similarities with IL-8. GROα designates a protein that was originally described as melanoma growth stimulatory activity (MGSA) on the basis of its effects on melanoma cell lines (56,57), and was later shown to be a powerful neutrophil chemoattractant (58). Two related molecules, GROß and GROγ, were identified subsequently (59). The three GRO proteins are very similar in structure (about 90% sequence identity) and activity toward neutrophils (Geiser et al., J. Biol. Chem., in press). ENA-78 was discovered in the culture supernatants of a type-II alveolar cell line (60), and appears to be produced preferentially by epithelial cells. NAP-2, another IL-8 analog, derives from the N-terminal processing of PBP or CTAP-III (61). In pathological conditions, it is likely that IL-8 and several of its related proteins are generated concomitantly. This is suggested by experiments with lung epithelial cells which, in addition to ENA-78, also release IL-8, GROα and GROγ (60), and by numerous reports showing that IL-8 and GROα are produced by the same cells under similar stimulatory conditions (62). The existence of so many structurally and functionally related cytokines may be taken to suggest that the function of IL-8 is so important that it must be "backed up" by several analogs. It cannot be excluded, however, that these cytokines have distinguishing biological properties of yet unknown nature in addition to their common effects on neutrophils.

IL-8 RECEPTORS

It was shown initially that neutrophils remain responsive to IL-8 after stimulation with fMet-Leu-Phe, C5a, PAF or LTB$_4$, and that IL-8 does not desensitize the cells toward structurally unrelated attractants, suggesting the existence of IL-8 selective receptors (36). Such receptors were demonstrated by binding studies with human neutrophils and myeloid cell lines (17,18). In general agreement with other reports, we found that human neutrophils possess on average 64,500 ± 14,000 receptors with an apparent K_D of 0.18 ± 0.07 nM (63). Radiolabelled IL-8 is displaced not only by cold IL-8, but also by NAP-2 and GROα. These experiments revealed the existence of two types of receptors on neutrophils: One with high affinity for all three ligands (K_D: 0.1-0.3 nM), and the other with high affinity for IL-8, but low affinity for NAP-2 and GROα (K_D: 100-130 nM) (63). Similar results were obtained in direct binding assays using radioiodinated, tyrosine-substituted NAP-2 and GROα (64). Two membrane proteins that bind IL-8, NAP-2 and GROα were also demonstrated by crosslinking experiments (63), and functional evidence for the sharing of receptors comes from the mutual desensitization between IL-8, NAP-2 and GROα (63), as well as IL-8 and ENA-78 (60) or GROß and GROγ (Geiser et al., J. Biol. Chem., in press).

Receptor Structure

Two IL-8 receptor cDNAs were cloned and shown to encode seven-transmembrane-domain receptors (19,20). The existence of two cDNAs is in agreement with the biochemical evidence for two IL-8 receptors (63,65). The cDNA-deduced sequence information adds conclusive evidence to the hypothesis (based on the pattern of biological activities and data on the mechanism of signal transduction) that IL-8 and its related CXC cytokines must be regarded primarily as chemotactic agonists.

Structural Requirements for Receptor Interaction

It was originally thought that IL-8 binds to its receptor through the C-terminal α—helix, which is a prominent feature of the three-dimensional structure. As recent studies show, however, the critical binding domain is at the N-terminus (35,66). Using chemically synthesized analogs, we have found that removal of the entire C-terminal helix decreases, but does not suppress biological activity. In contrast, receptor binding and neutrophil activation are abrogated when the N-terminal sequence Glu-Leu-Arg (ELR) that precedes the first cysteine is deleted (35). All three residues, Arg in particular, are highly sensitive to modification. Similar conclusions are drawn from a mutagenesis study showing that IL-8 activity is lost when residues of the ELR motif are replaced (66). It is interesting to note that the ELR motif is common to all chemotactic cytokines of the CXC subfamily, but is absent in PF4 and IP10, which are virtually inactive on neutrophils (61,67).

PROPERTIES OF CC CYTOKINES

Structure

As already pointed out, after the identification of a CC protein through cloning of the cDNA of the *LD78* gene (22), two murine inflammatory proteins, MIP-1α and MIP-1ß, were identified and found to activate leukocytes (23-25). MCP-1 isolated from different human sources (68-70), was later shown to attract monocytes, but not neutrophils (26-28). Subtractive hybridization screenings led to the identification of other proteins of this subfamily, I-309 (29) and RANTES (30), which, like MCP-1, attract mononuclear cells (31,32). On the basis of computer modelling studies, the folding of CC cytokines appears to be similar to that of CXC cytokines (71). MCP-1 is glycosylated (72-74), and the two forms of glioma- and leukocyte-derived monocyte chemoattractants described by Yoshimura et al. (26,27) are probably due to different degrees of glycosylation. By contrast, neither N- nor O-glycosylation has been reported so far for CXC proteins. The loops resulting from the disulfide bonds are virtually identical in size in both subfamilies. The CC cytokines have, like the CXC, a short N-terminal sequence preceding the first cysteine (6 to 10 residues) and a longer C-terminal domain following the fourth cysteine (18 to 24 residues). I-309, like its murine homolog TCA-3, has two additional cysteines, one about midway between Cys-2 and Cys-3, and the other near the C-terminus, which may influence the folding (29,75).

Biological Activities

MCP-1, the most thoroughly studied cytokine of the CC subfamily, was originally recognized as a chemoattractant for monocytes without activity toward neutrophils (26,27). I-309 (32) and RANTES (31) were later shown to share these properties. The situation is less clear for MIP-1α and MIP-1ß. The murine proteins were reported to stimulate neutrophils (23), and it is not known whether the same is true for the human homologs. MCP-1 also induces cytosolic free Ca^{2+} changes and the respiratory burst in human monocytes, but not in neutrophils (76). The $[Ca^{2+}]_i$ transient was reported to depend entirely on influx across the plasma membrane (77), which is unusual for receptor-dependent stimulation of leukocytes. Pretreatment with *B. pertussis* toxin, on the other hand, prevents the rise in $[Ca^{2+}]_i$ and chemotaxis suggesting the involvement of G-protein coupled receptors (77).

Most recently it was shown that CC cytokines activate basophil and eosinophil leukocytes (78-83) inducing chemotactic and release responses that are more pronounced than those observed with IL-8 (46,47,84). It is interesting that the relative activities and pattern of biological responses vary considerably for the single cytokines (80). These results are of obvious relevance for allergic inflammation, and I am sure that new information will be presented at the meeting. RANTES has also been reported to attract T-lymphocytes (31), and it will be important to study the effects of this cytokine *in vivo* and to identify the cells that are recruited at the site of application.

Receptors

Binding studies have demonstrated high affinity receptors for MCP-1 on human blood monocytes (85,86), for human MIP-1ß on activated peripheral blood lymphocytes, myeloid and lymphoid cell lines (87), and murine MIP-1α on T-cell and macrophage cell lines (88). Receptor numbers (1,000 to 2,000 for MCP-1, up to 10,000 for human MIP-1ß and 360 to 1200 for murine MIP-1α) are markedly lower than those of CXC cytokine receptors on human neutrophils, while the K_D values are similar, i.e. mostly around 1 nM. No information on receptor sequence and structure has been reported so far.

OPEN QUESTIONS

Although information about CXC and CC cytokines has accumulated at a breathtaking speed, many important questions are still open. I raise a few obvious ones that are worth our attention, and shall watch carefully for possible answers during the meeting.

Is chemotaxis the main effect? Chemotactic activity for neutrophils is well documented for several CXC cytokines. IL-8, the three GRO proteins, NAP-2 and ENA-78 induce, in addition to migration *in vitro and in vivo*, granule exocytosis and the respiratory burst, which are characteristic for activation by chemotactic agonists. Furthermore, like fMet-Leu-Phe and C5a, these cytokines act via heptahelical receptors coupled to *pertussis* toxin sensitive G-proteins. Other effects that have been reported, such as induction of cell proliferation, chemotaxis of lymphocytes, etc. must be studied beyond the present level of phenomenology. They must be substantiated with data on receptors and signal transduction, and validated by strict controls, using for instance unrelated chemotactic peptides like fMet-Leu-Phe, and chemotactically inactive CXC proteins. Much less is known about the CC subfamily, although present evidence indicates that the effects of these cytokines on their target leukocytes are not essentially different from those of IL-8 on neutrophils. We must wait, however, for a better characterization of the receptors and further information on signal transduction.

Are they important in pathology? The hypothesis that IL-8 is an inflammatory mediator is now supported by many studies showing its enhanced expression and release in inflammatory diseases. Of particular interest is the fact that inflammatory leukocytes as well as virtually all tissue cells are potential sources of IL-8, and that the main inducers are the notorious bad guys of inflammation, LPS, IL-1 and TNF. It has been shown that IL-8 is generated upon infection, ischemia, trauma, and other disturbances of tissue homeostasis (presumably as a consequence of enhanced IL-1 and/or TNF expression), and the local formation of IL-8 (and other chemotactic CXC cytokines) is considered as the main cause of the characteristic accumulation of neutrophils. IL-8 is very stable. It remains effective for a long time in the environment of the cells from which it is released, while classical chemoattractants (lipids and peptides) are rapidly inactivated. Of the CC cytokines, only MCP-1 has been studied in various pathologies. Expression appears to be frequent in inflamed tissues and often concomitant with that of CXC cytokines.

Therapy? It is obvious from what I just said that inhibiting the action of CXC and CC cytokines is desirable in pathological conditions that are aggravated by the influx of leukocytes. Some cytokines, like IL-4, IL-10 and gamma-interferon inhibit the expression of CXC and CC genes, and could thus be therapeutically useful, although their presumed lack of selectivity could be a draw-back. On the other hand, the observation that the CXC cytokines act on neutrophils through common receptors suggests that the search for antagonists would be worthwhile.

Why so many? In my opinion, the most remarkable feature of chemotactic cytokines is redundancy. There are no less than six CXC cytokines (including IL-8) that are chemotactic for neutrophils and act via common receptors. At least as many distinct proteins belong to the CC subfamily. They activate monocytes, and, to different degrees, basophil and eosinophil leukocytes. Redundancy clearly indicates that these cytokines have pathophysiologically important functions. They are likely to concur in the recruitment of leukocytes, but may also fulfill selective functions, that we still don't know.

REFERENCES

1. Deuel, T.F., P.S. Keim, M. Farmer and R.L. Heinrikson. 1977. Amino acid sequence of human platelet factor 4. *Proc. Natl. Acad. Sci. USA* 74: 2256-2258.
2. Hermodson, M., G. Schmer, and K. Kurachi. 1977. Isolation, crystallization, and primary amino acid sequence of human platelet factor 4. *J. Biol. Chem.* 252: 6276-6279.
3. Walz, D.A., V.Y. Wu, R. de Lamo, H. Dene, and L.E. McCoy. 1977. Primary structure of human platelet factor 4. *Thromb. Res.* 11: 893-898.
4. Morgan, F.J., F.S. Begg, and C.M. Chesterman. 1977. Primary structure of human platelet factor 4. *Thromb. Haemost.* 38: 231-235.
5. Schmid, J. and C. Weissmann. 1987. Induction of mRNA for a serine protease and a ß-thromboglobulin-like protein in mitogen-stimulated human leukocytes. *J. Immunol.* 139: 250-256.
6. Yoshimura, T., K. Matsushima, S. Tanaka, E.A. Robinson, E. Appella, J.J. Oppenheim, and E.J. Leonard. 1987. Purification of a human monocyte-derived neutrophil chemotactic factor that has peptide sequence similarity to other host defense cytokines. *Proc. Natl. Acad. Sci. USA* 84: 9233-9237.
7. Walz, A., P. Peveri, H. Aschauer, and M. Baggiolini. 1987. Purification and amino acid sequencing of NAF, a novel neutrophil-activating factor produced by monocytes. *Biochem. Biophys. Res. Commun.* 149: 755-761.
8. Holt, J.C., M.E. Harris, A.M. Holt, E. Lange, A. Henschen, and S. Niewiarowski. 1986. Characterization of human platelet basic protein, a precursor form of low-affinity platelet factor 4 and ß-thromboglobulin. *Biochemistry.* 25: 1988-1996.
9. Castor, C.W., J.W. Miller, and D.A. Walz. 1983. Structural and biological characteristics of connective tissue activating peptide (CTAP-III), a major human platelet-derived growth factor. *Proc. Natl. Acad. Sci. USA* 80: 765-769.
10. Begg, G.S., D.S. Pepper, C.N. Chesterman, and F.J. Morgan. 1978. Complete covalent structure of human beta-thromboglobulin. *Biochemistry.* 17: 1739-1744.
11. Luster, A.D., J.C. Unkeless, and J.V. Ravetch. 1985. τ-Interferon transcriptionally regulates an early-response gene containing homology to platelet proteins. *Nature* 315: 672-676.
12. Luster, A.D. and J.V. Ravetch. 1987. Biochemical characterization of a gamma interferon-inducible cytokine (IP-10). *J. Exp. Med.* 166: 1084-1097.
13. Baggiolini, M., A. Walz, and S.L. Kunkel. 1989. Neutrophil-activating peptide-1/interleukin 8, a novel cytokine that activates neutrophils. *J. Clin. Invest.* 84: 1045-1049.
14. Clore, G.M. and A.M. Gronenborn. NMR and X-ray analysis of the three-dimensional structure of interleukin-8. In: *Cytokine. Vol. 4. Interleukin-8 (NAP-1) and related chemotactic cytokines*, edited by Baggiolini, M. and Sorg, C. Basel: Karger, 1992, p. 18-40.
15. Mukaida, N., M. Shiroo, and K. Matsushima. 1989. Genomic structure of the human monocyte-derived neutrophil chemotactic factor IL-8. *J. Immunol.* 143: 1366-1371.
16. Modi, W.S., M. Dean, H.N. Seuanez, N. Mukaida, K. Matsushima, and S.J. O'Brien. 1990. Monocyte-derived neutrophil chemotactic factor (MDNCF/IL-8) resides in a gene cluster alon g with several other members of the platelet factor 4 gene superfamily. *Hum. Genet.* 84: 185-187.
17. Samanta, A.K., J.J. Oppenheim, and K. Matsushima. 1989. Identification and characterization of specific receptors for monocyte-derived neutrophil chemotactic factor (MDNCF) on human neutrophils. *J. Exp. Med.* 169: 1185-1189.
18. Besemer, J., A. Hujber, and B. Kuhn. 1989. Specific binding, internalization, and degradation of human neutrophil activating factor by human polymorphonuclear leukocytes. *J. Biol. Chem.* 264: 17409-17415.
19. Holmes, W.E., J. Lee, W.-J. Kuang, G.C. Rice, and W.I. Wood. 1991. Structure and functional expression of a human interleukin-8 receptor. *Science* 253: 1278-1280.
20. Murphy, P.M. and H.L. Tiffany. 1991. Cloning of complementary DNA encoding a functional human interleukin-8 receptor. *Science* 253: 1280-1283.
21. Baggiolini, M., P. Imboden, and P. Detmers. Neutrophil activation and the effects of interleukin-8/neutrophil-activating peptide 1 (IL-8/NAP-1). In: *Cytokines. Vol. 4. Interleukin-8 (NAP-1) and related chemotactic cytokines*, edited by Baggiolini, M. and Sorg, C. Basel: Karger, 1992, p. 1-17.

22. Obaru, K., M. Fukuda, S. Maeda, and K. Shimada. 1986. A cDNA clone used to study mRNA inducible in human tonsillar lymphocytes by a tumor promoter. *J. Biochem.* (Tokyo.) 99: 885-894.

23. Wolpe, S.D., G. Davatelis, B. Sherry, B. Beutler, D.G. Hesse, H.T. Nguyen, L.L. Moldawer, C.F. Nathan, S.F. Lowry, and A. Cerami. 1988. Macrophages secrete a novel heparin-binding protein with inflammatory and neutrophil chemokinetic properties. *J. Exp. Med.* 167: 570-581.

24. Davatelis, G., P. Tekamp Olson, S.D. Wolpe, K. Hermsen, C. Luedke, C. Gallegos, D. Coit, J. Merryweather, and A. Cerami. 1988. Cloning and characterization of a cDNA for murine macrophage inflammatory protein (MIP), a novel monokine with inflammatory and chemokinetic properties. *J. Exp. Med.* 167: 1939-1944.

25. Sherry, B., P. Tekamp Olson, C. Gallegos, D. Bauer, G. Davatelis, S.D. Wolpe, F. Masiarz, D. Coit, and A. Cerami. 1988. Resolution of the two components of macrophage inflammatory protein 1, and cloning and characterization of one of those components, macrophage inflammatory protein 1 beta. *J. Exp. Med.* 168: 2251-2259.

26. Yoshimura, T., E.A. Robinson, S. Tanaka, E. Appella, J. Kuratsu, and E.J. Leonard. 1989. Purification and amino acid analysis of two human glioma-derived monocyte chemoattractants. *J. Exp. Med.* 169: 1449-1459.

27. Yoshimura, T., E.A. Robinson, S. Tanaka, E. Appella, and E.J. Leonard. 1989. Purification and amino acid analysis of two human monocyte chemoattractants produced by phytohemagglutinin-stimulated human blood mononuclear leukocytes. *J. Immunol.* 142: 1956-1962.

28. Matsushima, K., C.G. Larsen, G.C. DuBois, and J.J. Oppenheim. 1989. Purification and characterization of a novel monocyte chemotactic and activating factor produced by a human myelomonocytic cell line. *J. Exp. Med.* 169: 1485-1490.

29. Miller, M.D., S. Hata, R. De Waal Malefyt, and M.S. Krangel. 1989. A novel polypeptide secreted by activated human T lymphocytes. *J. Immunol.* 143: 2907-2916.

30. Schall, T.J., J. Jongstra, B.J. Dyer, J. Jorgensen, C. Clayberger, M.M. Davis, and A.M. Krensky. 1988. A human T cell-specific molecule is a member of a new gene family. *J. Immunol.* 141: 1018-1025.

31. Schall, T.J., K. Bacon, K.J. Toy, and D.V. Goeddel. 1990. Selective attraction of monocytes and T lymphocytes of the memory phenotype by cytokine RANTES. *Nature* 347: 669-671.

32. Miller, M.D. and M.S. Krangel. 1992. The human cytokine I-309 is a monocyte chemoattractant. *Proc. Natl. Acad. Sci. USA* 89: 2950-2954.

33. Yoshimura, T., E.A. Robinson, E. Appella, K. Matsushima, S.D. Showalter, A. Skeel, and E.J. Leonard. 1989. Three forms of monocyte-derived neutrophil chemotactic factor (MDNCF) distinguished by different lengths of the amino-terminal sequence. *Mol. Immunol.* 26: 87-93.

34. Lindley, I., H. Aschauer, J.M. Seifert, C. Lam, W. Brunowsky, E. Kownatzki, M. Thelen, P. Peveri, B. Dewald, V. von Tscharner, A. Walz, and M. Baggiolini. 1988. Synthesis and expression in Escherichia coli of the gene encoding monocyte-derived neutrophil-activating factor: Biological equivalence between natural and recombinant neutrophil-activating factor. *Proc. Natl. Acad. Sci. USA* 85: 9199-9203.

35. Clark-Lewis, I., C. Schumacher, M. Baggiolini, and B. Moser. 1991. Structure-activity relationships of interleukin-8 determined using chemically synthesized analogs. Critical role of NH_2-terminal residues and evidence for uncoupling of neutrophil chemotaxis, exocytosis, and receptor binding activities. *J. Biol. Chem.* 266: 23128-23134.

36. Peveri, P., A. Walz, B. Dewald, and M. Baggiolini. 1988. A novel neutrophil-activating factor produced by human mononuclear phagocytes. *J. Exp. Med.* 167: 1547-1559.

37. Tanaka, S., E.A. Robinson, T. Yoshimura, K. Matsushima, E.J. Leonard, and E. Appella. 1988. Synthesis and biological characterization of monocyte-derived neutrophil chemotactic factor. *FEBS. Lett.* 236: 467-470.

38. Schroder, J.M., U. Mrowietz, E. Morita, and E. Christophers. 1987. Purification and partial biochemical characterization of a human monocyte-derived, neutrophil-activating peptide that lacks interleukin 1 activity. *J. Immunol.* 139: 3474-3483.

39. Thelen, M., P. Peveri, P. Kernen, V. von Tscharner, A. Walz, and M. Baggiolini. 1988. Mechanism of neutrophil activation by NAF, a novel monocyte-derived peptide agonist. *FASEB. J.* 2: 2702-2706.

40. Paccaud, J.-P., J.A. Schifferli, and M. Baggiolini. 1990. NAP-1/IL-8 induces upregulation of CR1 receptors in human neutrophil leukocytes. *Biochem. Biophys. Res. Commun.* 166: 187-192.

41. Detmers, P.A., S.K. Lo, E. Olsen-Egbert, A. Walz, M. Baggiolini, and Z.A. Cohn. 1990. Neutrophil-activating protein 1/interleukin 8 stimulates the binding activity of the leukocyte adhesion receptor CD11b/CD18 on human neutrophils. *J. Exp. Med.* 171: 1155-1162.

42. Detmers, P.A., D.E. Powell, A. Walz, I. Clark-Lewis, M. Baggiolini, and Z.A. Cohn. 1991.

Differential effects of neutrophil-activating peptide 1/IL-8 and its homologues on leukocyte adhesion and phagocytosis. *J. Immunol.* 147: 4211-4217.

43. Leonard, E.J., A. Skeel, T. Yoshimura, K. Noer, S. Kutvirt, and D. Van Epps. 1990. Leukocyte specificity and binding of human neutrophil attractant/activation protein-1. *J. Immunol.* 144: 1323-1330.

44. Walz, A., F. Meloni, I. Clark-Lewis, V. von Tscharner, and M. Baggiolini. 1991. $[Ca^{2+}]_i$ changes and respiratory burst in human neutrophils and monocytes induced by NAP-1/interleukin-8, NAP-2, and gro/MGSA. *J. Leukocyte Biol.* 50: 279-286.

45. Kernen, P., M.P. Wymann, V. von Tscharner, D.A. Deranleau, P.-C. Tai, C.J. Spry, C.A. Dahinden, and M. Baggiolini. 1991. Shape changes, exocytosis, and cytosolic free calcium changes in stimulated human eosinophils. *J. Clin. Invest.* 87: 2012-2017.

46. Krieger, M., T. Brunner, S.C. Bischoff, V. von Tscharner, A. Walz, B. Moser, M. Baggiolini, and C.A. Dahinden. 1992. Activation of human basophils through the IL-8 receptor. *J. Immunol.* 149: 2662-2667.

47. Dahinden, C.A., Y. Kurimoto, A.L. De Weck, I. Lindley, B. Dewald, and M. Baggiolini. 1989. The neutrophil-activating peptide NAF/NAP-1 induces histamine and leukotriene release by interleukin 3-primed basophils. *J. Exp. Med.* 170: 1787-1792.

48. Larsen, C.G., A.O. Anderson, E. Appella, J.J. Oppenheim, and K. Matsushima. 1989. The neutrophil-activating protein (NAP-1) is also chemotactic for T lymphocytes. *Science* 243: 1464-1466.

49. Bacon, K.B. and R.D.R. Camp. 1990. Interleukin (IL)-8-induced *in vitro* human lymphocyte migration is inhibited by cholera and pertussis toxins and inhibitors of protein kinase C. *Biochem. Biophys. Res. Commun.* 169: 1099-1104.

50. Wilkinson, P.C. and I. Newman. 1992. Identification of IL-8 as a locomotor attractant for activated human lymphocytes in mononuclear cell cultures with anti-CD3 or purified protein derivative of *Mycobacterium tuberculosis*. *J. Immunol.* 149: 2689-2694.

51. Grob, P.M., E. David, T.C. Warren, R.P. DeLeon, P.R. Farina, and C.A. Homon. 1990. Characterization of a receptor for human monocyte-derived neutrophil chemotactic factor/interleukin-8. *J. Biol. Chem.* 265: 8311-8316.

52. Colditz, I., R. Zwahlen, B. Dewald, and M. Baggiolini. 1989. In vivo inflammatory activity of neutrophil-activating factor, a novel chemotactic peptide derived from human monocytes. *Am. J. Pathol.* 134: 755-760.

53. Colditz, I.G., R.D. Zwahlen, and M. Baggiolini. 1990. Neutrophil accumulation and plasma leakage induced in vivo by neutrophil-activating peptide-1. *J. Leukocyte Biol.* 48: 129-137.

54. Swensson, O., C. Schubert, E. Christophers, and J.-M. Schröder. 1991. Inflammatory properties of neutrophil-activating protein-1/interleukin 8 (NAP-1/IL-8) in human skin: A light- and electronmicroscopic study. *J. Invest. Dermatol.* 96: 682-689.

55. Leonard, E.J., T. Yoshimura, S. Tanaka, and M. Raffeld. 1991. Neutrophil recruitment by intradermally injected neutrophil attractant/activation protein-1. *J. Invest. Dermatol.* 96: 690-694.

56. Derynck, R., E. Balentien, J.H. Han, H.G. Thomas, D. Wen, A.K. Samantha, C.O. Zachariae, P.R. Griffin, R. Brachmann, W.L. Wong, K. Matsushima, and A. Richmond. 1990. Recombinant expression, biochemical characterization, and biological activities of the human MGSA/gro protein. *Biochemistry* 29: 10225-10233.

57. Bordoni, R., R. Fine, D. Murray, and A. Richmond. 1990. Characterization of the role of melanoma growth stimulatory activity (MGSA) in the growth of normal melanocytes, nevocytes, and malignant melanocytes. *J. Cell. Biochem.* 44: 207-219.

58. Moser, B., I. Clark-Lewis, R. Zwahlen, and M. Baggiolini. 1990. Neutrophil-activating properties of the melanoma growth-stimulatory activity. *J. Exp. Med.* 171: 1797-1802.

59. Haskill, S., A. Peace, J. Morris, S.A. Sporn, A. Anisowicz, S.W. Lee, T. Smith, G. Martin, P. Ralph, and R. Sager. 1990. Identification of three related human *GRO* genes encoding cytokine functions. *Proc. Natl. Acad. Sci. USA* 87: 7732-7736.

60. Walz, A., R. Burgener, B. Car, M. Baggiolini, S.L. Kunkel, and R.M. Strieter. 1991. Structure and neutrophil-activating properties of a novel inflammatory peptide (ENA-78) with homology to interleukin 8. *J. Exp. Med.* 174: 1355-1362.

61. Walz, A., B. Dewald, V. von Tscharner, and M. Baggiolini. 1989. Effects of the neutrophil-activating peptide NAP-2, platelet basic protein, connective tissue-activating peptide III and platelet factor 4 on human neutrophils. *J. Exp. Med.* 170: 1745-1750.

62. Baggiolini, M., B. Dewald, and A. Walz. Interleukin-8 and related chemotactic cytokines. In: *Inflammation: Basic Principles and Clinical Correlates*, edited by Gallin, J.I., Goldstein, I.M. and Snyderman, R. New York: Raven Press, Ltd., 1992, p. 247-263.

9

63. Moser, B., C. Schumacher, V. von Tscharner, I. Clark-Lewis, and M. Baggiolini. 1991. Neutrophil-activating peptide 2 and *gro*/melanoma growth-stimulatory activity interact wi th neutrophil-activating peptide 1/interleukin 8 receptors on human neutrophils. *J. Biol. Chem.* 266: 10666-10671.

64. Schumacher, C., I. Clark-Lewis, M. Baggiolini, and B. Moser. 1992. High- and low-affinity binding of GROα and neutrophil-activating peptide 2 to interleukin 8 receptors on human neutrophils. *Proc. Natl. Acad. Sci. USA* 89: 10542-10546.

65. Lee, J., R. Horuk, G.C. Rice, G.L. Bennett, T. Camerato, and W.I. Wood. 1992. Characterization of two high affinity human interleukin-8 receptors. *J. Biol. Chem.* 267: 16283-16287.

66. Hébert, C.A., R.V. Vitangcol, and J.B. Baker. 1991. Scanning mutagenesis of interleukin-8 identifies a cluster of residues required for receptor binding. *J. Biol. Chem.* 266: 18989-18994.

67. Dewald, B., B. Moser, L. Barella, C. Schumacher, M. Baggiolini, and I. Clark-Lewis. 1992. IP-10, a gamma-interferon-inducible protein related to interleukin-8, lacks neutrophil activating properties. *Immunol. Lett.* 32: 81-84.

68. Robinson, E.A., T. Yoshimura, E.J. Leonard, S. Tanaka, P.R. Griffin, J. Shabanowitz, D.F. Hunt, and E. Appella. 1989. Complete amino acid sequence of a human monocyte chemoattractant, a putative mediator of cellular immune reactions. *Proc. Natl. Acad. Sci. USA* 86: 1850-1854.

69. Furutani, Y., H. Nomura, M. Notake, Y. Oyamada, T. Fukui, M. Yamada, C.G. Larsen, J.J. Oppenheim, and K. Matsushima. 1989. Cloning and sequencing of the cDNA for human monocyte chemotactic and activating factor (MCAF). *Biochem. Biophys. Res. Commun.* 159: 249-255.

70. Yoshimura, T., N. Yuhki, S.K. Moore, E. Appella, M.I. Lerman, and E.J. Leonard. 1989. Human monocyte chemoattractant protein-1 (MCP-1). Full-length cDNA cloning, expression in mitogen-stimulated blood mononuclear leukocytes, and sequence similarity to mouse competence gene JE. *FEBS. Lett.* 244: 487-493.

71. Gronenborn, A.M. and G.M. Clore. 1991. Modeling the three-dimensional structure of the monocyte chemo-attractant and activating protein MCAF/MCP-1 on the basis of the solution structure of interleukin-8. *Protein Eng.* 4: 263-269.

72. Jiang, Y., A.J. Valente, M.J. Williamson, L. Zhang, and D.T. Graves. 1990. Post-translational modification of a monocyte-specific chemoattractant synthesized by glioma, osteosarcoma, and vascular smooth muscle cells. *J. Biol. Chem.* 265: 18318-18321.

73. Jiang, Y., L.A. Tabak, A.J. Valente, and D.T. Graves. 1991. Initial characterization of the carbohydrate structure of MCP-1. *Biochem. Biophys. Res. Commun.* 178: 1400-1404.

74. Rollins, B.J., P. Stier, T. Ernst, and G.W. Wong. 1989. The human homolog of the JE gene encodes a monocyte secretory protein. *Mol. Cell Biol.* 9: 4687-4695.

75. Burd, P.R., G.J. Freeman, S.D. Wilson, M. Berman, R. DeKruyff, P.R. Billings, and M.E. Dorf. 1987. Cloning and characterization of a novel T cell activation gene. *J. Immunol.* 139: 3126-3131.

76. Rollins, B.J., A. Walz, and M. Baggiolini. 1991. Recombinant human MCP-1/JE induces chemotaxis, calcium flux, and the respiratory burst in human monocytes. *Blood* 78: 1112-1116.

77. Sozzani, S., W. Luini, M. Molino, P. Jílek, B. Bottazzi, C. Cerletti, K. Matsushima, and A. Mantovani. 1991. The signal transduction pathway involved in the migration induced by a monocyte chemotactic cytokine. *J. Immunol.* 147: 2215-2221.

78. Kuna, P., S.R. Reddigari, D. Rucinski, J.J. Oppenheim, and A.P. Kaplan. 1992. Monocyte chemotactic and activating factor is a potent histamine-releasing factor for human basophils. *J. Exp. Med.* 175: 489-493.

79. Alam, R., M.A. Lett-Brown, P.A. Forsythe, D.J. Anderson-Walters, C. Kenamore, C. Kormos, and J.A. Grant. 1992. Monocyte chemotactic and activating factor is a potent histamine-releasing factor for basophils. *J. Clin. Invest.* 89: 723-728.

80. Bischoff, S.C., M. Krieger, T. Brunner, and C.A. Dahinden. 1992. Monocyte chemotactic protein 1 is a potent activator of human basophils. *J. Exp. Med.* 175: 1271-1275.

81. Kuna, P., S.R. Reddigari, T.J. Schall, D. Rucinski, M.Y. Viksman, and A.P. Kaplan. 1992. Rantes, a monocyte and T lymphocyte chemotactic cytokine releases histamine from human basophils. *J. Immunol.* 149: 636-642.

82. Alam, R., P.A. Forsythe, S. Stafford, M.A. Lett-Brown, and J.A. Grant. 1992. Macrophage inflammatory protein-1α activates basophils and mast cells. *J. Exp. Med.* 176: 781-786.

83. Rot, A., M. Krieger, T. Brunner, S.C. Bischoff, T.J. Schall, and C.A. Dahinden. 1992. RANTES and macrophage inflammatory protein 1α induce the migration and activation of normal human eosinophil granulocytes. *J. Exp. Med.* 176: 1489-1495.

84. Schultz, D.R., J.E. Volanakis, P.I. Arnold, N.L. Gottlieb, K. Sakai, and R.M. Stroud. 1974. Inactivation of C1 in rheumatoid synovial fluid, purfied C1 and C1 esterase, by gold compounds. *Clin. Exp. Immunol.* 17: 395-406.

85.   Yoshimura, T. and E.J. Leonard. 1990. Identification of high affinity receptors for human monocyte chemoattractant protein-1 on human monocytes. *J. Immunol.* 145: 292-297.

86.   Valente, A.J., M.M. Rozek, C.J. Schwartz, and D.T. Graves. 1991. Characterization of monocyte chemotactic protein-1 binding to human monocytes. *Biochem. Biophys. Res. Commun.* 176: 309-314.

87.   Napolitano, M., K.B. Seamon, and W.J. Leonard. 1990. Identification of cell surface receptors for the Act-2 cytokine. *J. Exp. Med.* 172: 285-289.

88.   Oh, K.-O., Z. Zhou, K.-K. Kim, H. Samanta, M. Fraser, Y.-J. Kim, H.E. Broxmeyer, and B.S. Kwon. 1991. Identification of cell surface receptors for murine macrophage inflammatory protein-1α. *J. Immunol.* 147: 2978-2983.

INDUCTION OF CHEMOTACTIC CYTOKINES BY MINIMALLY OXIDIZED LDL

Judith A. Berliner[1], David S. Schwartz[1], Mary C. Territo[2], Ali Andalibi[2], Laura Almada[2], Aldons J. Lusis[2], D. Quismorio[2], Zhuang P. Fang[1] and Alan M. Fogelman[2]

Department of Pathology[1]
Department of Medicine[2]
13-186 Center for Health Sciences
University of California
Los Angeles, CA 90024, USA

INTRODUCTION

An important initiating event in atherogenesis is the entry of monocytes into the vessel wall. In cholesterol fed animals after approximately one week of feeding, monocytes begin to bind to the vessel wall and migrate into the subendothelial space (1,2). Here they take up lipid and form foam cells. Monocyte binding occurs only in certain areas of the vessel referred to as areas of prediliction (3). In these areas liproteins accumulate (4), and several investigators have shown that these lipoproteins are oxidized (5,6). Our group has studied the effects of minimally oxidized LDL (MM-LDL) on the interactions of monocytes with endothelial cells (7,8).

METHODS

For studies of effects of lipoproteins, rabbit aortic endothelial cells (RAEC) and human aortic endothelial and smooth muscle cells (HAEC and HASMC), cultured as described previously (7,8) have been employed. MM-LDL was obtained by storage of native LDL or was synthesized by treatment of native LDL with lipoxidase and phospholipase A2 by a modification of the method of Sparrow et al (9). Monocyte chemotaxis and binding studies were performed as described previously (7,8). Murine MIP 2, a recombinant protein made in yeast, was obtained from Chiron. Antibody to MCP-1 was a polyclonal antibody obtained from Dr. T Valente.

Two antibodies to Gro were employed: one antibody was a polyclonal antibody to Gro beta obtained from Chiron, the other was a monoclonal antibody reacting only with fixed Gro homologues obtained from Anne Richmond, Vanderbilt University.

RESULTS

The MM-LDL employed in these studies exhibited a 2-3 fold increase in conjugated dienes but showed little degradation of Apo B detectable by Comassie blue staining. Therefore it is likely that MM-LDL is taken up by the LDL and not the scavenger receptor. Treatment of RAEC or HAEC with 100 µg/ml of MM-LDL caused a 5 fold increase in the monocyte chemotactic activity in the medium; no neutrophil chemotactic activity was observed (Fig 1).

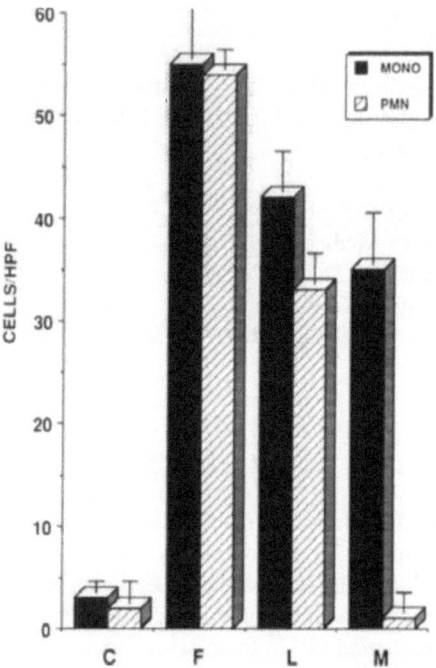

Figure 1. Chemotactic activity produced by endothelial cells treated with MM-LDL and LPS. Medium was collected from RAEC treated for 6 hours with 1ng/ml of LPS (L), 100ug/ml MM-LDL (M) or medium without additives (C); 10-⁹M FMLP (F) served as a positive control for these studies. Chemotactic activity for monocytes (mono) or neutrophils (PMN) was tested in a neuroprobe chemotactic chamber. Values are given as mean + SE, N=8 fields.

Synthesis of MCSF, a known monocyte chemotactic factor, was increased both at the level of mRNA and protein by treating RAEC for 4-24 hours with MM-LDL (10). The monocyte chemotactic activity in medium from RAEC (Table 1) or HAEC (8) was completely neutralized by incubation with antibody to MCP-1.

Incubation of co-cultures of human aortic endothelial cells and HASMC with native LDL also caused the formation of MM-LDL which increased the migration of monocytes into co-cultures (11). For these studies one set of co-cultures was treated for 24 hours with 500ug/ml of native LDL. Then this LDL was refloated and naive co-cultures were exposed to the refloated LDL. After 4 hours monocytes were added to the co-cultures and migration across the endothelium was measured. Migration was increased three fold and this migration was blocked by pretreatment of co-cultures with antibody to MCP-1 (11).

We have also shown that treating endothelial cells with MM-LDL increases monocyte but not neutrophil binding to the monolayers of RAEC (7) and HAEC (Fig 2).

Table 1. Inhibition of monocyte chemotactic activity by antibody to MCP-1.

	-AB	+AB
C	3 + 1	3 + 2
F	64 + 5	61 + 6
M	47 + 6	2 + 1

Medium from untreated cells (C) and cells treated with MM-LDL (M) was collected as described for Figure 1. Aliquots of these media and medium containing 10^{-9} M FMLP were treated with a polyclonal antibody to MCP-1 and chemotactic activity of antibody containing medium (+ AB) and medium without antibody (-AB) were tested in a neuroprobe chamber. Data are reported as mean cells/high power field + SE. N = 8.

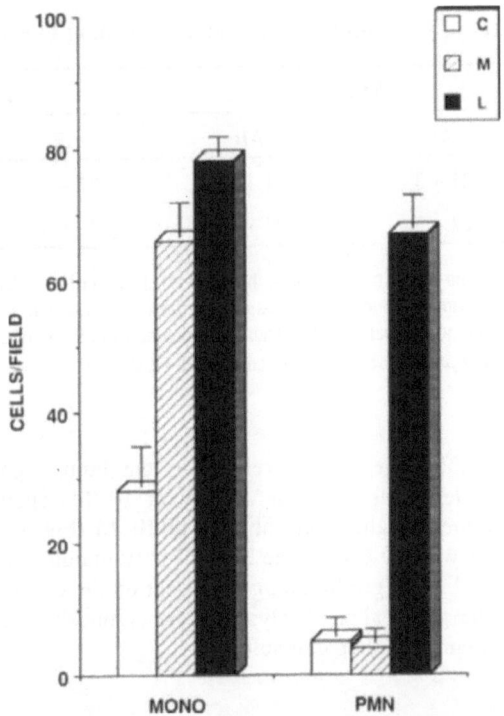

Figure 2. Binding of leukocytes to endothelial cells treated with LPS and MM-LDL. Human aortic endothelial cells (HAEC) were treated for 4 hours with 3ng/ml LPS (L), 125ug/ml MM-LDL (M) or medium containing no additives (C). Binding of human monocytes (Mono) or neutrophils (PMN) to these monolayers was then measured and is given as cells/field. Values are mean + SE. N = 8.

This binding to HAEC was not blocked by FAB 2 fragments of antibodies to VCAM, ELAM or ICAM. The identity of the binding molecule(s) induced by MM-LDL is not known, but one molecule that contributes to binding in both HAEC and RAEC is a gro homologue. This homologue has a 70 % homology to human gro proteins in the translated region and a lower homology to murine MIP-2. Gro protein is increased on the surface of HAEC treated with LPS or MM-LDL as detected by ELISA (Table 2).

Treatment of HAEC with antibody to Gro blocked monocyte binding by 50-100% (Table 3).

Table 2. Surface expression of Gro homologue on treated endothelial cells.

	4H	24H
C	1 + .1	.5 + .2
L	2.5 + .2	2.4 + .3
M	3.3 + .2	3.1 + .3

Human aortic endothelial cells were treated for 4 hours (4H) or 24 hours (24H) with 3ng/ml of LPS (L), 125 ug/ml MM-LDL (M) or medium containing no additives (C). Cells were rinsed and the binding to the cell surface of a monoclonal antibody to Gro alpha (but which reacted with all Gro homologues) was determined using an ELISA assay. Values are given as mean OD units of color produced in the assay + SE. N = 6.

Table 3. Effect of Gro antibody on monocyte binding to endothelial cells.

	EXPT 1		EXPT 2	
	-AB	+AB	-AB	+AB
C	24 + 3	21 + 2	27 + 2	22 + 4
M	67 + 5	23 + 4	76 + 4	43 + 3

Human aortic endothelial cells were treated for 4 hours with 125ug/ml of MM-LDL (M) or medium containing no additives (C). Human monocytes were suspended for 10 minutes in medium (-AB) or medium containing polyclonal antibody to gro beta (+AB). These monocytes were then added to the endothelial cells and binding measured. Values are reported as mean cells/field + SE. N=8.

Murine MIP-2, a molecule closely related to the human gro homologues, was chemotactic for monocytes with maximal activity at 10^{-9}M; chemotactic activity for neutrophils peaked at a much higher concentration of 10^{-7}M (Fig 3).

Treating monocytes with 10^{-9}M murine MIP-2 for 10 minutes increased their binding to untreated HAEC 2 fold . The gro homologue present on the cell surface may be bound to heparan sulfate of the glycocalyx. Analysis of the composition of the glycocalyx of RAEC showed it to contain 90% heparan sulfate .

DISCUSSION

Treatment of endothelial cells with MM-LDL increases the formation of 2 chemotactic cytokines: MCP-1 and a Gro homologue. From our studies it appears that both MCP-1 and Gro facilitate monocyte migration across the endothelium after treatment with MM-LDL. The gro homologue seems to be mainly present on the cell surface while MCP-1 is in the medium surrounding the cells. This is shown by the fact that antibody to MCP-1 completely inhibited the monocyte chemotactic activity in the medium of MM-LDL treated endothelial cells. Several others at this meeting have shown that other members of the CXC family are more active as leukocyte chemotactic or activating factors when bound to heparan sulfate. We hypothesize that MCP-1 and gro homologye molecules probably act, together with a selectin, to stabilize monocyte binding and migration as components of the 3 step mechanism previously described (12). The finding that an endothelial cell activator induces both surface bound and soluble chemotactic cytokines suggests a role for both types of molecules in monocyte adhesion and migration.

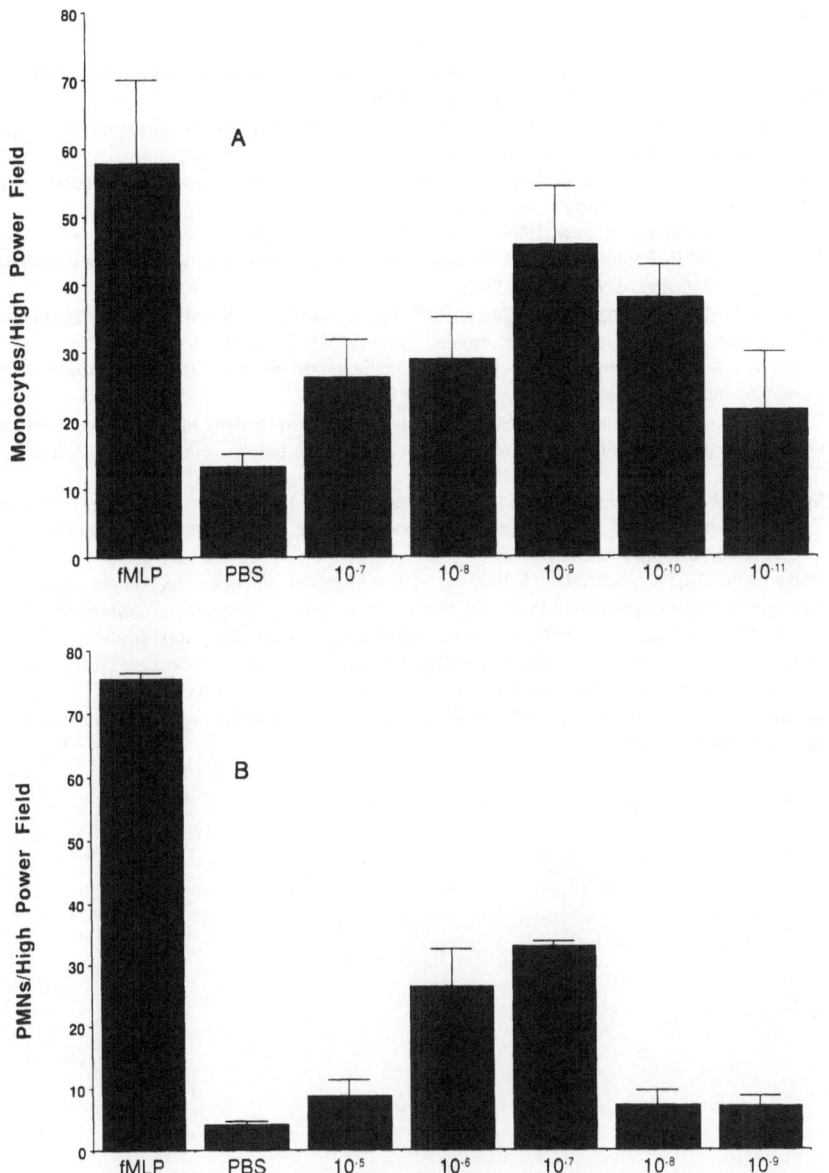

Figure 3. Chemotactic activity of murine MIP-2 for monocytes and neutrophils. The chemotactic activity of murine MIP-2 for monocytes (Panel A) and neutrophils (Panel B) was compared using a neuroprobe chamber. Values are given as mean cells/HPF + SE. N=6.

ACKNOWLEDGEMENTS

This research was supported by NIH grants HL 30568, TRDRP grant RT 372 and the Laubish Fund. Salary support for D. Schwartz was provided by a Fellowship from the American Heart Association.

REFERENCES

1. Gerrity, R.G, H.K. Naito et al. 1979. Dietary induced atherogenesis in swine. Morphology of the intima in prelesion stages. *Am J Pathol* 95: 775-792.
2. Faggiotto, A., R. Ross, R. Harker. 1984. Studies of hypercholesterolemia in the nonhuman primate. I. Changes that lead to fatty streak formation. *Arterioclerosis* 4: 323-340.
3. Gerrity, R.G. 1981. The role of the monocyte in atherogenesis: I. Transition of blood-borne monocytes into foam cells in fatty lesions. *Am J Pathol.* 103: 181-190.
4. Schwenke, D. and T.D. Carew. 1989. *Arteriosclerosis* 9: 220-226.
5. Palinski, W., M.E. Rosenfeld et al. 1989. Low density lipoprotein undergoes oxidative modification in vivo. *Proc Natl Acad Sci USA*, 86: 1372-76.
6. Haberland, M.E., D. Fong, and L. Cheng. 1988. Malondialdehyde-altered protein occurs in atheroma of Watanabe heritable hyperlipidemic rabbits. *Science* 241: 215-218.
7. Berliner, J.A., M.C. Territo et al. 1990. Minimally modified low density lipoprotein stimulates monocyte endothelial interactions. *J Clin Invest* 85: 1260-66.
8. Cushing, S.D., J.A. Berliner et al. 1990. Minimally modified low density lipoprotein induces monocyte chemotactic protein 1 in human endothelial cells and smooth muscle cells. *Proc Natl Acad Sci USA* 87: 5134-8.
9. Sparrow, C.P., S. Parthasarathy, and D. Steinberg. 1988. Enzymatic modification of low density lipoprotein by purified lipoxygenase plus phospholipase A2 mimics cell-mediated oxidative modification. *J Lipid Res* 29: 745-53.
10. Rajavasisth, T.B., A. Andalibi et al. 1990. Induction of endothelial cell expression of granulocyte and macrophage colony-stimulating factors by modified low-density lipoproteins. *Nature* 344: 254-7.
11. Navab, M., S.S. Imes et al. 1991. Monocyte transmigration induced by modification of low density lipoprotein in cocultures of human aortic wall cells is due to induction of monocyte chemotactic protein 1 synthesis and is abolished by high density lipoprotein. *J Clin Inves* 88: 2039-46.
12. Butcher, E.C. 1991. Leukocyte-endothelial cell recognition: three (or more) steps to specificity and diversity. *Cell* 67: 1033-36.

THE IMMUNOPATHOLOGY OF CHEMOTACTIC CYTOKINES

Robert M. Strieter[1] and Steven L. Kunkel[2]

Departments of [1]Internal Medicine and [2]Pathology
The University of Michigan Medical School
Ann Arbor
Michigan 48109-0602, USA

INTRODUCTION

One of the hallmarks of acute and chronic inflammation is the accumulation of blood-born leukocytes to inflamed tissue. The recruitment of specific leukocyte populations to these areas is one of the most fundamental of all mechanisms associated with an inflammatory response. In order for this process to successfully deliver leukocytes to a site of inflammation, a number of specific steps must occur in an absolute faithful manner. This cascade of events is initiated with the binding of the leukocyte to an activated or "inflamed" endothelium via induced expression of various adhesion molecules on both endothelial cells and leukocytes. Interestingly, this initial process usually takes place in the small venules under a large shear stress. Once the leukocyte populations bind to the endothelium, these cells migrate from the vascular space, through the basement membrane matrix, to the area of interstitial inflammation. Recent studies have dramatically increased our knowledge concerning leukocyte extravasation. This is especially true of the research directed at understanding the role of leukocyte-endothelial cell interactions and adhesion molecule expression. In addition, recent advances in understanding leukocyte chemotaxis have revealed a large supergene family of chemotactic polypeptides (1-5). Information gained from studying these novel chemotactic cytokines have generated exciting new concepts regarding the inflammatory response.

The supergene family of chemotactic cytokines is comprised of two major groups of polypeptides which are identified by the position of 4 cysteine amino acid residues that form 2 internal disulfide bonds (Table 1). In one division of this polypeptide family (C-C chemotactic cytokines) 2 of the 4 cysteine amino acid residues are found in juxtaposition to each other and is exemplified by the protein monocyte chemotactic protein-1 (MCP-1). The other family division (C-X-C chemotactic cytokines) possesses 2 of the 4 cysteines separated by 1 different amino acid. This group is structurally typified by interleukin-8 (IL-8). Much of the interest in these chemotactic polypeptides arises from the observations that theses molecules possess a degree of specificity for the movement of a given leukocyte population. For example, it appears that members of the C-C family have specificity for the elicitation of mononuclear cells, either monocytes or lymphocytes, and members of the C-X-C family possess some degree of specificity for primarily the movement of neutrophils. While research in this area has generated basic information and set forth new mechanisms which may explain how leukocytes are attracted to an area of inflammation, few studies have identified the expression of these chemotactic cytokines in association with specific disease states.

Table 1. Supergene family of chemotactic cytokines and the general specificity for leukocyte populations.

C-X-C (Neutrophil chemotaxis)		C-C (monnuclear cell chemotaxis)	
Name	Cell Source	Name	Cell Source
IL-8	Variety of Cells	MCP-1	Variety of Cells
ENA-78	Non-immune cells	Rantes	Variety of Cells
GRO/MGSA	Variety of Cells	MIP-1	Macrophages
NAP-2	Platelets		
MIP-2	Macrophages		

An interesting aspect regarding the expression of these chemotactic cytokines is that their production has been observed from a number of immune and non-immune cells. For example, IL-8 is a product of many cells (6-9). Historically, monocytes were the first cell type to be identified as a rich source of IL-8; however, it is now known that most nucleated cells can synthesize IL-8. Hepatocytes, endothelial cells, fibroblasts, epithelial cells, neutrophils, and a host of cell lines have been shown to produce IL-8. The wide cellular expression of IL-8 also appears to be true for other members of this supergene family, including MCP-1. Interleukin-1 (IL-1) and tumor necrosis factor (TNF) appear to be important cytokines which can induce the expression of either IL-8 or MCP-1, at nM concentrations. The expression of either IL-8 or MCP-1 by certain non-immune cells demonstrates a degree of stimulus specificity, in that lipopolysaccharide (LPS) does not serve as an effective stimulus, while both IL-1 or TNF stimulation are extremely efficacous. This paradigm is not true for mononuclear phagocytes or neutrophils, as either LPS, IL-1 or TNF are effective in causing the expression of IL-8. The apparent selectivity of LPS stimulation for mononuclear phagocytes may be important for the initiation of cytokine networking between tissue macrophages and surrounding structural cells. During inflammation, the macrophage serves as the first line of defense and can respond to a LPS challenge and rapidly secrete both IL-1 and TNF. These two early response mediators can in turn stimulate the stromal and parenchymal cells, which comprise the involved tissue, resulting in the synthesis of distal cytokines. Thus, stromal and parenchymal cells can serve as effector cells during the inflammatory response by expressing chemotactic cytokines; all under the influence of activated fixed tissue macrophages. The information contained in Table 2 demonstrates that the expression of IL-8 and MCP-1 is not limited to only mononuclear cells and the production is stimulus specific.

Much of the above information provides circumstantial evidence that chemotactic cytokines are likely to be important in various inflammatory responses; however, few investigations have provided an actual causal relationship between the production of chemotactic cytokines and specific diseases. In this chapter, we provide evidence that demonstrates the expression of specific chemotactic cytokines in association with human disease.

Table 2. Interleukin-8 and monocyte chemotactic protein-1 are products of many cells and are generated in response to a number of stimuli.

Cell Type	Interleukin-8				Monocyte Chemotactic Protein-1			
	LPS	IL-1	TNF	IL-6	LPS	IL-1	TNF	IL-6
Macrophages	+	+	+	--	+/-	+/-	+/-	+/-
Fibroblasts	--	+	+	--	--	+	+	+
Epithelial Cells	--	+	+	--	--	+	+	+
Endothelial Cells	+	+	+	--	+	+	+	+
Neutrophils	+	+	+	--	--	--	--	--

ROLE OF CHEMOTACTIC CYTOKINES IN LUNG DISEASE

Neutrophil accumulation is often the hallmark of a number of disease states many of which are infectious in nature (10). This is particularly true of pulmonary infectious diseases. Histologically, it is not unusual for bacteria pneumonia to be characterized by the elicitation of a nearly pure population of neutrophils. This observation is interesting from the standpoint of neutrophil chemotaxis, since many of the traditional chemotactic factors, such as C5a or fMLP, lack specificity for the movement of only neutrophils. Therefore, a factor with predominately chemotactic bioactivity for neutrophils must be generated during the infectious process. We have initiated studies to address the mechanism responsible for the elicitation of neutrophils into the airways of human patients. Studying a population of Cystic Fibrosis patients with Pseudomonas-induced airway inflammation, we have found significantly elevated levels of IL-8 in both the unconcentrated bronchoalveolar lavage (BAL) fluid, as well as in sputum specimens. The levels of IL-8 correlated statistically with the magnitude of pulmonary disease activity. Interestingly, the levels of IL-8 dropped precipitously in patients placed on an effective course of antibiotics. Immunohistologic assessment of the cells recovered by BAL demonstrated that alveolar macrophages were the major cellular source of antigenic IL-8 during active airway inflammation. Although alveolar macrophages appear to be the major cellular source of airway-derived IL-8, it must be remembered that epithelial cells and cells of the interstitium are capable of synthesizing IL-8 and may in addition be a likely source of IL-8 in the airway.

While pulmonary bacterial infection can definitely be characterized by the accumulation of neutrophils, other lung diseases of unknown etiology can also have stages characterized by the presence of neutrophils in the distal airspace. For example, idiopathic pulmonary fibrosis (IPF) is an immunologically-mediated lung disorder in which activated macrophages and neutrophils may play a pathogenic role. Clinically, an increase in neutrophils in both the bronchoalveolar lavage (BAL) fluid and lung tissue have often been demonstrated in IPF patients (11). In addition, a decline in BAL neutrophils typically occur amoung patients exhibiting favorable response to therapy. As IPF is an inflammatory disease characterized by the sequestration of activated neutrophils and mononuclear cells, we have examined the cellular pellets from BAL fluid of patients with IPF for the expression of steady-state levels of IL-8 mRNA and cell-associated protein. Approximately 50% of patients diagnosed with active IPF demonstrated significant levels of constitutive IL-8 mRNA expression in the BAL fluid pellet. Healthy nonsmoking volunteers were found not to express any detectable steady-state levels of IL-8 mRNA. Having first demonstrated that there is a positive correlation between IPF active disease and IL-8 mRNA expression, we next demonstrated a positive correlation (r=0.95) between levels of steady state IL-8 mRNA and the percentage of neutrophils found in the BAL fluid. In contrast to the relationship between IL-8 mRNA levels and percent BAL neutrophils, no such relationship was noted between constitutive IL-8 mRNA and percent of BAL lymphocytes or eosinophils. Further studies were directed at determining if increased IL-8 mRNA levels corresponded with an increase in IL-8 antigen found in the cells recovered by BAL. Immunohistolochemistry demonstrated a strong correlation between IL-8 antigen expression by BAL cells and patients with active IPF disease. No antigenic IL-8 was identified in BAL cells from normal nonsmoking nondisease volunteers. Interleukin-8 is not the only cytokine which has been found in association with IPF. Previous studies have found elevated levels of transforming growth factors (alpha and beta), platelet-derived growth factor, IL-1, and TNF in the BAL fluid of patients with IPF. While the cellular events that result in the alveolitis of IPF remains unknown, several soluble mediators may play an important role. Circulating immune complexes have been demonstrated in the serum and BAL of patients with IPF, and these complexes were shown to stimulate the generation of neutrophil chemotactic activity from alveolar macrophages. Preliminary studies from our laboratory suggest the immune complexes can serve as a primary stimulus for the generation of IL-8 mRNA and antigen from mononuclear phagocytic cells. Also, we have found that fibronectin and collagen, proteins present in increased levels in IPF, can stimulate the expression of steady-state levels of IL-8 mRNA from both monocytes and alveolar macrophages. Furthermore, an increase in IL-1 and TNF by resident macrophages from IPF patients may function in an autocrine manner to induce the expression of IL-8 in a cytokine network fashion. Although IL-8 has previously been shown to be chemotactic for lymphocytes in vitro, we observed no correlation between IL-8 mRNA levels and lymphocytic alveolitis. The above investigations suggest that IL-8 may be an important mediator of neutrophil recruitment during the cellular phase of IPF. Further investigations assessing the expression of IL-8 mRNA and antigenic IL-8 protein must be conducted in order to determine the utility of IL-8 as a clinical indicator of IPF disease activity and potential response to therapy.

CHEMOTACTIC CYTOKINES (IL-8 AND MCP-1) IN RHEUMATOID ARTHRITIS

An additional disease state, in which neutrophils have been shown to play a pathologic role, is rheumatoid arthritis; however, the mechanism for neutrophil recruitment is not entirely clear (12). To determine whether IL-8 was present in the synovial fluids from patients with rheumatoid arthritis, synovial fluid specimens from these patients were assessed and compared to samples drawn from osteoarthritis patients. Using a sensitive ELISA for measuring IL-8 protein, we observed that patients with rheumatoid arthritis had the highest levels of IL-8 (14.37±5.8 ng/ml). These levels were significantly greater than that found in the synovial fluid of osteoarthritis patients (0.13±0.017 ng/ml). Additional studies determined that peripheral blood from patients with rheumatoid arthritis contained elevated levels of IL-8, as compared to normal controls (13). Serum IL-8 levels in the rheumatoid arthritis patients were found to have a mean of 8.44 ng/ml, with a range of 0.05 to 31.31 ng/ml. Linear regression analysis of the serum IL-8 and synovial fluid IL-8 levels demonstrated a strong positive correlation (r=0.93). Further analysis of the neutrophil chemotactic bioactivity demonstrated that the majority of the bioactivity could be neutralized with antibodies to human IL-8.

To explore the potential role of nonimmune cellular constituents of the synovium as a potential source of IL-8, isolated synovial fibroblasts were assessed for IL-8 production. As shown in Figure 1, isolated synovial fibroblasts stimulated with different agents demonstrated the production of IL-8 in a time- and dose-dependent manner. Upon stimulation with either IL-1, TNF, or LPS, IL-8 levels were first detectable at 4 hours and steadily increased over the next 20 hour time period. By 24 hours post stimulation, the levels of IL-8 produced by IL-1, TNF, or LPS triggered synovial fibroblasts were 9.6±2.2 ng/ml, 7.6±2.6 ng/ml, and 5.1±1.7 ng/ml, respectively. IL-8 levels from unstimulated fibroblasts were low throughout these studies and was maximal at 24 hours (0.3±0.2 ng/ml). The production of IL-8 by synovial fibroblasts was also dose-dependent.

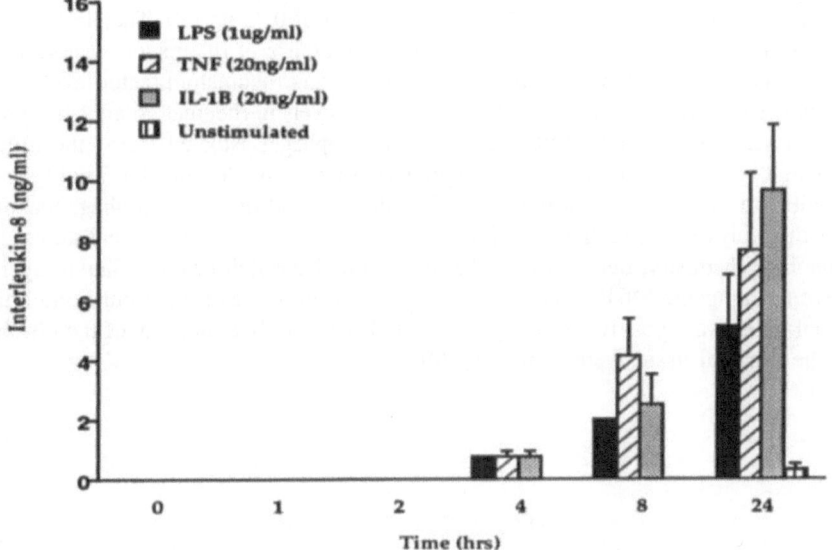

Figure 1. Time-dependent generation of IL-8 by synovial fibroblasts in response to LPS, TNF, and IL-1B.

Stimulatory concentrations of 100 ng/ml of LPS and 1 ng/ml of TNF were found to be half-maximal concentrations on the dose-response curve. Maximum stimulatory concentrations of LPS and TNF were 1 ug/ml and 20 ng/ml, respectively. Since synovial fibroblasts were shown to synthesis IL-8 we next assessed synovial tissue macrophages for the production of IL-8. Previous studies from our laboratory had identified that fixed tissue macrophages such as alveolar macrophages and Kupffer cells, are significant contributors to the pool of IL-8. Thus, the synovial macrophage, which is found in the invasive pannus of rheumatoid synovial tissue, were studied. Macrophages isolated from synovial tissue from patients with active disease were treated with either IL-1, TNF, or LPS and assessed for the production of antigenic IL-8. Under conditions of stimulation or no stimulation, the synovial macrophages produced similar levels of IL-8 (Figure 2). Therefore, constitutive levels of IL-8 released by the synovial macrophages could not be enhanced by the addition of a stimulus known to increase the production of IL-8. These data are in support of the theory that macrophages isolated from an active inflammatory lesion are already exhibiting signs of in vivo stimulation.

While neutrophils are clearly an important inflammatory cell found in the synovial joint fluid of patients with rheumatoid arthritis, the mononuclear phagocyte is an important and characteristic inflammatory cell found in the synovial tissue and developing pannus.

Blood-born monocytes, which are elicited to the inflammatory synovial tissue, become activated and mediate rheumatoid joint destruction mainly due to their ability to present antigen, and release reactive oxygen metabolites, proteolytic enzymes, and a variety of cytokines (14-16). These latter polypeptide mediators serve as communication links and orchestrate the evolving inflammatory response. The synovial tissue macrophage is truly an important cellular component of the developing lesion, however, little is known regarding the mechanism of elicitation of its precusor cell, monocytes. Recent investigations from our laboratory have identified that MCP-1 is a likely candidate responsible for the recruitment of monocytes into the inflamed synovium. Antigenic MCP-1 was found in both the synovial fluid (mean of 25.5 ng/ml) and the serum (mean of 22.4 ng/ml) of patients with rheumatoid arthritis. Patients with osteoarthritis had significantly less MCP-1 in the synovial fluid (mean of 0.92 ng/ml) and serum (mean of 0.72 ng/ml). The elevated levels of MCP-1 present in the serum was not due to the presence of rheumatoid factor, as there was no significant correlation between MCP-1 levels and rheumatoid factor levels. There was a positive correlation between MCP-1 and IL-8 levels in rheumatoid arthritic synovial fliud and sera. Both synovial fibroblasts and macrophages isolated from the inflamed rheumatoid synovial tissue were found to produce MCP-1. While synovial fibroblasts were susceptible to stimulation with IL-1 and TNF, the synovial tissue macrophage was found to constitutively express high levels of MCP-1 mRNA and protein. Additional studies using immunohistochemistry, demonstrated that the macrophage rich synovial lining layer was positive for antigenic MCP-1. The majority of the macrophages from rheumatoid arthritis synovial tissue were positive for antigenic MCP-1, while only a minority of the fibroblasts from the synovial tissue were expressing MCP-1.

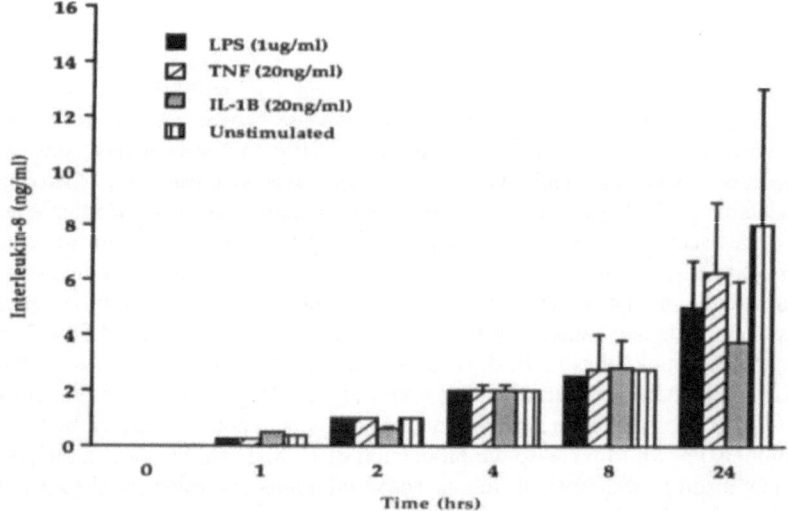

Figure 2. Time-dependent generation of IL-8 by synovial macrophages in response to LPS, TNF, and IL-1B.

The initiation and maintenance of synovial inflammation results from the influx of various inflammatory cells in response to specific mediators. In the above studies we identified that synovial fluids and sera contain chemotactic factors for monocytes (MCP-1), as well as lymphocytes and neutrophils (IL-8). Actually, a strong correlation was found between MCP-1 and IL-8 levels in both the sera and synovial fluid of patients with rheumatoid arthritis. This information suggests that signals which may upregulated the expression of MCP-1 may also influence IL-8 in the context of the same disease. Indeed, IL-1 and TNF are potent stimuli for the production of fibroblast-derived IL-8 and MCP-1. Since IL-1 and TNF are produced in abundance by tissue macrophages, it is not unreasonable to speculate that cytokine networking is occurring between mononuclear phagocytes and cells which comprise the surrounding tissue. This complex cascade of cell-to-cell communication is ultimately responsible for events which both initiate the movement of cells from the peripherial blood to the inflamed area and allow a continued influx of leukocytes to the inflamed tissue.

INTERLEUKIN-8 IN ISCHEMIA/REPERFUSION INJURY

A striking observation which is routinely made upon histologic examination of certain pathologic disorders is the characteristic infiltrates of specific leukocyte populations. This is particularly true of ischemia/reperfusion injury, which possesses a significant neutrophil component to the overall pathology (17). Interestingly, the mechanism(s) that results in neutrophil recruitment to areas of ischemia/reperfusion injury have yet to be elucidated. Our laboratory has employed a cellular model to simulate ischemia/reperfusion by determining whether anoxia-hyperoxia induced stress upon monocytes could cause the production of IL-8. Initial studies demonstrated that anoxic pre-conditioning of monocytes would augment the production of IL-8 when challenged with a subsequent hyperoxic event. Monocytes cultured in the absence of oxygen for 6 hours and then incubated in either normoxic (room air) or hyperoxic (95% oxygen) conditions demonstrated a significant increase in mean levels of IL-8, 11.4 ng/ml and 26.3 ng/ml, respectively, as compared to monocytes cultured continuously in normoxia (5.9 ng/ml). Northern blot analysis of steady-state levels of IL-8 mRNA isolated from monocytes cultured under conditions of anoxia-hyperoxia, demonstrated an increase in IL-8 gene expression. A molecular assessment of this observation demonsrtated that the increase in oxidative stress was inducing nuclear (gene) transcription of IL-8, as determined by nuclear transcriptional analysis. These latter studies are important since they identify that the signal responsible for the gene expression of IL-8 is likely tied to the generation of oxygen radical metabolites. In addition, the promoter region of the IL-8 gene is apparently under the influence of NF kappa-beta, as this cis-acting nuclear binding protein has been implicated in oxygen radical-dependent polypeptide expression.

Since ischemia/reperfusion injury may be associated with conditions favoring the presence of Gram negative bacteria or LPS, we initiated studies to assess the effects of an oxygen stress in the presence of LPS. Monocyte cultures were stimulated with 100 ng/ml of LPS at time 0 and incubated under anoxia/hyperoxic conditions. When human monocytes were challenged with LPS under normoxic conditions for 18 hours, a mean of 38.4 ng/ml of IL-8 was produced. However, when LPS-challenged monocytes were cultured under anoxic conditions for 6 hours followed by 12 hours of hyperoxia, a 300% increase in IL-8 production was found. Conditions of anoxia/hyperoxia in the presence of LPS resulted in the production of a mean of 95.7 ng/ml of IL-8. To determine whether the above LPS study was due to elevated steady-state IL-8 mRNA transcripts, we performed Northern blot analyses of mRNA isolated from monocytes cultured in the presence of LPS under conditions of anoxia/hyperoxia. A time-dependent increase in the expression of IL-8 mRNA was observed in both the LPS normoxia cultured group and the LPS anoxia-hyperoxia group. However, a significant elevation in IL-8 mRNA was found in the monocyte culture group which were incubated under conditions of augmented oxygen stress.

A number of early studies have identified a close association between oxygen toxicity/hyperoxia-induced tissue injury and the infiltration of neutrophils. Classical studies by Fox and coworkers identified the presence of enhanced chemotactic activity in the BAL fluid of experimental animals exposed to hyperoxia (18). Other investigators reported that hyperoxia could cause an augmentation in endotoxemia-mediated neutrophil alveolitis in an experimental animal model, and this increase in neutrophil numbers was associated with the presence of a neutrophil chemotaxin (19,20). Furthermore, neutrophil depletion studies in an experimental model of myocardial ischemia attenuates the extent of subsequent reperfusion injury. These studies demonstrate the key role of the neutrophil in anoxia/hyperoxia related tissue injury and IL-8 is a likely candidate neutrophil chemotaxin to serve in the recruitment of neutrophils to the area of reperfusion injury. Recent information from our laboratory has further increased the excitement in this area, as endothelial cells also appear to be susceptible to oxidative stress with a subsequent increase in the expression of IL-8. Given the apparent role that neutrophils play in ischemia/reperfusion related tissue injury, the evidence for hyperoxic induction of neutrophil chemotactic factors, and the demonstration of anoxia/hyperoxia-induced IL-8, we hypothesize that a variety of cells exposed to an oxygen stress are candidates for expressing IL-8. The importance of a variety of cells contributing to the in vivo pool of IL-8 is important, such that non-immune cells can perform effector cell activity and aid in driving the inflammatory response.

CONCLUSION

The expression in vivo of leukocyte specific chemotactic factors and the subsequent recruitment of these cells by chemotaxins are essential elements of an inflammatory reaction in response to a number of stimuli. For example, physical trauma, infection, cancer, and shock are different states of inflammation which may possess a common leukocyte histology. The actual type of leukocyte population that will predominate at the lesion depends upon the nature of the stimulus and the resulting spectrum of chemotactic factors. While many antigen specific stimuli lead to the recruitment of lymphocytes and monocytes, acute inflammation may result in the elicitation of only neutrophils. Thus, the phenomenon of leukocyte chemotaxis is dependent upon a redundancy of chemotaxins, involving a host of different leukocyte chemotactic factors. This concept addresses the importance of a successful chemotactic response to host defense.

Our understanding of the leukocyte recruitment has been greatly enhanced in the last few years with the discovery of the chemotactic cytokine supergene family. These polypeptide mediators of inflammation exert an important role during the initiation and maintenance of an inflammatory lesion, as they are likely responsible in both the induction of adhesion molecules resulting in the promotion of leukocyte-endothelial cell interaction and the subsequent extravascular extravasation. With regard to this latter mechanism, there is some evidence that the migration response through the interstitium may be governed by haptotaxis. This directed migration in response to an insoluble gradient may actually be the final mechanism that allows specific leukocyte populations to pin-point the area of tissue injury.

Our data provides evidence that different members of the family of chemotactic cytokines are important in human diseases. Our initial studies have targeted the lung, as this organ can be can be easily studied by BAL fluid and recovered cells. Using a cellular and molecular approach we have identified IL-8 as a major mediator in both pulmonary airway inflammation (Cystic Fibrosis patients) and IPF patients. In many of these studies the levels of IL-8 correlated with the activity of the disease. Interestingly, the levels of IL-8 were highly elevated in Pseudomonas airway inflammation, IL-8 could be identified in both unconcentrated BAL fluid and unconcentrated sputum. Both IL-8 and MCP-1 were found in elevated levels in synovial fliud and sera of patients with rheumatoid arthritis. This disease has historically been found to possess a mixture of leukocyte populations in the synovial fluid or the invading pannus. We found a high correlation between the presence of IL-8 and MCP-1. In tissue specimens taken from rheumatoid arthritis patients, immunohistochemistry routinely found the tissue macrophage expressing antigen for these chemotactic cytokines. For years the treatment of choice for this and other joint disease has been the use of nonsteroidial antiinflammatory medications, which have never been demonstrated to possess remittive efficacy. The increase in our understanding of the mechanism(s) responsible for initiating and maintaining these enigmatic disease states may lead to more reliable therapeutic intervention. Targeting the chemotactic cytokines as a therapeutic approach to limit disease could prove to be an innovative and useful pharmacologic direction.

ACKNOWLEDGEMENTS

This work was supported in part by National Institutes of Health grants 1P50HL46487, HL02401, HL31693, HL35276, and DK38149. Dr. Strieter is an RJR Nabisco Research Scholar.

REFERENCES

1. Peveri, P., A. Walz, B. DeWald, and M. Baggiolini. 1988. A novel neutrophil-activating factor produced by human mononuclear phagocytes. *J. Exp. Med.* 167: 1547-1559.
2. Matsushima, K., T. Morishita, T. Yoshimura, S. Lavu Y. Obayashi, W. Lew, E. Appella, E.J. Leonard, and J.J. Oppenheim. 1988. Molecular cloning of human monocyte-derived neutrophil chemotactic factor (MDNCF) and induction of MDNCF by interleukin-1 and tumor necrosis factor. *J. Exp. Med.* 167: 1883-1893.
3. Baggiolini, M., A. Walz, and S.L. Kunkel. 1989. Neutrophil-activating peptide-1/interleukin-8, a novel cytokine that activates neutrophils. *J. Clin. Invest.* 84: 1045-1049.
4. Yoshimura, T., E.A. Robinson, S. Tanaka, E. Appella, and E.J. Leonard. 1989. Purification and amino acid analysis of two human glioma-derived cytokines. *J. Exp. Med.* 169: 1449-1459.
5. Matsushima, K., C.G. Larson, G.C. DuBois, and J.J. Oppenheim. 1989. Purification and characterization of a novel monocyte chemotactic and activating factor produced by a human myelomonocyte cell line. *J. Exp. Med.* 169; 1484-1490.
6. Standiford, T.J., S.L. Kunkel, M.A. Basha, S.W. Chensue, J.P. Lynch, G.B. Toews, and R.M. Strieter. 1990. Interleukin-8 gene expression by a pulmonary epithelial cell line: a model for cytokine networks in the lung. *J. Clin. Invest.* 86: 1945-1953.
7. Strieter, R.M., S.H. Phan, H.J. Showell, J.P. Lynch and S.L. Kunkel. 1989. Monokine-induced neutrophil chemotactic factor gene expression in human fibroblasts. *J. Biol. Chem.* 264: 10621-10626.
8. Strieter, R.M., S.L. Kunkel, H.J. Showell, S.H. Phan, P.A. Ward, and R.M. Marks. 1989. Endothelial cell gene expression of a neutrophil chemotactic factor by TNF, LPS, and IL-l. *Science* 243: 1467-1469.
9. Thornton, A.J., R.M. Strieter, I. Lindley, M. Baggiolini, and S.L. Kunkel. 1989. Cytokin-induced gene expression of a neutrophil chemotactic factor/IL-8 in human hepatocytes. *J. Immunol.* 144: 2609-2613.
10. Weiss, S.J. 1989. Tissue destruction by neutrophils. *N. Eng. J. Med.* 320: 365-376.
11. Hunninghake, G.W., O. Kawanami, V.J. Ferrans, W.C. Roberts, and R.G. Crystal. 1981. Characterization of the inflammatory and immune effector cells in the lung parenchyma of patients with interstitial lung disease. *Am. Rev. Respir. Dis.* 123: 407-421.
12. Harris, E.D. 1990. Rheumatoid arthritis: pathophysiology and implications for therapy. *N. Eng. J. Med.* 322: 1277-1280.
13. Koch, A.E., S.L. Kunkel, J.C. Burrows, H.L. Evanoff, G.K. Haines, R.M. Pope, and R.M. Strieter. 1991. Synovial tissue macrophages as a source of the chemotactic cytokine IL-8. *J. Immunol.* 147: 2187-2195.

14. Koch, A.E., J.C. Burrows, A. Skoutelis, R. Marder, P. Domer, and S.J. Leibovich. 1991. Monoclonal antibodies detecting monocyte/macrophage activation and differentiation antigens and identifying functionally distinct subpopulatoins of human rheumatoid synovial tissue macrophages. *Am. J. Pathol.* 138: 165-173.

15. Arend, W.P. and J.M. Dayer. 1990. Cytokines and cytokine inhibitors or antagonists in rheumatoid arthritis. *Arthritis Rheum.* 33: 305-313.

16. Husby, G. and R.C. Williams. 1988. Synovial localization of tumor necrosis factor in patients with rheumatoid arthritis. *J. Autoimmun.* 1: 363-371.

17. Linas, S.L., P.F. Shanley, D. Whittenburg, E. Berger, and J.E. Repine. 1988. Neutrophils accentuate ischemia-reperfusion injury in isolated perfused rat kidneys. *Am. J. Physiol.* 255: 728-735.

18. Fox, R.B., J.R. Hoidal, D.M. Brown, and J.E. Repine. 1981. Pulmonary inflammation due to oxygen toxicity: involvement of chemotactic factors and polymorphonuclear leukocytes. *Am. Rev. Respir. Dis.* 123: 521-523.

19. Rinaldo, J.E., J.H. Dauber, J. Christman, and R.M. Rogers. 1984. Neutrophil alveolitis following endotoxemia. *Am. Rev. Respir. Dis.* 130: 1065-1071.

20. Christman, J.W., J.E. Rinaldo, J.E. Henson, S.A. Moore, and J.H. Dauber. 1985. Modification by hyperoxia in vivo of endotoxin-induced neutrophil alveolitis in rats. *Am. Rev. Respir. Dis.* 132: 152-158.

RECEPTOR/LIGAND INTERACTIONS IN THE C-C CHEMOKINE FAMILY

Thomas J. Schall, John Y. Mak, David DiGregorio, Kuldeep Neote

Department of Immunology
Genentech, Inc.
460 Point San Bruno Blvd.
South San Francisco, CA 94080
USA

INTRODUCTION

The past few years have been particularly enriching for investigators interested in basic immunoregulatory and inflammatory processes. Receptors for such central regulators as IL-1, TNF, and IFN-g have been cloned by molecular techniques, and progress is being made in the understanding of their signal transduction mechanisms. In addition, recent immunological history has witnessed the appreciation of entirely new areas in immunobiology, some of which are beginning to fundamentally impact on our understanding of basic inflammatory processes. Nowhere is this better exemplified than in the area of the recently discovered superfamily of immune cytokines which we have come to call the *chemokines*.

The chemokines are intriguing in many regards, not the least of which is their ability to effect different actions on unique subsets of immune effector cells. While the chemokines, like all immune cytokines, do exhibit some pleiotropy and redundancy, they also seem to divide their duties in immune regulation along definable boundaries. For example, studies have shown that generally C-X-C chemokines act on neutrophils, and C-C chemokines act on monocytes (1-9). Indeed all of the C-C chemokines tested apparently have some effect on monocytes, exhibiting what might appear to be an unnecessary level of redundancy. In point of fact, the specific functions of the chemokines may be more subtle. As discussed below, although overlap does exist, the *in vivo* functions of the chemokines may be much more specific. One central question then of chemokine biology is how these differing and common activities are mediated at the level of the cell surface via receptor/ligand interactions.

This report attempts to examine these interactions with particular emphasis on the C-C branch of the chemokine superfamily. We have undertaken to understand chemokine receptor/ligand interactions by examining four areas: 1) defining the spectrum of target cells acted upon by C-C chemokines; 2) understanding early signal transduction events in those target cells; 3) understanding the binding interaction of the protein ligands with their receptors, and 4) by cloning the various receptors using molecular techniques. Some of the results of these four approaches are presented here, and, taken together, they indicate that the C-C chemokines act via a set of complex interactions with a variety of shared and specific receptors on a diverse group of target cells.

PRIMARY TARGETS FOR C-C CHEMOKINES

As determined mostly through *in vitro* studies, C-C chemokines as a class have the capacity to act on 4 of the 5 major classes of blood leukocytes: monocytes, lymphocytes, basophils and eosinophils. Most of the activities described are chemotactic. For example, RANTES is a chemoattractant for memory T cells *in vitro* (9) and we have shown that HuMIP-1α and HuMIP-1β are chemoattractant for distinct subpopulations of lymphocytes including naive T cells and B cells. MCP-1 and RANTES are potent direct mediators of the release of histamine by human basophils (10-12), while RANTES, and to a lesser extent HuMIP-1a, are chemoattractants and activators of eosinophils (13, 14). These properties suggest that the C-C chemokines are an important link between monocytes, lymphocytes, basophils, and eosinophils during immune and inflammatory processes. The regulatory signals imparted by C-C chemokines can also be negative: MIP-1α has been reported to be a unique and specific inhibitor of the proliferation of hematopoietic stem cells (15, 16).

While these activities may seem to be somewhat confusing at first glance, it is likely that the chemokines serve specialized functions *in vivo*. For example, as shown schematically in figure 1, MCP-1 is widely expressed by many cells and tissues, but has only two known targets: monocytes and basophils (6, 7, 10, 12). MIP-1α, however, has a much more restricted pattern of cellular production. Although it too will act on monocytes, (and possibly basophils) it clearly will act on a larger range of targets: eosinophils, lymphocytes, hematopoietic stem cells, and perhaps even the hypothalamus in fever induction (14-17). RANTES is expressed by an overlapping but distinct set of cells and tissues as compared to the other chemokines, and it too has its own specific array of targets: eosinophils and memory T cells for example (9, 13, 14). Moreover, these molecules are regulated temporally as well as spatially: MCP-1 and MIP-1α are apparently not expressed with the same kinetics as RANTES. Lastly, the relative potency of a given chemokine for a common target may vary widely. MCP-1 is a more potent attractant of monocytes *in vitro* than is RANTES, and RANTES attracts eosinophils much better than does MIP-1α (14). Thus, it is possible that the actions of the chemokines can "fine tune" an immune response to a challenge via three simple mechanisms: spatial regulation, time of production, and differences in activities for different target cells.

Ca^{++} MOBILIZATION IN TARGET CELLS IN RESPONSE TO C-C CHEMOKINES

The measurement of the mobilization of intracellular Ca^{++} can provide a straightforward indication of early signal transduction in leukocytes. In addition, it has been shown to have a predictive value in ascertaining whether various ligands signal through the same or different cell-surface receptors. For example, rapid, successive exposure to the same ligand is known to desensitize the signalling capacity of G-protein linked receptors (18). Desensitization can also occur when the two different agonists signal through the same receptor. We therefore used Ca^{++} flux assays to address whether the C-C chemokines were using the same or distinct receptors on panels of target cells. DMSO-differentiated U937 cells loaded with the Ca^{++} indicator Indo-1-AM transiently mobilize Ca^{++} in response to E-coli derived recombinant RANTES, HuMIP-1α, HuMIP-1β, and MCP-1 chemokines (used at a final concentration of 100 nM), thus showing that these materials were biologically active (data not shown). In order to determine their activities on a more physiologically relevant non-transformed cell type, we isolated monocytes from whole blood of normal donors. These monocytes were purified from other leukocytes by adhering the monocytes to culture dishes for 2-3 days, during which time the non-adhering lymphocytes were removed. The "cultured" monocytes, presumably mirroring an *in vivo* differentiated state, de-adhere after 48-72 hours in culture and are assayed immediately thereafter.

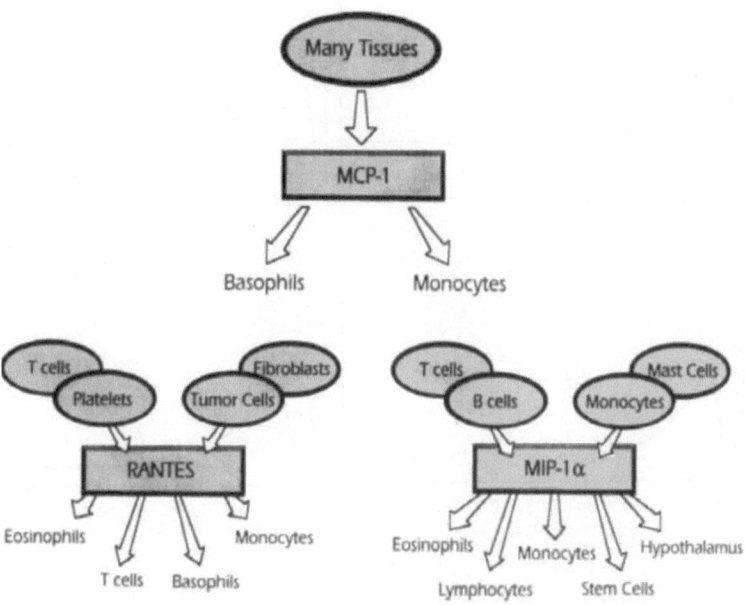

Figure 1. Schematic representation depicting the differences in cells and tissues which are capable of producing C-C chemokines (ovals) as well as their distinct sets of target cells and tissues (bottom arrows). C-C chemokines clearly exhibit overlapping but distinct spatial regulations as well as target cell specificities.

The use of these cells reveals an unusual pattern of Ca^{++} signalling desensitization in chemokine-challenged cells. HuMIP-1α clearly blocks the ability of the cultured monocytes to transmit a second Ca^{++} signal when HuMIP-1α is added to Indo-1-AM loaded cells twice in succession (not shown). Similarly, RANTES blocks the response to a second challenge at the same concentration (not shown). When HuMIP-1α is added first, it blocks the response to RANTES, indicating that complete desensitization has occurred (Fig 2). However, challenging the cells first RANTES did not prevent a subsequent Ca^{++} flux by the addition of HuMIP-1α (Fig 2), indicating that receptor desensitization had not occurred. Interestingly, this pattern of desensitization is dependent on what type of target cell is being challenged with the chemokines. If either basophils or eosinophils (14) are challenged with the identical combination of chemokines, the pattern of desensitization is exactly the opposite of that seen in cultured monocytes. In these cells RANTES clearly causes signalling desensitization to subsequent HuMIP-1α challenge, and not vice versa (Fig 2).

These phenomena can be explicated in terms of a two receptor model (Fig 3). In monocytes, for example, one can envision a receptor shared by both RANTES and HuMIP-1α as well as a HuMIP-1α-specific receptor. If RANTES is presented first to the cells, they flux Ca⁺⁺ via the shared receptor. Subsequent exposure to HuMIP-1α results in

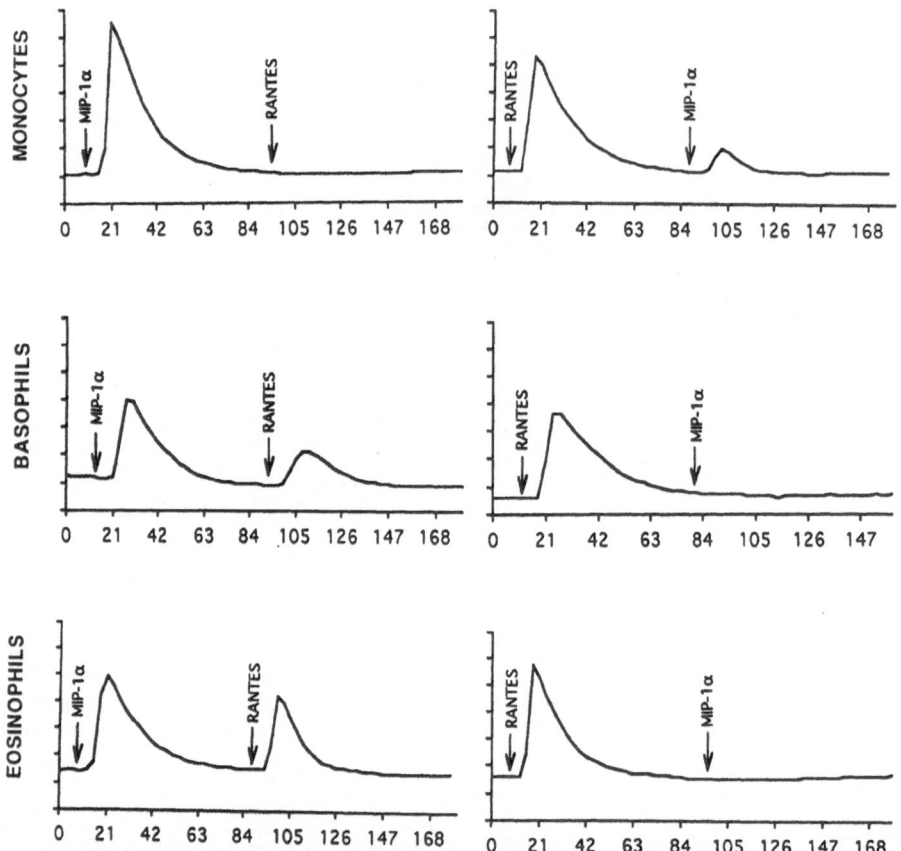

Figure 2. Measurements of intracellular calcium levels in human cultured monocytes, blood basophils, and blood eosinophils in response to HuMIP-1α and RANTES. Measurements were done on a spectrofluoremeter using cells loaded with a Ca⁺⁺ indicator. Peaks represent the increase in intracellular Ca⁺⁺ content after the addition of the agonist (time of addition denoted by the arrows), as measured by an increase in the relative fluorescence intensity (Y-axis). The X-axis depicts time in seconds that the assay was run.

a second Ca⁺⁺ flux via the specific receptor. To reverse the order of ligand challenge, however, results in only a single flux as HuMIP-1α occupies all of the available binding sites. This model can be extended to basophils and eosinophils by postulating the existence of a RANTES-specific receptor rather than a HuMIP-1α-specific receptor in these cells.

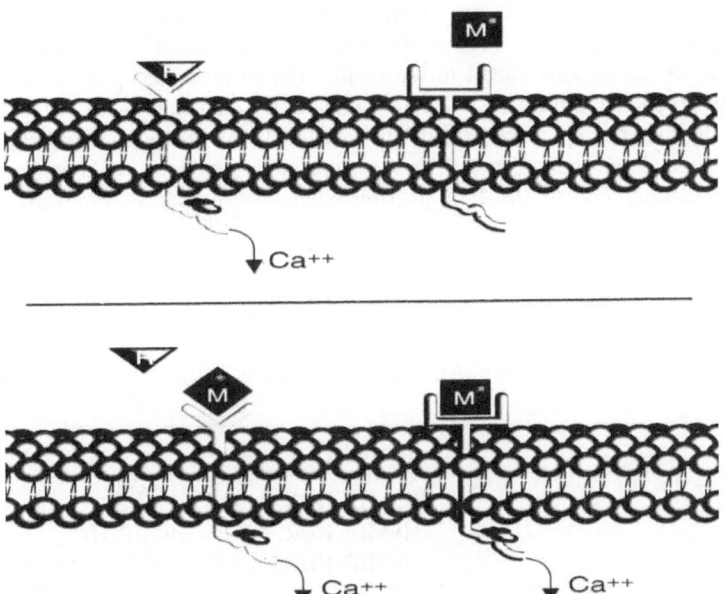

Figure 3. Two receptor model for RANTES (R) and HuMIP-1α (M) on human cultured monocytes. A shared receptor for R and M, and a specific receptor for M would explain the Ca^{++} flux data observed for monocytes in Fig. 2.

The picture becomes even more complex when four C-C chemokines are used in various combinations to challenge cultured monocytes. If HuMIP-1α is the first chemokine added to the cells, it will induce an appreciable transient Ca^{++} increase, but no other chemokine added after HuMIP-1α is capable of Ca^{++} mobilization in these cells (Fig. 4). This indicates that HuMIP-1α is desensitizing the cell to all other chemokines tested. By contrast, if MCP-1 is added first, followed by HuMIP-1β, then RANTES, then HuMIP-1α, four Ca^{++} fluxes are clearly induced (Fig. 4). In fact, depending solely on the order in which the same ligands are added to the cells, all combinations of 1, 2, 3 and 4 fluxes are observed.

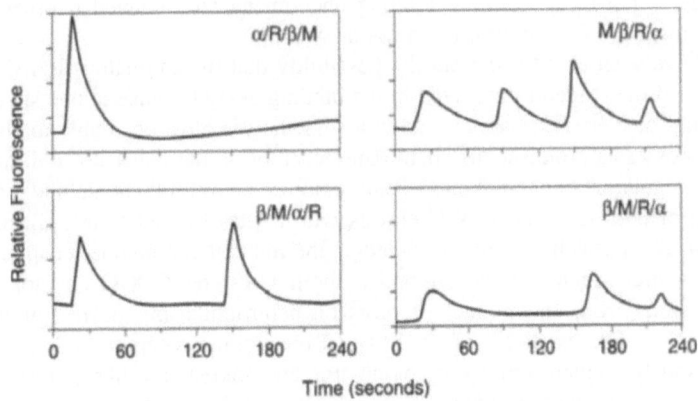

Figure 4. Representation of intracellular Ca^{++} fluxes in human cultured monocytes in response to 4 chemokine challenges. Experiments were done as in Fig. 2, with a=HuMIP-1α; R=RANTES; b=HuMIP-1β; M=MCP-1. Agonists (100 nM each) were added in the order indicated, with the first added at approximately 15 seconds and subsequent challenges at intervals of approximately 70 seconds.

Again this asymmetrical pattern of signalling desensitization suggests an array of shared and specific receptors on these cells which are differentially engaged dependent upon the order of the agonist added to the cells. The array of receptors which would explain the observed calcium fluxes in the four chemokine challenge experiments is given in Table I. In this model, since HuMIP-1α is capable of interacting with all of the receptors, it alone desensitizes the cell to any subsequent challenge. Conversely, since MCP-1 binds to only one of the four possible receptors, it induces desensitization to nothing except itself.

Table 1. Postulated array of chemokine receptor types on cultured human monocytes.

Specific	Shared
HuMIP-1α	HuMIP-1α/RANTES
	HuMIP-1α/RANTES/HuMIP-1β
	HuMIP-1α/RANTES/HuMIP-1β/MCP-1

ANALYSIS OF CHEMOKINE BINDING TO TARGET CELLS

In order to further investigate the interaction of the C-C chemokines and their target cells, direct binding experiments with ^{125}I-labelled ligands were attempted. Cultured monocytes were incubated with 0.5 nM ^{125}I-labelled HuMIP-1α or RANTES and increasing concentrations of the same unlabelled chemokine. Interestingly, analysis of the direct binding of RANTES to the cells could not be accomplished. As the amount of unlabelled RANTES was increased in the binding assay, a concomitant increase of ^{125}I-RANTES bound to cells was observed, as shown schematically in Figure 5. Displaceable binding of ^{125}I-HuMIP-1α to the cells could be observed, however (Fig 5). Scatchard analysis of HuMIP-1α binding to cultured monocytes showed a dissociation constant (K_d) of about 5 nM and ~50,000 sites/cell. The reason for the unusual binding profile observed for ^{125}I-RANTES is not clear. Similar binding phenomena are obtained if other target cells responding to RANTES are used (data not shown).

We have been able to rule out the possibility that the anomalous RANTES binding is an artifact of the reagents employed in the binding assay because in one special case the direct binding of RANTES can be readily assessed. We have been able to show that red blood cells possess a promiscuous chemokine receptor on their surface, and may therefore be acting as regulators of chemokines in circulation. The red cell chemokine receptor binds chemokines of both C-C and C-X-C classes with a Kd of about 5 nM, and is present at about 5,000 sites per cell. To our knowledge, the red cell chemokine receptor is the only binding structure which will accommodate both C-C and C-X-C chemokines, and is therefore distinct from chemokine receptors on neutrophils, monocytes, or lymphocytes (1-4, 6, 19-22). The existence of the red cell chemokine receptor further suggests that unlike previously studied immune cytokines the chemokines are likely to interact with a complex array of receptors *in vivo*.

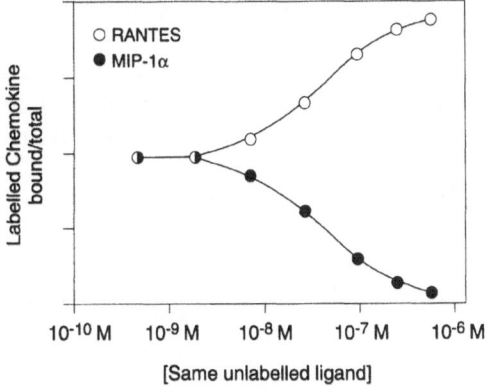

Figure 5. Schematic representation of competition binding curves for RANTES and HuMIP-1b on cultured human monocytes. RANTES exhibits anomalous enhanced binding of radiolabelled ligand in the presence of an increasing concentration of unlabelled competitor. HuMIP-1a displays a more usual competition profile.

CHEMOKINE RECEPTOR CLONING STRATEGIES

The appreciation of three separate factors has led to a strategy to attempt to clone C-C chemokine receptors by molecular techniques. First, all chemoattractant receptors cloned to date which have a role in leukocyte motility: the receptor for C5a (23), the bacterial tripeptide fMLP (24), and the IL8 receptors A and B (20, 21); have been shown to belong to the superfamily of seven-transmembrane-spanning (7-TM) proteins. Second, we have observed that many of the effects of C-C chemokines are sensitive to pertussis toxin, and therefore the receptors for the C-C chemokines are also likely to be G-protein linked and share the seven transmembrane architecture. Third, there are domains of conserved amino acids in the known chemoattractant 7-TM receptors which are not generally conserved in other 7-TM molecules. Accordingly, two degenerate oligonucleotides corresponding to conserved amino acid sequences in two transmembrane regions of the IL8rA, the C5a and the fMLP receptors were synthesized: LNLA(L/V)AD(L/F)(L/G) from a region in transmembrane domain 2 (TM2), and NP(I/M)(I/L)Y(A/V)(F/V)(I/M/A)GQ from TM7. These were then used as primers in RT-PCR experiments using cDNA substrates from different hematopoietic cell types known to respond to C-C chemokines. Subcloning and sequencing of the PCR products revealed the presence of IL8rA, IL8rB, C5a receptor, as well as some novel clones having characteristics of seven-transmembrane-segment receptors and marked similarity to the two IL8 receptors. Partial functional characterization of this first set of orphan receptors revealed that they do not bind to C-C chemokines. However, it was noted that some of these clones, which were more related to the IL-8 receptors than to other seven-transmembrane-spanning receptors, possessed a new conserved amino acid motif which seemed to define a subfamily of IL8 receptor-related seven-transmembrane-spanning molecules that excluded the C5a and the fMLP receptors. A second round of RT-PCR/orphan cloning was therefore carried out exploiting the new motif. Cloning and sequencing of these PCR products revealed several additional unique seven-transmembrane-spanning receptors. Table 2 is a roster of the types of orphan cDNA clones encoding novel 7-TM proteins that have been obtained in these experiments. A cDNA corresponding to one clone isolated in the second round of RT-PCR was transiently expressed and shown indeed to be a functional C-C chemokine receptor, the details of which are to be presented elsewhere (22).

Table 2. Types of clones encoding novel 7-TM proteins obtained by chemokine "orphan" receptor cloning strategy.

Most closely related known molecule	Number of different cDNA clones obtained
Neuropeptide-Y receptor	2
mas oncogene product	2
IL8R subfamily	6
Undefined 7-TM	2

CONCLUDING REMARKS

The information summarized in this report suggests that the C-C chemokines are likely to be important links between monocytes, lymphocytes, basophils, and eosinophils during inflammatory and other immunoregulatory processes. We present evidence that the specific and differential actions of these molecules on their target cells is mediated through a complex array of shared and specific receptors. The evidence for this comes mostly from signalling data derived from calcium mobilization studies in various cell types. At least two things are noteworthy from these studies. First, the same cell type yields different Ca^{++} flux patterns dependent on the order of the addition of the C-C chemokine challenge. Second, different cell types yield different Ca^{++} flux patterns to the identical C-C chemokine challenge.

Some of the C-C chemokines, such as RANTES, display unusual competition and dissociation profiles on nucleated cells. This fact has made direct biochemical characterization of chemokine receptors difficult. However, direct binding for RANTES and other chemokines can be done on red blood cells, and we have accordingly identified a promiscuous chemokine receptor on the surface of those cells. The red cell chemokine receptor is the only binding site to date which will accommodate chemokines of C-X-C and C-C classes, and its existence suggests that the red cell is playing a role as a regulator of inflammatory peptides in circulation.

Lastly we have established various approaches to cloning C-C chemokine receptors, and have used an "orphan cloning" strategy to isolate the first identified C-C chemokine receptor.

The characterization of some of the receptor/ligand interactions in the C-C chemokine family may provide clues into the actions of these important immune regulators. A deeper understanding of the interaction of the C-C chemokines with their receptors will no doubt lead to a better understanding of cell recruitment during infection and inflammation as well as providing useful therapeutic insights.

ACKNOWLEDGMENTS

We thank Louis Tamayo for his help with the figures and Evelyn Berry for her assistance with typing the manuscript.

REFERENCES

1. Schall, T.J. 1991. Biology of the RANTES/SIS cytokine family. *Cytokine.* 3: 165-183.
2. Oppenheim, J.J., C.O. Zachariae, N. Mukaida, and K. Matsushima. 1991. Properties of the novel

proinflammatory supergene "intercrine" cytokine family. *Ann. Rev. Immunol.* 9: 617-648.

3. Baggiolini, M., A. Walz, and S.L. Kunkel. 1989. Neutrophil-activating peptide-1/interleukin 8, a novel cytokine that activates neutrophils. *J. Clin. Invest.* 84: 1045-1049.

4. Matsushima, K. and J.J. Oppenheim. 1989. Interleukin 8 and MCAF: novel inflammatory cytokines inducible by IL 1 and TNF. *Cytokine.* 1: 2-12.

5. Yoshimura, T., K. Matsushima, S. Tanaka, E.A. Robinson, E. Appella, J.J. Oppenheim, and E.J. Leonard. 1987. Purification of a human monocyte-derived neutrophil chemotactic factor that has peptide sequene similarity to other host defence cytokines. *Proc. Natl. Acad. Sci. USA.* 84: 9233-9237.

6. Leonard, E.J. and T. Yoshimura. 1990. Human monocyte chemoattractant protein-1 (MCP-1). *Immunol Today.* 11: 97-101.

7. Matsushima, K., C.G. Larsen, G.C. DuBois, and J.J. Oppenheim. 1989. Purification and characterization of a novel monocyte chemotactic and activating factor produced by a human myelomonocytic cell line. *J. Exp. Med.* 169: 1485-1490.

8. Walz, A., R. Burgener, B. Car, M. Baggiolini, S.L. Kunkel, and R.M. Strieter. 1992. Structure and neutrophil-activating properties of a novel inflammatory peptide (ENA-78) with homology to Interleukin 8. *J. Exp. Med.* 174: 1355-1362.

9. Schall, T.J., K. Bacon, K.J. Toy, and D.V. Goeddel. 1990. Selective attraction of monocytes and T lymphocytes of the memory phenotype by cytokine RANTES. *Nature.* 347: 669-671.

10. Kuna, P., S.R. Reddigari, D. Rucinski, J.J. Oppenheim, and A.P. Kaplan. 1992. Monocyte chemotactic and activating factor is a potent histamine-releasing factor for human basophils. *J. Exp. Med.* 175: 489-493.

11. Kuna, P., S.R. Reddigari, T.J. Schall, D. Rucinski, M.Y. Viksman, and A.P. Kaplan. 1992. RANTES, a monocyte and T lymphocyte chemotactic cytokine, releases histamine from human basophils. *J. Immunol.* 149: 636-642.

12. Bischoff, S.C., M. Krieger, T. Brunner, and C.A. Dahinden. 1992. Monocyte chemotactic protein 1 is a potent activator of human basophils. *J. Exp. Med.* 175: 1271-1275.

13. Kameyoshi, Y., A. Dorschner, A.I. Mallet, C. Christophers, and J.-M. Schroder. 1992. Cytokine RANTES released by thrombin-stimulated platelets is a potent attractant for human eosinophils. *J. Exp. Med.* 176: 587-592.

14. Rot, A., M. Krieger, S. Brunner, T.J. Schall, and C.A. Dahinden. 1992. RANTES and macrophage inflammatory protein 1a induce the migration and activation of normal human eosinophil granulocytes. *J. Exp. Med.* 176: 1489-1495.

15. Graham, G.J., E.G. Wright, R. Hewick, S.D. Wolpe, N.M. Wilkie, D. Donaldson, S. Lorimore, and I.B. Pragnell. 1990. Identification and characterization of an inhibitor of haemopoietic stem cell proliferation. *Nature.* 344: 442-444.

16. Dunlop, D.J., E.G. Wright, S. Lorimore, G.J. Graham, T. Holyoake, D.J. Kerr, S.D. Wolpe, and I.B. Pragnell. 1992. Demonstration of stem cell inhibition and myeloprotective effects of SCI/rhMIP1 alpha *in vivo*. *Blood.* 79: 2221-2225.

17. Davatelis, G., S.D. Wolpe, B. Sherry, J.-M. Dayer, R. Chicheportiche, and A. Cerami. 1989. Macrophage inflammatory protein-1: a prostaglandin-independent endogenous pyrogen. *Science.* 243: 1066-1068.

18. Schild, H.O. 1973. Receptor classification with special reference to b-adrenergic receptors. In Drug Receptors. H.P. Rang, ed. (University Press). pp. 29-36.

19. Yoshimura, T. and E.J. Leonard. 1990. Identification of high affinity receptors for human monocyte chemoattractant protein-1 on human monocytes. *J. Immunol.* 145: 292-297.

20. Holmes, W.E., J. Lee, W.J. Kuang, G.C. Rice, and W.I. Wood. 1991. Structure and functional expression of a human interleukin-8 receptor. *Science.* 253: 1278-1280.

21. Murphy, P.M. and H.L. Tiffany. 1991. Cloning of complementary DNA encoding a functional human interleukin-8 receptor. *Science.* 253: 1280-1283.

22. Neote, K., D. DiGregorio, J.Y. Mak, R. Horuk, and T.J. Schall. 1993. Molecular cloning and expression, and signalling characteristics of a C-C chemokine receptor. *Cell.* 72: (in press).

23. Gerard, N.P. and C. Gerard. 1991. The chemotactic receptor for human C5a anaphylatoxin. *Nature.* 349: 614-617.

24. Boulay, F., M. Tardif, L. Brouchon, and P. Vignais. 1990. The human N-formylpeptide receptor. Characterization of two cDNA isolates and evidence for a new subfamily of G-protein-coupled receptors. *Biochemistry.* 29: 11123-11133.

ADHESION MOLECULES IN ACUTE AND CHRONIC LUNG INFLAMMATION

L. Gordon Letts, Craig. D. Wegner and Robert H. Gundel

Department of Pharmacology
Boehringer Ingelheim Pharmaceuticals Inc.
900 Ridgebury Road
Ridgefield, CT 06877, USA

INTRODUCTION

In the quest to understand the events involved with an inflammatory response, research has focused on understanding the role of single mediators such as leukotrienes, thromboxane, platelet activating factor, histamine, etc. Whereas each of these single mediators undoubtedly contributes to the complexity of disease, illustrating individual pathophysiologic or therapeutic modalities by selectively antagonizing or inhibiting the activity of each has proven difficult. In light of these findings attention has focused on understanding the types of cells and their pattern of migration into inflamed tissue. It is thought that by inhibiting either the presence or function of pro inflammatory cells in the lungs it is possible to eliminate a whole range of potentially adverse mediators and promote homeostasis.

It has become clear that the induction of pulmonary inflammation requires the presence of inflammatory cells. This is readily shown in animal models of allergic inflammation where the migration of cells into the lungs has been induced by exposing the lungs of a sensitized animal to a challenge of the allergen. Evaluation of the response typically encompasses an immediate bronchoconstriction (acute phase response) and six to eight hours later, a second obstruction of airway function which in animals, unlike man, is usually of greater severity (late phase response). Although the initiation of this late response is signaled during the immediate response, its severity is due solely to the influx of cells. The degree of the late obstruction relates directly to the time course and number of invading cells which have been activated. It is also felt that the induction of airway hyperresponsiveness is associated with airway inflammation.

Of particular interest to us has been the modulation of trafficking and inhibition of specific functions of the different cell types which migrate into the lungs during an inflammatory response. In particular, we have been interested in the eosinophil and its role in the onset of airway inflammation and hyperresponsiveness. To achieve this we have used monoclonal antibodies to specific human adhesion glycoproteins and administered them to primates which possess a naturally occurring sensitivity to inhaled Ascaris suum antigen. In this article we describe the effects of anti-adhesion antibodies in primate models of acute and chronic lung inflammation.

METHODS AND MATERIALS

Adult, male cynomolgus monkeys with a naturally occurring sensitivity to inhaled Ascaris suum antigen have been used. Pulmonary function was measured by oscillatory mechanics, airway responsiveness by construction of dose responses to inhaled methacholine and airway cellular composition by bronchoalveolar lavage (BAL) and biopsy. The methods have been described in detail elsewhere (1,3). The monoclonal antibodies used were murine anti-human ICAM-1 (6,7) and anti-human E-Selectin (4,5).

ACUTE AND CHRONIC LUNG INFLAMMATION MODELS

The chronic lung inflammation model utilizes sub-human primates which exhibit an intrinsic and persistent airway eosinophilia; i.e. animals with an underlying inflammation characterized by increased eosinophils and levels of eosinophil peroxidase (EPO) in the bronchoalveolar lavage fluids (BAL). This chronic inflammation does not require an antigenic stimulus and reflects a small percentage of the Ascaris- sensitive animals. The animals exhibit a profound sensitivity to inhaled methacholine and are classified as hyperresponsive. Characterization of these animals has also involved immunohistochemical staining of expressed and upregulated ICAM-1 on airway epithelium and vascular endothelium. The numbers of neutrophils found in the BAL is small and not much different from normal animals and E-Selectin staining is not evident on the vascular endothelium.

The acute lung inflammation model uses primates which respond to a single exposure to antigen with a dual response (i.e. an early and late phase airway obstruction). These dual responding animals have an intrinsic pulmonary eosinophilia as described above. They comprise a small percentage of primates as the majority of animals respond to antigen with a single immediate response only. The single responders are distinguishable from the dual responders in that their cellular composition differs significantly. In comparison and as already described, dual responders have significantly more eosinophils and EPO activity in their airways (recovered by BAL) prior to antigen challenge. After antigen inhalation, airway eosinophils rapidly decrease and the degree of the late phase airway obstruction significantly correlates with the quantity of neutrophils invading and the level of the increase in BAL myeloperoxidase activity (6 hours post antigen). Dual responding animals have an acute inflammatory event superimposed on an underlying (but milder) chronic inflammation. During the late phase obstruction, ELAM-1 staining is markedly upregulated on vascular endothelium (it is not detectable prior to antigen inhalation). As mentioned, ICAM-1 staining is expressed prior to antigen inhalation. Similar levels of expression are observed during the late phase response (6 hours).

These models of acute and chronic lung inflammation have been described in detail before (3, 8, 9) and their properties are summarized in Table 1.

PROTOCOL

Illustrated below are the protocols in the two models which have been used to evaluate the effects of the murine anti-human adhesion monoclonal antibodies. In each case studies were done to confirm cross reactivity of the antibodies in the cynomolgus monkey.

Table 1. Summary of the properties of the acute and chronic models of pulmonary inflammation.

	CHRONIC MODEL	ACUTE MODEL	
		Early Response	Late Response
BAL LUNG CELL COMPOSITION			
eosinophils	++++	++++	++
neutrophils	(+)	(+)	+++
BAL MEDIATOR LEVELS			
EPO	++	++	+++
MPO	-	--	+++
AIRWAY RESPONSIVENESS	+++	+++	+++
IMMUNOHISTOCHEMICAL STAINING			
ICAM-1			
- epithelium	+++	+++	+++
- endothelium	+++	+++	+++
E-SELECTIN			
- epithelium	--	--	--
- endothelium	--	--	+++

Also the doses were selected on the basis of providing excess circulating antibody as determined by in vitro cell-adhesion assays (5,10). In the chronic model the antibodies were administered once per day for seven days and blood was withdrawn to monitor for anti-idiotypic responses.

In the acute model protocol the anti-human E-Selectin and ICAM-1 antibodies were administered as a single dose of 2 mg/kg iv. Lung function was monitored as indicated (Rrs). In the chronic model protocol the anti-human ICAM-1 antibody was administered as a daily single intravenous injection of 2 mg/kg. Airway responsiveness to inhaled methacholine (PC100)was determined on days 0 and 7.

RESULTS

The administration of the murine anti-human E-Selectin antibody to dual responder monkeys dramatically reduced the numbers of neutrophils migrating into the airways during the late phase response. There was also a marked reduction in the levels of MPO in the BAL and a significant decrease in the late phase airway obstruction. On the other hand, the murine anti-human ICAM-1 antibody did not effect any of the late phase variables. The results are summarized in Table 2.

Acute model protocol

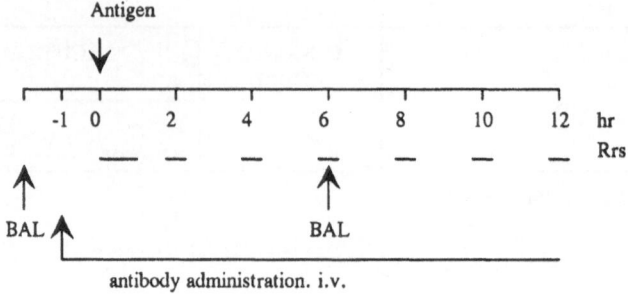

Chronic model protocol

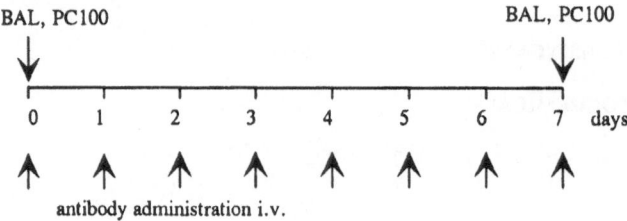

The daily administration for 7 days of the ICAM-1 antibody to the animals exhibiting a chronic lung inflammation did not provide any measurable benefit in the monitored variables throughout the study. The effects of dexamethasone are also included and show significant reductions in BAL eosinophil count, MPO and EPO levels as well as airway responsiveness to inhaled methacholine. The results are summarized in Table 3.

Table 2. Effects of Monoclonal Antibody Treatment on Antigen-induced Acute Airway Inflammation.

Treatment	Nr. of Infiltrating cells[1]			BAL Fluid Levels of[2]		% Inhibit.
	Total Cells	Eos.	Neut.	MPO	EPO	LPR
Control	7.6±3.6	-2.4±1.7	3.3±2.1	423±87	590±52	----
Anti-E Selectin	-2.4±0.9*	-1.2±0.3	0.6±0.9*	101±20	315±110	72±8*
Anti-ICAM-1	3.9±3.7	-3.1±1.1	3.6±2.3	388±92	409±97	19±15

[1]Number of cells (x10^5/ml)) recovered in BAL fluid 6 hr after antigen challenge
[2]Optional Density
LPR Late Phase Response
*p<0.05, compared to pre-challenge values

Table 3. Effects of Monoclonal Antibody treatment on Chronic Airway Inflammation. The variables were measured after 7 consecutive days of treatment.

Treatment	Nr. of Cells in BAL Fluid[1]			BAL Fluid Levels of[2]		AHR[3]
	Total Cells	Eos.	Neut.	MPO	EPO	PC$_{100}$
Control	7.1±1.1	3.3±0.8	0.1±2.4	192±41	389±53	0.32±1.5
Anti-E Selectin	7.8±1.9	4.1±1.3	0.4±1.9	165±49	240±78	0.25±1.6
Anti-ICAM-1	7.9±1.7	4.6±1.0	0.4±0.9	178±37	310±105	0.79±2.3
Dexamethasone	3.3±0.6	0.6±0.2*	0.1±0.2	43±10*	52±14*	6.31±2.0*

[1]Number of cells (x10^5) recovered in BAL Fluid.
[2]Optical Density
[3]Airway Hyperresponsiveness
*$p < 0.05$, compared to day 7

DISCUSSION

In earlier publications, we have reported that multiple antigen inhalations induce changes in the airways including a prolonged and selective eosinophilia temporally associated with epithelial damage and the onset of airway hyperresponsiveness (1,3,9). We also showed that intravenous pretreatment with an anti-ICAM-1 antibody inhibited the influx of eosinophils and subsequently reduced the BAL levels of EPO, epithelial damage and increase in airway responsiveness. We proposed from these studies that ICAM-1 was pivotal for the trafficking of eosinophils into the airways and their alignment near the basolateral surface of epithelium. The importance of this close association of eosinophils with the epithelium was highlighted by several studies. Firstly, we showed that inhaled anti-ICAM-1 was effective in this model (11) and secondly, treatment with an anti-CD11b/CD18 (mac-1) antibody did not inhibit eosinophil migration into the lungs, but blocked the activation events (as determined by BAL EPO levels and the increase in airway responsiveness) (12). Thirdly we have reported that instilled major basic protein, one of the cationic proteins secreted by eosinophils, is capable of inducing increases in airway responsiveness (13). The dominant role of ICAM-1 for the entry of eosinophils into the airways is illustrated by the failure of the E-Selectin antibody to affect this process.

In the chronic model the failure of the ICAM-1 antibody to reverse any of the variables, particularly the numbers of eosinophils in the BAL, is presumably due to the failure to provide an adhesive pathway for cellular exodus. This study reaffirms the importance of ICAM-1 in the events associated with movement of eosinophils in the airways. It is therefore possible to either prohibit cells entering the lungs or alternatively, once present (as in the chronic model), prevent their passage out. The results obtained with dexamethasone in the chronic model are included to further emphasize the importance of the eosinophil and its products in the maintenance of the persistent inflammation. Daily treatment with dexamethasone is able to reduce the numbers of cells in the BAL, reduce the levels of EPO and decrease the airway responsiveness to inhaled methacholine (10).

The late phase obstruction observed in the acute model of airway inflammation is primarily correlated with the influx of neutrophils and increases in BAL MPO activity. At the same time there is a slight decrease in BAL eosinophil count and increase in EPO levels, indicating the activation of resident (or newly recruited) eosinophils. We believe this observation with the eosinophil is important since it may reflect that activation of this resident cell is paramount to the cascade of events which ultimately leads to the late phase response. Immunohistochemical staining has shown the upregulation of E-Selectin during the late response. The time course of E-Selectin is relatively transient in that its expression on endothelial cells peaks 4-6 hours after stimulation and is gone by approximately 12 hours. It is not expressed on epithelial cells and is undetectable on resting endothelium. Interestingly, the rolling and initial adhesion of neutrophils to endothelial surfaces has been shown to be Selectin dependent (14, 15).

The results showing the E-Selectin antibody to inhibit the airway obstruction and block the migration of neutrophils into the airways during the late phase response supports the role of the neutrophil in the acute inflammatory response. It is thought that during the immediate events following antigen stimulation the activation of resident lung cells (such as mast cells, eosinophils, macrophages) releases mediators such as cytokines which are capable of stimulating endothelial cells to express E-Selectin. The release of mediators into the BAL immediately following antigen inhalation has been previously reported (16). The failure of anti-ICAM-1 antibody to affect these responses illustrates the divergence and relative importance of different adhesive ligands for different cell types during an inflammatory reaction. These studies illustrate not only the relative effects of different cell types but emphasize the different kinetics of adhesion glycoprotein expression during an acute versus a chronic (sustained) inflammatory event.

In summary, in the acute inflammation model we have monitored airway cellular composition, biochemical markers of cell activation, lung function and adhesion protein expression. In these dual responder primates we conclude that the presence of activated eosinophils results in a "priming" of adjacent cells so that following antigen stimulation the lung reacts with an "acute" response which involves the E-Selectin dependent influx of neutrophils. The severity of the late phase response correlates with the numbers of cells infiltrating and the status of their activation.

In the chronic model of pulmonary inflammation we have identified a selective eosinophilia, biochemical evidence of eosinophil activation, upregulation of cell surface expression of ICAM-1 on endothelium and epithelium and a persistent hyper-responsiveness to inhaled methacholine. E-Selectin is not detected in this model. The dominant functional role of ICAM-1 is illustrated by its failure to decrease the numbers of eosinophils (and markers of eosinophil activation) recovered in BAL fluid and to modify the underlying inflammation. This lack of decrease in the numbers of activated eosinophils is likely caused by the anti-ICAM-1 antibodies ability to block the exodus of eosinophils from the lungs.

In conclusion we believe that by modulating cellular adhesive events it is possible to obtain a better perspective on the role of different cell types, both resident and non-resident, in an "acute" and chronic inflammatory setting. It is hoped that this will permit a better understanding of the science and eventually lead to the development of new therapeutics.

ACKNOWLEDGEMENTS

The authors wish to thank the excellent technical assistance of Ms. C. Torcellini, Mr. C. Clarke, Ms. A. LaPlante, Ms. N. Haynes, Ms. C. Stearns. Also Dr. R. Rothlein and Dr. W. Smith for supply of antibodies.

REFERENCES

1. Gundel, R.H., M.E. Gerritsen, G.J. Gleich, and C.D. Wegner. 1990. Repeated antigen inhalation results in a prolonged airway eosinophilia and airway hyperresponsiveness in primates. *J. Appl. Physiol.* 68: 779-786.

2. Gundel, R.H., M.E. Gerritsen, and C.D. Wegner. 1989. Antigen-coated sepharose beads induce airway eosinophilia and airway responsiveness in cynomolgus monkeys. *Am. Rev. Respir. Dis.* 140: 629-633.

3. Gundel R.H., C.D. Wegner, C.A. Torcellini, and L.G. Letts. 1992. Antigen induced acute and late-phase responses in primates. *Am. Rev. respir. Dis.* 146: 369-373.

4. Rothlein, R., M.L. Dustin, S.D. Marlin, and T.A. Springer. 1986. A human intercellular adhesion molecule (ICAM-1) distinct from LFA-1. *J. Immunol.* 137: 1270-1274.

5. Wegner C.D., R.H. Gundel, P. Reilly, N. Haynes, L.G. Letts, and R. Rothlein. 1990. Intercellular molecules in the pathogenesis of asthma. *Science* 247: 456-459.

6. Kishimoto, T.K., R.A. Warnock, M.A. Jutlia, E.C. Butcher, C.L. Lane, D.C. Anderson, and C.W. Smith. Antibodies against human neutrophil (LECAM-1) and endothelial cell ELAM-1 inhibits a common CD18 indpendent adhesion pathway in vitro.

7. Gundel, R.H., C.D. Wegner, C.A. Torcellini, C.C. Clarke, N. Haynes, R. Rothlein, C.W. Smith, and L.G. Letts. 1991. *J. Clin. Invest.* 88: 1407-1411.

8. Gundel, R.H., C.D. Wegner, C.A. Torcellini, and L.G. Letts. 1992. The role of intercellular adhesion molecule-1 in chronic airway inflammation. *Clin. Exp. Allergy* 22: 569-575.

9. Wegner, C.D., R.N. Gundel, R. Rothlein, and L.G. Letts. 1992. Expression and probable role of cell adhesion molecules in lung inflammation. *Chest* 101: 34s-39s.

10. Wegner, C.D., C.W. Smith, and R. Rothlein. 1989. CD18 dependence of primate eosinophils adherence in vitro. In: Springer, T.A., D.C. Anderson, A.S. Rosenthal, and R. Rothlein, eds. Leukocyte adhesion molecules: structure, function and regulation. New York Springer - Verlag, 208-214.

11. Wegner, C.D., R. Rothlein, C.C. Clarke, N. Haynes, C.A. Torcellini, A.M. LaPlante, D.R. Averill, L.G. Letts, and R.H. Gundel. 1991. Inhaled anti-intercellular adhesion molecule-1 (ICAM-1) reduces antigen-induced airway hyperresponsiveness in monkeys. (Abstract) *Am. Rev. Respir. Dis.* 143: 418.

12. Wegner, C.D., R.H. Gundel, L. Churchill, D. Souza, C. Stearns, and L.G. Letts. 1992. Mac-1 (CD11b/CD18) mediates antigen-induced eosinophil activation and airway hyperresponsiveness in monkeys. (Abstract) *Am. Rev. Respir. Dis.* 145: 461

13. Gundel, R.H., L.G. Letts, and G.J. Gleich. 1991. Human eosinophil major basic protein induces acute airway contriction and airway hyperresponsiveness in primates. *J. Clin. Invest.* 87: 1470-1473.

14. Kishimoto, T.K., M.A. Jutlin, E.L. Berg, E.C. Butcher. 1989. Neutrophil Mac-1 and Mel-14 adhesion proteins inversely regulated by chemotactic factors. *Science* 245: 1238-1241.

15. Lasky, L.A. 1991. Lectin cell adhesion molecules (LEC-CAM's): a new family of cell adhesion proteins involved with inflammation. *J. Cell. Biochem.* 45: 139-146.

16. Gundel, R.H., P. Kinkade, C.A. Torcellini, C.C. Clarke, J. Watrous, S. Desai, C.A. Homon, P.R. Farina, and C.D. Wegner. 1991. Antigen-induced mediator release in primates. *Am. Rev. Respir. Dis.* 144: 76-82.

MONOCYTE CHEMOTACTIC PROTEIN-1 (MCP-1): SIGNAL TRANSDUCTION AND INVOLVEMENT IN THE REGULATION OF MACROPHAGE TRAFFIC IN NORMAL AND NEOPLASTIC TISSUES

Alberto Mantovani, Silvano Sozzani, Barbara Bottazzi, Giuseppe Peri, Francesca Luisa Sciacca, Massimo Locati, and Francesco Colotta

Istituto di Ricerche Farmacologiche "Mario Negri"
Via Eritrea 62
20157 Milan, Italy

INTRODUCTION

The identification of tumor-derived chemotactic factors (TDCF) specific for monocytes (1) is one of the pathways that led to the identification of MCP-1 (2-6). Macrophages are a major component of the lymphoreticular infiltrate of rodent and human tumors (7). Since these cells are situated at the very interface between tumor and host, they may represent a strategically located target for therapeutic intervention. Interest in these cells is stimulated by the knowledge that macrophages have the potential to kill neoplastic cells including drug-resistant variants surviving conventional chemotherapy. Tumor-associated macrophages (TAM) derive from circulating monocytic precursors. The functional properties of macrophages infiltrating murine and human metastatic tumors have been characterized in an effort to obtain indications as to the role played by these cells in the immunobiology of neoplastic tissues (7). This analysis has indicated how TAM can contribute to important aspects of tumor tissue biology, such as fibrin deposition and angiogenesis.

Moreover, TAM in certain tumors are a source of growth factors which actually provide the optimal conditions for tumor growth. More in general, this type of analysis has shown how TAM, within the mononuclear phagocyte system, represent a population with peculiar phenotypic and functional properties. Here we will summarize the role of MCP-1 in regulating macrophage infiltration *in vivo* and we will discuss recent progress in the signal transduction pathway used by this cytokine.

The Chemokines, Edited by I.J.D. Lindley
et al., Plenum Press, New York 1993

MCP-1 AS A DETERMINANT OF TAM LEVELS

Early work on TDCF indicated some correlation between macrophage infiltration *in vivo* and release of monocyte chemotactic activity. The molecular identification of TDCF as MCP-1 provided the tools to revisit the role of MCP-1 in regulating TAM levels. In a murine tumor model, a correlation was found between the percentage of TAM and MCP-1 expression (8). In addition MCP-1 gene transfer in a non-expressing mouse line and in CHO cells resulted in higher levels of TAM in vivo (9,10). More recently, the metastatic capacity of clones expressing the human MCP-1 gene was studied by injecting tumor cells i.v. MCP-1 expressing clones were more metastatic than control cells in terms of lung involvement (number or weight) and of the occurrence of extrapulmonary lesions (Table 1). This observation is in line with a higher tumorigenicity at low tumor inocula in spite of a slower *in vivo* growth rate (9). Mononuclear phagocytes recruited by MCP-1 may help the initial implantation and outgrowth of the small number of cells that ultimately give rise to metastasis.

Table 1. Metastatic capacity of melanoma cells after MCP-1 gene transfer[a]

Exp.	Clone	MCP-1	Mice with met.	Lung colonies (total number)	Weight (mg±SE)	Extra-pulmonary lesions[b]
1	V14	-	7/8	20	210±21	0/8
	V12	+	5/5	<300	862±156	1/5
2	V14	-	7/10	16	138±66	1/10
	V16	-	9/9	23	76/17	0/9
	L12	+	8/8	14	296±116	8/8
	L4	+	7/7	15	174/80	0/7

a) The human MCP-1 gene under the control of a retroviral promoter was transferred in the B78/H subline of the B16 melanoma. The obtainement and properties of MCP-1 expressing (L12-L4) and control (V14-V16) clones are described by Bottazzi et al (9), 5 x 10^5 cells of *in vitro* cultured cells were injected in the tail vein and mice were autopsied on day 30 (experiment 1) or 34 (experiment 2).

b) Extrapulmonary metastases involved lymph nodes and liver.

THE SIGNAL TRANSDUCTION PATHWAY OF MCP-1: REGULATION OF Ca^{2+} INFLUX

The signal transduction pathway(s) of members of the Cys-Cys chemokine family is still unknown, and their receptors have not been characterized. In the effort to elucidate monocyte activation by these cytokines, we started to investigate the early events following receptor activation. One of the most common signalling pathway, although not unique, of many "classical" chemotactic agents, such as FMLP, C5a, LTB$_4$ and PAF, involves the activation of a pertussis toxin (PT)-sensitive GTP-binding protein and the activation of phosphatidylinositol 4, 5-biphosphate (PIP$_2$)-specific phospholipase C (PLC). The two products of this process are diacylglycerol which will activate protein kinase C (PKC) and inositol triphosphate (IP$_3$), which will rise the cytosolic concentration of free Ca^{2+} ([Ca^{2+}]i) inducing the discharge of intracellular stores (11).

MCP-1, RANTES and MIP-1α, three members of the Cys-Cys family, induced a rapid and transient increase of [Ca^{2+}]i in human monocytes. Monocyte migration toward these three proteins was inhibited in a dose-dependent manner by PT and by PKC/cAMP-dependent protein kinase inhibitors, such as C-I, H-7, HA1004, and staurosporine (Table 2). Also genistein, and erbstatin, two tyrosine kinase inhibitors, strongly reduced monocyte migration to these agonists. Interestingly, while neutrophil migration in response to IL-8 (a member of the Cys-X-Cys family) was inhibited by tyrosine kinase inhibitors, C-I and staurosporine were able to increase neutrophil chemotaxis. These results suggest that members of the two chemokine families may differ in their activation pathway.

Table 2. Effect of Protein Kinase Inhibitors on chemokine-induced chemotaxis.

Inhibitor	Specificity	Cys-Cys			Cys-X-Cys
		MCP-1	RANTES	MIP-1α	IL-8
C-I	PKC/PKA[a]	+[b]	+	+	-[c]
H-7	PKC/PKA	+	N.D.	N.D.	-[c]
Staurosporine	PKC/PKA/TyrPK	+	N.D.	N.D.	-[c]
HA1004	PKC/PKA	+	N.D.	N.D.	+
KT5823	PKG	-	N.D.	N.D.	N.D.
KT5926	MLCK	-	N.D.	N.D.	N.D.
KT5720	PKA	+	N.D.	N.D.	N.D.
Genistein	TyrPK	+	+	+	+
Erbstatin	TyrPK	+	+	+	+

a: PKA, cAMP-dependent protein kinase; PKC, protein kinase C; PKG, cGMP-dependent protein kinase; MLCK, myosin L chain kinase, TyrPK, tyrosine protein kinase.

b: + means inhibition, - no effect.

c: increase of chemotaxis was observed.

Thus, at least for some of their actions, MCP-1, RANTES and MIP-1α appeared to fit the general activation process described above. However, when we investigated in more detail the mechanisms of monocyte activation by MCP-1, a different picture was obtained. MCP-1 failed to activate the breakdown of PIP$_2$ and the production of detectable levels of IP$_3$. In agreement with these results, MCP-1 did not induce the release of Ca^{2+} from intracellular stores. The rise in [Ca^{2+}]i was completely dependent on the presence of extracellular Ca^{2+} and could be blocked by plasma membrane Ca^{2+} channels inhibitors, such as 5mM Ni^{2+} and the imidazole derivative SC 38249. Thus, the increase in [Ca^{2+}]i has to be ascribed only to Ca^{2+} influx across the plasma membrane (Table 3). Ca^{2+} influx was directly evaluated measuring Mn^{2+}-quenching of Fura-2 fluorescence. In these experiments Mn^{2+} is assumed to be transported via the same plasma membrane channels that are permeable to Ca^{2+}. Once penetrated into the cells, Mn^{2+} binds to Fura-2 with high affinity and produces a quench of the fluorescence signal. MCP-1 activation of human monocytes markedly enhanced the rate of Mn^{2+} quenching. The effect was dose-dependent and was blocked by both Ni^{2+} and SC 38249. Inhibitors of voltage-operated Ca^{2+} channels (VOCs) had no effect on MCP-1-induced Ca^{2+} influx. Collectively these data indicate that MCP-1 stimulation results in the opening of plasma membrane Ca^{2+} channels that are different from VOCs and that possibly belong to the class of receptor-activated Ca^{2+} channels (12).

The role on Ca^{2+} influx in MCP-1-activated monocytes remains to be established. We have observed that monocytes exposed to Ni^{2+} or EGTA show a marked reduction in chemotaxis to MCP-1. However, the exact nature of the second messengers generated by MCP-1 and their role in monocyte activation needs further evaluation.

Table 3. Evidence that MCP-1-induced increase in $[Ca^{2+}]i$ depends on influx

---

1.   The rise in $[Ca^{2+}]i$ is dependent on the presence of extracellular Ca^{2+}

2.   MCP-1 induces dose-dependent Mn^{2+} quenching of Fura-2 fluorescence

3. Ni^{2+} and SC 38249 block both Mn^{2+} influx and the rise in $[Ca^{2+}]i$

4.   MCP-1 does not induce IP_3 production

MCP-1-INDUCED INCREASE IN $[Ca^{2+}]i$: CROSS DESENSITIZATION WITH OTHER CYS-CYS CHEMOKINES

Monocytes exposed to 10 ng/ml MCP-1 become deactivated to a second stimulation with the same agonist assessed in terms of monocyte migration or $[Ca^{2+}]i$ transients (13). This is a property shared by a number of different receptors including those of the Cys-X-Cys chemokine family (14). Therefore, it was of interest to investigate possible interactions (in terms of Ca^{2+} fluxes) between MCP-1 and two other members of the Cys-Cys family, RANTES and MIP-1α Desensitization of monocytes was always observed whenever the same agonist was applied twice. As shown in Fig. 1, a first stimulation with either RANTES or MIP-1α did not affect the rise in $[Ca^{2+}]i$ induced by a subsequent stimulation with MCP-1. By contrast, prestimulation with MCP-1 abolished the response to a subsequent challenge with either RANTES or MIP-1α Further, RANTES and MIP-1α desensitized monocytes reciprocally. A possible explanation for this pattern of cross desensitization is that these chemokines interact with distinct receptors, which are all recognized by MCP-1. However, it is also possible that a common mechanism of receptor deactivation is shared by closely related receptors.

MACROPHAGE INFILTRATION IN MCP-1-NEGATIVE TUMORS

The results summarized above suggest that MCP-1 can indeed play a role in regulating macrophage infiltration in tumors. However, there is evidence that chemoattract-ants other than MCP-1 are produced by tumor cells. Van Damme et al. (15) recently identified MCP-2 and 3 in tumor supernatants. Our attention was focused on human ovarian carcinoma, a source of monocyte chemotactic activity related to macrophage infiltration *in vivo* (16). The SW626 ovarian carcinoma line releases a monocyte-specific chemoattractant distinct from MCP-1, and has a conspicuous infiltrate (34%) when injected in nude mice. By Northern blot analysis we found no evidence of expression of MCP-1, LD78/MIP-1α, Act-2, RANTES, groß. Purification (in collaboration with Drs. J. Van Damme and R. Bertini) resulted in a single 24 KDa protein chemotactic for monocytes and inactive on neutrophils. Molecular identification of this novel TDCF is in progress.

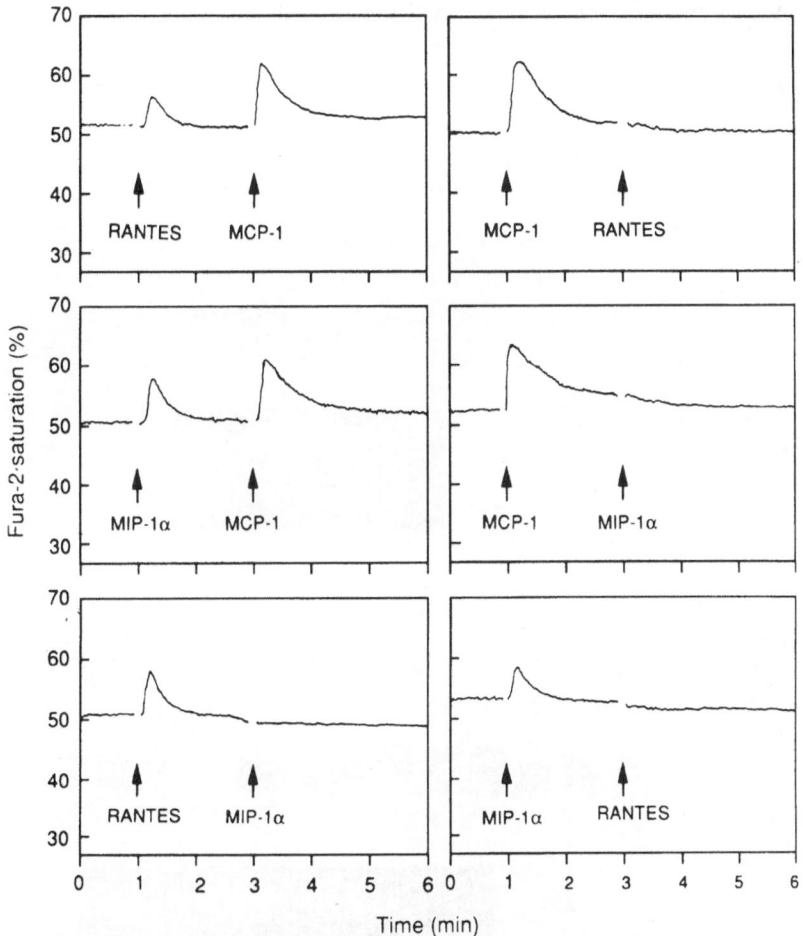

Figure 1. Cross deactivation of Ca^{2+} transients among MCP-1, RANTES and MIP-1α. MCP-1 (10 ng/ml), RANTES (200 ng/ml) and MIP-1α(200 ng/ml) were added to cell suspension as indicated in the figure.

MCP-1 ARMING OF MONOCYTES FOR MIGRATION: INDUCTION OF UROKINASE-TYPE PLASMINOGEN ACTIVATOR (u-PA) AND u-PA RECEPTOR (uPA-R)

Migratory behaviour of cells is frequently associated with production of proteolytic enzymes, u-PA in particular, which are important for progression in the context of the extracellular matrix. This consideration prompted us to investigate whether MCP-1-induced expression of u-PA and uPA-R. As shown in Fig. 2, MCP-1 (as well as classical chemoattractants, C5a and FMLP) induced expression of u-PA mRNA. Similar results were obtained when uPA-R transcripts were examined (Fig. 3). In order to get more insight into the molecular mechanisms of the induction of u-PA and uPA-R by chemotactic stimuli, we examined the effects of metabolic inhibitors (cycloheximide, CH, and actinomycin D, ActD).

CH (a protein synthesis inhibitor) superinduced both constitutive and inducible expression of u-PA and uPA-R mRNAs. Transcriptional block, operated by ActD, inhibited the inducible expression of uPA-R and u-PA. We also found that chemotactic stimuli prolonged the half-life of u-PA and uPA-R mRNA (data not shown).

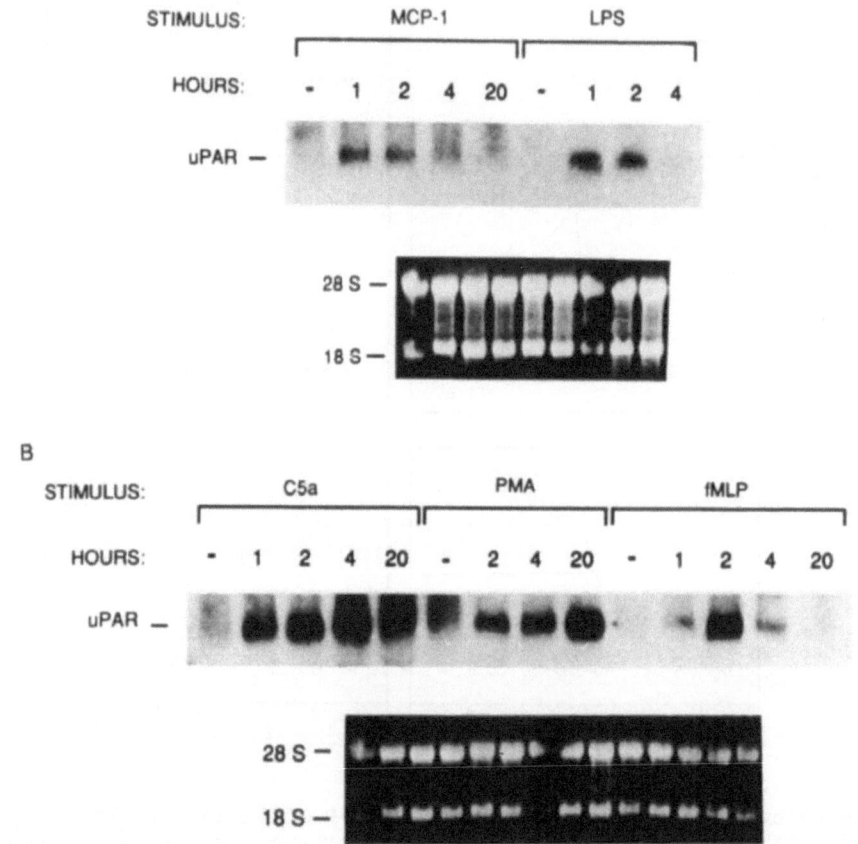

Figure 2. Expression of u-PA mRNA after stimulation with MCP-1 and LPS (Panel A) and with C5a, PMA and fMLP (Panel B).

CONCLUDING REMARKS

The results summarized here indicate that MCP-1 is indeed an important, though by no means unique (7), determinant of monocyte recruitment in tumors. Tumors provide at this time the only pathophysiological conditions in which an actual *in vivo* role of MCP-1 has been determined. The evidence for an involvement of MCP-1 in regulating monocyte traffic in non-neoplastic conditions rests mainly on *in vitro* studies on cells or tissue sections. In this context it is particularly noteworthy that cells strategically located to gate leukocyte traffic, typically endothelial cells (17), produce conspicuous amounts of MCP-1. In this context, we recently found that mesothelial cells, exposed to inflammatory signals, express MCP-1 (18). This finding, as well as the expression by these cells of a distinct set of adhesion molecules, suggest that the mesothelial lining plays an active role in regulating monocyte traffic between the circulation, the peritoneal fluid and the peritoneal wall, and provide a basis for the long known macrophage appearance and disappearance reactions (18).

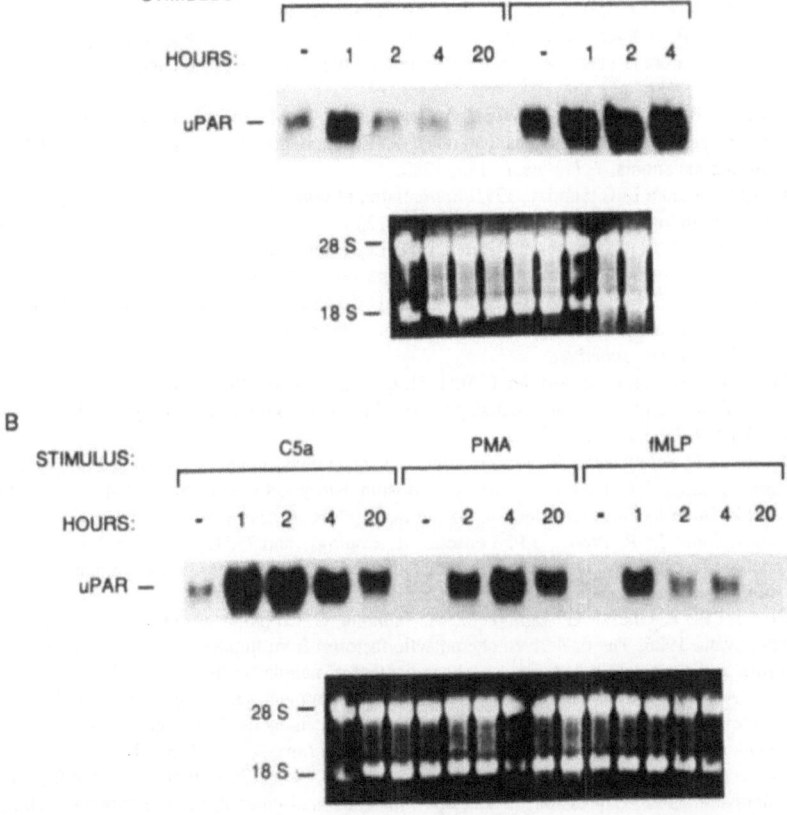

Figure 3. Expression of uPA-R mRNA after stimulation with MCP-1 and LPS (Panel A) and with C5a, PMA and fMLP (Panel B).

REFERENCES

1. Bottazzi, B., N. Polentarutti, R. Acero, A. Balsari, D. Boraschi, P. Ghezzi, M. Salmona, and A. Mantovani. 1983. Regulation of the macrophage content of neoplasms by chemoattractants. *Science*, 220: 210.
2. Furutani, Y., H. Nomura, M. Notake, Y. Oyamada, T. Fukui, M. Yamada, C.G. Larsen, J.J. Oppenheim, and K. Matsushima. 1989. Cloning and sequencing of the cDNA for human monocyte chemotactic and activating factor (MCAF). *Biochem. Biophys. Res. Commun.*, 159: 248.

3.   Yoshimura, T., N. Yuhki, S.K. Moore, E. Appella, M.I. Lerman, and E.J. Leonard. 1989. Human monocyte chemoattractant protein-1 (MCP-1). Full-lenght cDNA cloning, expression in mitogen-stimulation blood mononuclear leukocytes, and sequence similarity to mouse competence gene JE. *FEBS Letters*, 244: 487.

4.   Van Damme, J., B. Decock, J.P. Lenaerts, R. Conings, R. Bertini, A. Mantovani, and A. Billiau. 1989. Identification by sequence analysis of chemotactic factors for monocytes produced by normal and transformed cells stimulated with virus, double stranded RNA or IL-1. *Europ. J. Immunol.*, 19: 2367.

5.   Bottazzi, B., F. Colotta, A. Sica, N. Nobili, and A. Mantovani. 1990. A chemoattractant expressed in human sarcoma cells (Tumor-derived chemotactic factor, TCDF) is identical to monocyte chemoattractant protein-1/monocyte chemotactic and activating factor (MCP-1/MCAF). *Int. J. Cancer*, 45: 795.

6.   Matsushima, K., C.G. Larsen, G.C. DuBois, and J.J. Oppenheim. 1989. Purification and characterization of a novel monocyte chemotactic and activating factor produced by a human myelomonocytic cell line. *J. Exp. Med.*, 169: 1485.

7.   Mantovani, A., B. Bottazzi, F. Colotta, S. Sozzani, and L. Ruco. 1992. Origin and function of tumor-associated macrophages. *Immunology Today*, 13: 265.

8.   Walter, S., B. Bottazzi, D. Govoni, F. Colotta, and A. Mantovani. 1991. Macrophage infiltration and growth of sarcoma clones expressing different amounts of monocyte chemotactic protein/JE. *Int. J. Cancer*, 49: 431.

9.   Bottazzi, B., S. Walter, D. Govoni, F. Colotta, and A. Mantovani. 1992. Monocyte chemotactic cytokine gene transfer modulates macrophage infiltration, growth and susceptibility to IL-2 therapy of a murine melanoma. *J. Immunol.*, 148: 1280.

10. Rollins B.J., and M.E. Sunday. 1991. Suppression of tumor formation in vivo by expression of the JE gene in malignant cells. *Mol. Cell. Biol.* 11: 3125.

11. Di Virgilio F., O. Stendhal, D. Pittet, P.D. Lew, and T. Pozzan. 1990. Cytoplasmic calcium in phagocyte activation. *Curr. Topics Membr. Transport.* 35: 303.

12. Sozzani S., M. Molino, M. Locati, W. Luini, C. Cerletti, A. Vecchi, and A. Mantovani. 1992. Receptor-activated calcium influx in human monocytes exposed to monocyte chemotactic protein-1 and related cytokines, *submitted*.

13. Sozzani S., W. Luini, M. Molino, P. Jìlek, B. Bottazzi, C. Cerletti, K. Matsushima, and A. Mantovani. 1991. The signal transduction pathway involved in the migration induced by a monocyte chemotactic cytokine. *J. Immunol.* 147: 2215.

14. Moser B., C. Schumacher, V. von Tscharner, I. Clark-Lewis, and M. Baggiolini. 1991. Neutrophil-activating peptide 2 and gro/melanoma growth-stimulatory activity interact with neutrophil-activating peptide 1/interleukin 8 receptors on human neutrophils. *J. Biol. Chem.* 266: 1066.

15. Van Damme J., P. Proost, J.P. Lenaerts, R. Conings, and G. Opdenakker. 1992. Structural and functional identification of two human tumor-derived monocyte chemotactic proteins (MCP-2, MCP-3) belonging to the chemokine family. *J. Exp. Med.* 176: 59.

16. Bottazzi B., P. Ghezzi, G. Taraboletti, M. Salmona, C. Colombo, C. Bonazzi, C. Mangioni, and A. Mantovani. 1985. Tumor-derived chemotactic factor(s) from human ovarian carcinoma: evidence for a role in the regulation of the macrophage content of neoplastic tissues. *Int. J. Cancer* 36: 167.

17. Sica, A., J.M. Wang, F. Colotta, E. Dejana, A. Mantovani, J.J. Oppenheim, C.G. Larsen, C.O. Zachariae, and K. Matsushima. 1990. Monocyte chemotactic factor gene expression induced in endothelial cells by IL-1 and tumor necrosis factor. *J. Immunol.*, 144: 3034.

18. Jonjic N., G. Peri, S. Bernasconi, F.L. Sciacca, F. Colotta, P.G. Pelicci, L. Lanfranconi, and A. Mantovani. 1992. Expression of adhesion molecules and chemotactic cytokines in cultured human mesothelial cells. *J. Exp. Med.*, *in press*.

SECRETION OF MONOCYTE CHEMOATTRACTANT PROTEIN-1 (MCP-1) BY HUMAN MONONUCLEAR PHAGOCYTES

Edward J. Leonard[1], Alison Skeel[1], Teizo Yoshimura[1], and John Rankin[2]

[1]Immunopathology Section
Laboratory of Immunobiology
National Cancer Institute
Frederick, Maryland, USA

[2]Pulmonary and Critical Care Section
Veteran's Administration Hospital
Department of Medicine
Yale University School of Medicine
New Haven, Connecticut, USA

INTRODUCTION

The stage for the work that is presented here was set in a 1973 paper by Altman et al., entitled "A human mononuclear leukocyte chemotactic factor: characterization, specificity and kinetics of production by homologous leukocytes" (1). They reported that stimulation by tuberculin (PPD) of blood mononuclear leukocytes (PBMC's) from individuals with a positive skin test to tuberculin caused production of a chemotactic factor for human monocytes. Since the factor was not produced by PPD-stimulated leukocytes from PPD-negative subjects, the authors suggested that it could account for recruitment of macrophages in delayed cutaneous hypersensitivity reactions. The factor was a heat stable macromolecule with a molecular mass, estimated by gel filtration, of about 12,500 daltons. Altman later referred to this molecule as leukocyte-derived chemotactic factor, or LDCF (2). In 1989, we reported the purification to homogeneity of the predominant chemotactic activity for monocytes in culture fluids of PHA-stimulated human mononuclear leukocytes (3). We suggested that this 8700 dalton protein, which we call monocyte chemoattractant protein-1 (MCP-1), is the attractant described by Altman et al. In the interval since the last international meeting on chemotactic cytokines, we have developed a sandwich ELISA that can quantify the amount of MCP-1 in biological fluids (4). Using this assay, we find that PPD-stimulated PBMC's from a tuberculin-positive human subject secrete large amounts

The Chemokines, Edited by I.J.D. Lindley
et al., Plenum Press, New York 1993

of MCP-1 within 48 hrs. The amount is too much to be accounted for by secretion by the number of lymphocytes that react with specific antigen in this system. Therefore, MCP-1 must be secreted by non-immune cells -- lymphocytes or monocytes -- stimulated by the very low concentrations of a mediator released by PPD-reactive lymphocytes. The focus of this communication is on the MCP-1 secretory capacity of the monocyte.

SECRETION OF MCP-1 BY PPD-STIMULATED HUMAN PBMC's

MCP-1 and NAP-1 (neutrophil attractant protein-1/IL-8) secretion by PBMC's in response to PPD is illustrated in Figure 1. We measured high concentrations of both MCP-1 and NAP-1 in 48 hr culture fluid of PPD-stimulated PBMC's from a PPD-positive human subject. There was no response by PBMC's from the tuberculin-negative donor.

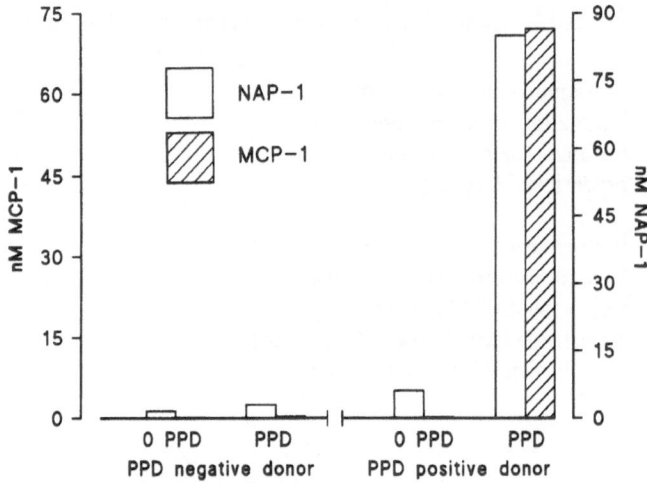

Figure 1. Response of human PBMC's to 10 TU PPD/ml. Culture fluids were collected 48 hours after addition of the stimulus.

Several aspects of this result are of interest. [1] If we assume that MCP-1 was produced only by the PPD-reactive lymphocytes in the culture, and that these comprise the order of 1 in 10,000 lymphocytes, secretion of MCP-1 would be about 1 ng/cell -- an impossibly high number when compared to stimulated T-cell lymphokine secretion of about 10^{-15} gm/cell (5) or the monocyte MCP-1 secretion of 0.16 pg/cell described later in the present communication. This argues for recruitment of non-immune cells in the culture. [2] Contrary to the implication of the name "purified protein derivative," PPD is a crude ammonium sulfate precipitate from a tubercle bacillus culture. Although the preparation did not stimulate PBMC's from an tuberculin-negative subject, it is possible that there are co-stimulatory actions by other substances in this crude material that could act on PBMC's from tuberculin-positive subjects. We did observe similar stimulation with PPD that was passed through an LPS-binding column. [3] PPD also induced secretion of NAP-1. Since neutrophil infiltration is not a prominent feature of a 48 hr tuberculin skin reaction, our PBMC culture system is not a faithful model of the DCH reaction. The PPD dose-response curve in Figure 2 is illustrative of 2 of 3 subjects, in which there is a PPD concentration that induced preferential secretion of MCP-1. Furthermore, as suggested by data shown later in this communication, the enhanced secretory capacity of blood monocytes cultured under conditions that promote maturation is much greater for MCP-1 than NAP-1.

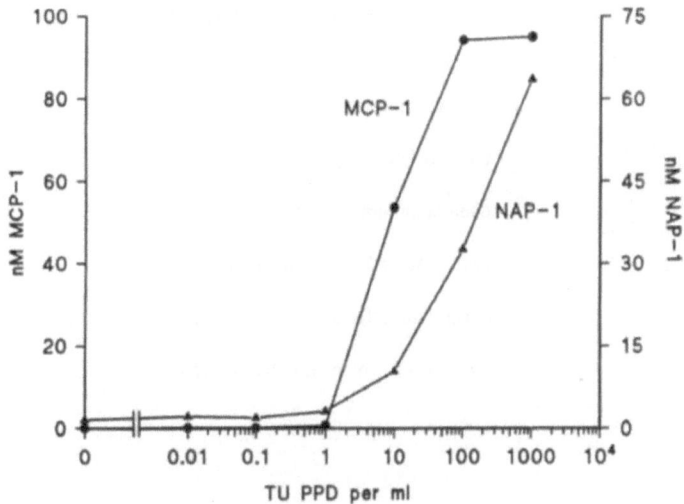

Figure 2. Stimulation of PBMC's from a tuberculin-positive subject: PPD dose-response curves. Culture fluid was collected 48 hours after addition of the stimulus.

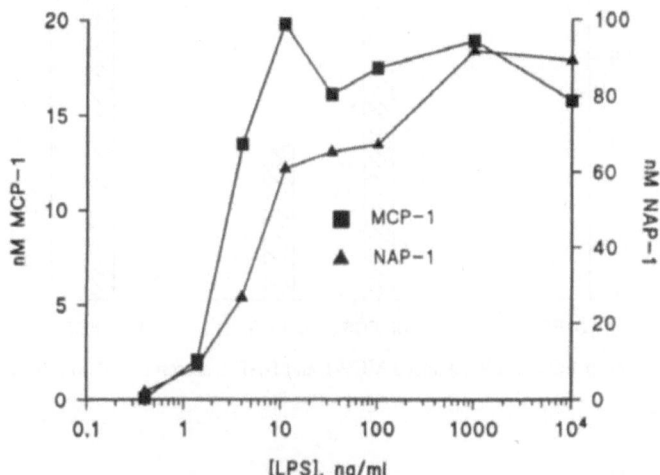

Figure 3. LPS stimulation of monocyte MCP-1 and NAP-1 secretion in Iscove's DMEM with 10% FCS.

LPS-INDUCED SECRETION OF MCP-1 BY HUMAN BLOOD MONOCYTES

The essentials of the experimental conditions for our studies on monocyte MCP-1 and NAP-1 secretion are summarized in Table 1. Figure 3 shows that freshly isolated human monocytes do not secrete detectable amounts of MCP-1 or NAP-1. LPS induced secretion of both. Figures 4 and 5 show that the magnitude of the response to LPS is critically dependent on culture conditions. Figure 4 shows that heat-inactivated fetal calf serum enhances the secretory response. Figure 5 shows that secretion of MCP-1 is much greater in Iscove's DMEM-10% FCS than in RPMI-1640-10% FCS. The LPS-induced MCP-1 secretory response is much more fastidious than the NAP-1 response. This is summarized in Table 2.

Table 1. Experimental methods

Monocytes purified by elutriation.

About 90% monocytes, no NK cells.

Adherent culture, 10^6 cells/ml.

RPMI 1640 or Iscove's DMEM.

Collect culture fluid at 24 or 36 hrs.

Sandwich ELISA for MCP-1 and NAP-1.

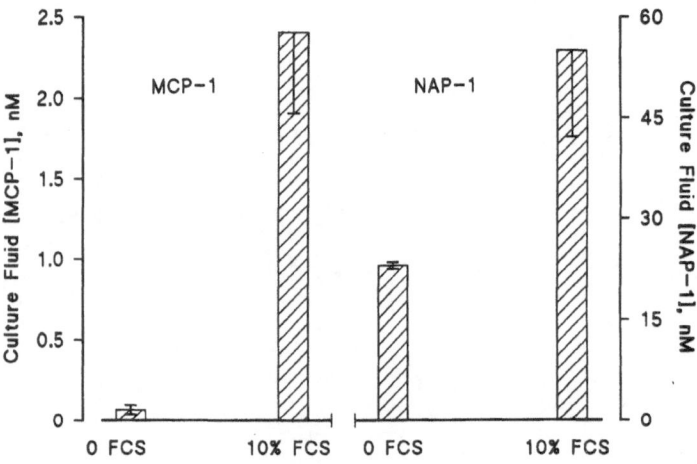

Figure 4. Effect of FCS on LPS-induced MCP-1 and NAP-1 secretion. SEM's for 6 experiments.

These data underline the need for careful definition of the *in vitro* system, and may account for the disparate results in the literature on human monocyte MCP-1. Rollins et al. found MCP-1 by immunoprecipitation in unstimulated human monocytes, which decreased in response to LPS (6). Difficulty in defining factors that affect monocyte MCP-1 gene expression was illustrated in a recent report by Colotta et al on blood monocyte preparations from 15 normal human subjects (7). Eight monocyte preparations had no or low MCP-1 mRNA, and expression was induced by LPS; monocytes from the other 7 subjects had high levels of MCP-1 mRNA, with little or no response to LPS. The authors failed to find the cause for the two MCP-1 patterns among their 15 subjects. In contrast to the results of Colotta et al., elutriated monocytes from all of our 15 subjects produced undetectable or negligible MCP-1 in control medium, and secreted MCP-1 in response to LPS stimulation. These differences might be explained if a human platelet release product causes monocyte MCP-1 secretion, either directly or indirectly by activation of a plasma component. Whereas elutriated monocytes are free of platelets, monocytes isolated from buffy coats by density centrifugation may be accompanied by platelets, the number of which may vary in different preparations (8).

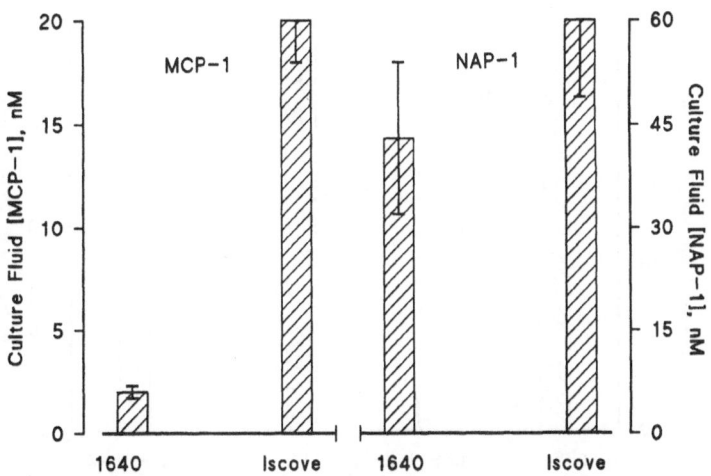

Figure 5. LPS-induced MCP-1 and NAP-1 secretion in 2 different media containing 10% heat inactivated FCS. SEM's for 9 experiments.

Table 2. LPS-induced concentration ratios in different media

	MCP-1	NAP-1
1640: FCS/0 FCS	24	2.4
Iscove/1640	10	1.4

In the comparison between Iscove's DMEM and RPMI-1640, both media contained 10% heat-inactivated FCS.

Is the enhancement of LPS-induced secretion of MCP-1 by FCS or Iscove's medium at the level of transcription or translation? Northern blots of our preparations should answer this question. Colotta et al reported that the addition of FCS had no effect on MCP-1 transcription. If our preparations are comparable, it is possible that the effect of FCS is at the level of translation or secretion. Colotta et al also also reported that cycloheximide inhibited MCP-1 transcription. The enhancing action of Iscove's medium could be on the LPS-induced production of a transcription-regulating protein.

In view of the inhibition by IL-4 of LPS-induced monocyte production of TNFα (9,10), IL-1α (9,10) and NAP-1 (11), we determined if this inhibitory effect of IL-4 also extended to MCP-1 secretion. Figure 6 shows that IL-4 inhibits LPS-induced secretion of MCP-1 as well as NAP-1. The data of Colotta et al on NAP-1 mRNA suggest that IL-4 may act by producing a protein inhibitor of mRNA synthesis, since the suppressive effect of IL-4 on mRNA expression is abolished by cycloheximide.

CYTOKINE PATHWAYS FOR INDUCTION OF MONOCYTE MCP-1 SECRETION

As noted above, PPD-induced stimulation of MCP-1 and NAP-1 secretion in PBMC cultures suggests that recruitment of non-immune cells in the culture must occur. Because of the potential for stimulation by IL-1, TNF or IL-2 of MCP-1 or NAP-1 secretion by PBMC's, we tested effects of these cytokines on monocytes, one of the two abundant cell types in the PBMC population.

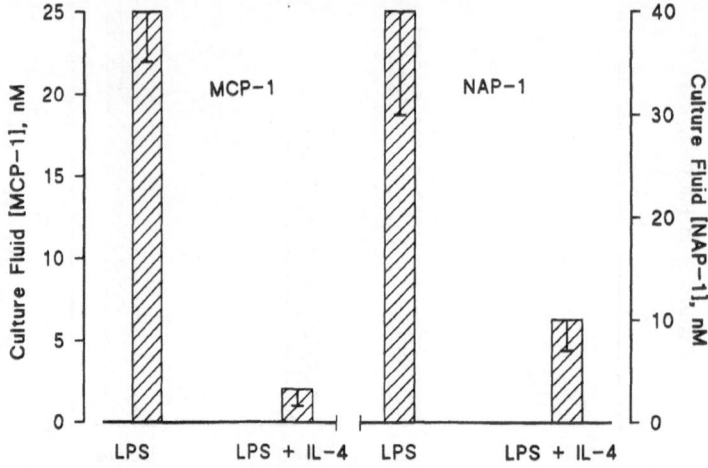

Figure 6. Inhibition by IL-4 of LPS-induced secretion of MCP-1 and NAP-1. SEM's for 3 experiments.

Figure 7 shows that IL-2, in combination with IFNγ, stimulates MCP-1 secretion by elutriated human monocytes. The response is about 1/10 of that induced by LPS. IFNγ alone is a minimal stimulus of MCP-1 secretion. This is in contrast to the stimulation of MCP-1 mRNA expression, reported by two laboratories (7,12).

Dose-response curves for IL-1α and TNFα are shown in Figure 8. The data are presented as a percentage of the LPS-response for the corresponding monocyte preparations.

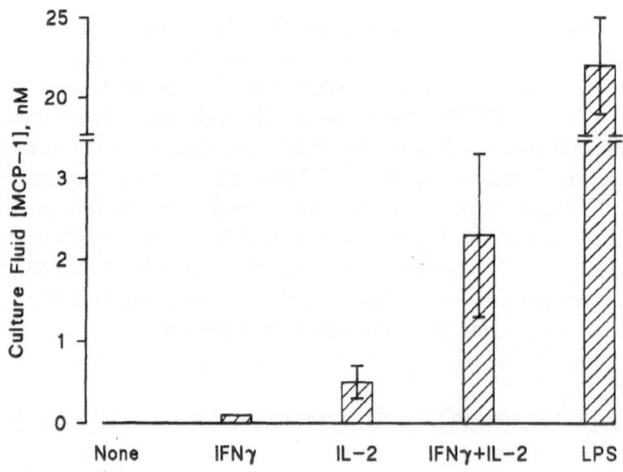

Figure 7. Effect of IFNγ and IL-2 on monocyte MCP-1 secretion. SEM's for 4 experiments.

Table 3 summarizes MCP-1 secretion responses to these various stimuli, with NAP-1 secretion for comparison. As shown in Figure 1, the concentration of MCP-1 in cultures of 5×10^6 PPD-stimulated PBMC's/ml (containing about 1.5×10^6 monocytes/ml) is considerably higher than the values shown in Table 3. Thus, the cell and mediator interactions that account for the PPD response remain to be elucidated.

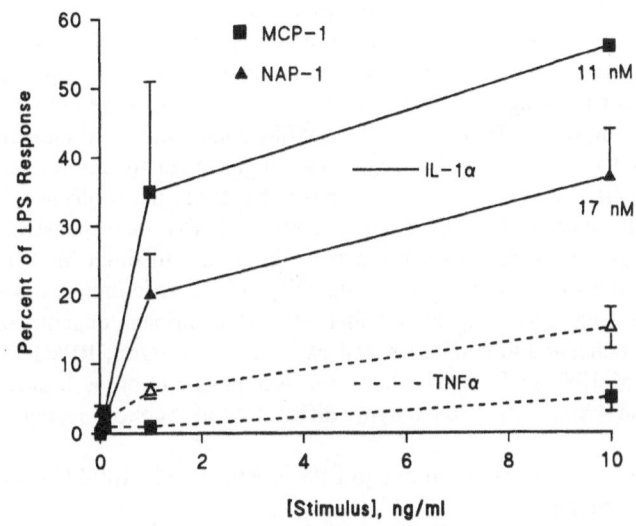

Figure 8. Monocyte secretory responses in Iscove's DMEM-10% FCS to IL-1α and TNFα, as a percentage of the response to LPS. SEM's for 6 experiments.

Table 3. Summary of monocyte secretory responses in Iscove's DMEM with 10% FCS

	nM MCP-1	nM NAP-1
No stimulus	0	0
LPS	20 ± 2	60 ± 11
IL-1α	11 ± 3	17 ± 4
TNFα	0	6 ± 2
IFNγ	0.1 ± 0	0.5 ± 0.2
IL-2	0.5 ± 0.2	2 ± 0.6
IFNγ + IL-2	2.3 ± 1	2 ± 1

COMPARISON OF LPS-INDUCED MCP-1 AND NAP-1 SECRETION IN ELUTRIATED MONOCYTES, CULTURED MONOCYTES AND BAL MACROPHAGES

This part of the study was designed to determine if phenotypically different mononuclear phagocytes had different MCP-1 and NAP-1 secretion responses to LPS. We identified 3 distinct patterns of secretion in response to LPS (in RPMI-1640-10% FCS), illustrated in Figures 9 and 10. The first pattern is shown by the reference population, monocytes stimulated with LPS shortly after elutriation. These cells responded with secretion of both MCP-1 and NAP-1. Mean concentrations of MCP-1 and NAP-1 in the culture fluids were 2 nM and 50 nM respectively, which are optimal chemotactic concentrations for these agonists. When monocytes were transferred to Costar-24 wells after 3-7 days culture in teflon beakers in DMEM containing 1000 units/ml CSF-1 (13), MCP-1 secretion in response to LPS was much higher than for the reference population (Fig. 9). The third secretory pattern is shown by BAL macrophages. Like reference monocytes and cultured monocytes, they secreted NAP-1 in response to LPS. But in contrast to monocytes, LPS caused no increase in culture medium MCP-1 concentration above the 0-stimulus level of 0.5 nM. The inability of BAL macrophages to secrete MCP-1 in response to LPS was confirmed under several additional experimental conditions including LPS concentrations of 20 ng/ml as well as 10 µg/ml, RPMI-1640-10% FCS, Iscove's DMEM-10% FCS, and culture for 4-5 days in teflon beakers before LPS challenge. Table 4 summarizes phenotypic differences in secretory responses to LPS.

Table 4. Secretory responses to LPS in RPMI-1640-10%FCS depend on phenotype

Monocytes:

> secrete both MCP-1 (2 nM) and NAP-1 (50 nM).

Cultured monocytes:

> [MCP-1] cultured/[MCP-1] fresh: 23 $\pm$ 15
> [NAP-1] cultured/[NAP-1] fresh: 3 $\pm$ 1

Bronchoalveolar lavage macrophages:

> secrete large amounts of NAP-1 (150 nM).
> no MCP-1 response above baseline (0.5 nM).

These experiments identify one example of stimulus-induced selective secretion of a leukocyte-specific chemoattractant. Comparison of Figures 9 and 10 shows that LPS caused accumulation of high concentrations of NAP-1 in BAL macrophage cultures, whereas there was no change in the low level of MCP-1. Thus, the programming of the BAL macrophage in this respect is different from the relatively undifferentiated blood monocyte, which responds to LPS with secretion of both attractants. This finding is consistent with the results of Strieter et al., who detected NAP-1 but not MCP-1 mRNA in LPS-stimulated human BAL macrophages (14). When blood monocytes were put into culture for 3-7 days, their secretory responses to LPS had a relative shift in favor of MCP-1 (Table 4). The secretion by cultured monocytes of a specific attractant for monocytes may reflect an in vivo pattern in which tissue macrophages -- as part of an amplification loop in cellular immunity -- are the cells that produce MCP-1 to attract blood monocytes to the site.

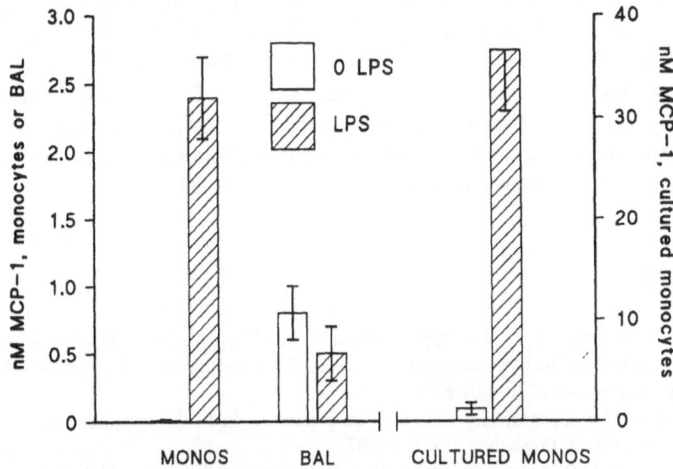

Figure 9. Secretion of MCP-1 by different types of mononuclear phagocytes. Note the different scale for cultured monocytes. SEM's for 5 experiments.

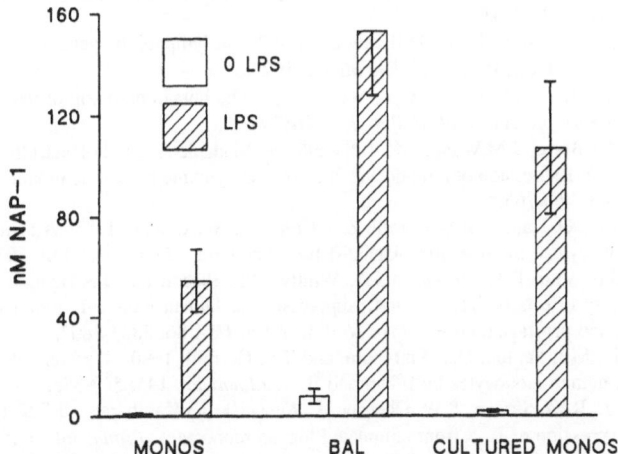

Figure 10. Secretion of NAP-1 by different types of mononuclear phagocytes.

SUMMARY

Concentrations of MCP-1 and NAP-1 in culture fluids of human leukocytes were measured by sandwich ELISA. PPD caused PBMC's from tuberculin-sensitive subjects to secrete MCP-1 and NAP-1. PPD did not stimulate secretion by cells from a tuberculin-negative subject. Since the amounts secreted were more than could be produced by the few PPD-sensitized lymphocytes in the culture, we postulate that other cells were stimulated to secrete these chemoattractants. This study evaluated secretory capacity of one of the cell types in the PBMC culture. Unstimulated monocytes did not secrete MCP-1 or NAP-1. In order of increasing effect, IL-2 + IFNγ, IL-1α, and LPS caused monocyte secretion of MCP-1. The rank order for NAP-1 secretion was the same. TNFα did not cause secretion of MCP-1, but caused about the same amount of NAP-1 secretion as IL-2 + IFNγ. Composition of the culture medium was especially critical for LPS-induced secretion of MCP-1, which was greatly enhanced by FCS and by Iscove's DMEM compared to RPMI 1640. IL-4 inhibited LPS-induced secretion of both MCP-1 and NAP-1. Secretory patterns were also a function of mononuclear phagocyte phenotype. LPS-induced secretion of MCP-1 was much greater for monocytes cultured several days in CSF-1 than for freshly isolated monocytes. LPS stimulation of bronchoalveolar macrophages caused NAP-1 secretion, but no secretion of MCP-1 above a relatively low baseline level.

REFERENCES

1. Altman, L.C., R. Snyderman, J.J. Oppenheim, and S.E. Mergenhagen. 1973. A human mononuclear leukocyte chemotactic factor: characterization, specificity and kinetics of production by homologous leukocytes. *J. Immunol.* 110: 801-810.

2. Altman, L.C. 1978. Chemotactic lymphokines: a review. In *Leukocyte Chemotaxis.* J.I. Gallin and P.G. Quie. eds. Raven Press. New York. p 267.

3. Yoshimura, T., E.A. Robinson, S. Tanaka, E. Appella, and E.J. Leonard, 1989. Purification and amino acid analysis of two human monocyte chemoattractants produced by phytohemagglutinin-stimulated human blood mononuclear leukocytes. *J. Immunol.* 142: 1956-1962.

4. Yoshimura, T., M. Takeya, K. Takahashi, J. Kuratsu, and E.J. Leonard. 1991. Production and characterization of mouse monoclonal antibodies against human monocyte chemoattractant protein-1. *J. Immunol.* 147: 2229-2233.

5. Ehlers, S. and K.A. Smith. 1991. Differentiation of T cell lymphokine gene expression: the in vitro acquisition of T cell memory. *J.Exp.Med* 173: 25-36.

6. Rollins, B.J., P. Stier, T. Ernst, and G.G. Wong. 1989. The human homolog of the JE gene encodes a monocyte secretory protein. *Mol. Cell.Biol.* 9: 4687-4695.

7. Colotta, F., A. Borre, J.M.Wang, M. Tattanelli, F. Maddalena, N. Polentarutti, G. Peri, and A. Mantovani. 1992. Expression of a monocyte chemotactic cytokine by human mononuclear phagocytes. *J. Immunol.* 148: 760-765.

8. Pawlowski, N.A., G. Kaplan, A.L. Hamill, Z.A. Cohn, and W.A. Scott. 1983. Arachidonic metabolism by human monocytes. Studies with platelet-depleted cultures. *J.Exp.Med.* 158: 393-412.

9. Hart, P.H., G.F. Vitti, D.R. Burgess, G.A. Whitty, D.S. Piccoli and J.A.Hamilton. 1989. Potential antiinflammatory effects of interleukin-4: Suppression of human monocyte tumor necrosis factor α, interleukin 1, and prostaglandin E$_2$. *Proc.Natl.Acad.Sci. USA* 86: 3803-3807.

10. Donnelly, R.P., M.J. Fenton, D.S. Finbloom, and T.L. Gerrard. 1990. Differential regulation of IL-1 production in human monocytes by IFN-γ and IL-4. *J.Immunol.* 145: 569-575.

11. Standiford, T.J., R.M. Strieter, S.W. Chensue, J. Westwick, K. Kasahara, and S.L. Kunkel. 1990. Il-4 inhibits the expression of IL-8 from stimulated human monocytes. *J.Immunol.* 145: 1435-1439.

12. Chang, H.C., F. Hsu, G.J. Freeman, J.D. Griffin, and E.L. Reinherz. 1989. Cloning and expression of a γ-interferon-inducible gene in monocytes: a new member of a cytokine gene family. *International Immunol.* 1: 388-397.

13. Becker, S., M.K. Warren, and S. Haskill. 1987. Colony-stimulating factor-induced monocyte survival and differentiation into macrophages in serum-free cultures. *J. Immunol.* 139:3703-3709.

14. Strieter, R.M., S.W. Chensue, T.J. Standiford, M.A. Basha, H.J. Showell, and S.L. Kunkel. 1990. Disparate gene expression of chemotactic cytokines by human mononuclear phagocytes. *Biochem. Biophys.Res.Comm.* 166: 886-891.

L-ARGININE/NITRIC OXIDE PATHWAY: A POSSIBLE SIGNAL
TRANSDUCTION MECHANISM FOR THE REGULATION OF THE
CHEMOKINE IL-8 IN HUMAN MESANGIAL CELLS

Zarin Brown, Rachel L. Robson and John Westwick

Department of Pharmacology
University of Bath
Claverton Down
Bath, UK.

INTRODUCTION

The central function of the mesangial cell within the kidney is the control of glomerular filtration rate primarily via the regulation of mesangial cell tone. A secondary function of this specialised pericyte is to maintain the integrity of the glomerulus via matrix synthesis and provide support for the capillary loops. The elicitation of neutrophils and monocytes from the circulation into the inflamed glomerulus accompanied by proliferation of resident mesangial cells and expansion of mesangial matrix are prominent features of immunologically mediated glomerulonephritis (1).

The migration of inflammatory cells into the glomeruli requires a series of orchestrated signals, which include the generation of a chemotactic gradient by the cells of the extravascular compartment. Recently a growing family of target cell specific chemotactic polypeptides have been identified and structurally characterised by the location of four cysteine residues (2). The released peptides of this family are generally less than 10 KDa and belong either to the C-X-C family of which the neutrophil chemoattractant IL-8 is the best characterised or the C-C subfamily of which the monocyte chemoattractant peptide (MCP-1) is the prototype (3, 4).

Glomerular mesangial cells upon activation by cytokines such as IL-1 or TNF, display a host of pro-inflammatory functions, including mesangial cell proliferation, synthesis and secretion of collagenase, prostaglandins, growth factors, oxygen radicals and cytokines (5, 6). Mesangial cells in culture do not constitutively express the chemokines IL-8 or MCP-1 but upon activation with the pro-inflammatory cytokines IL-1 or TNFα, express mRNA transcripts and secrete peptide for both the chemokines IL-8 and MCP-1, as well as the mesangioproliferative cytokine IL-6 (7, 8, 9, 10). Treatment of mesangial

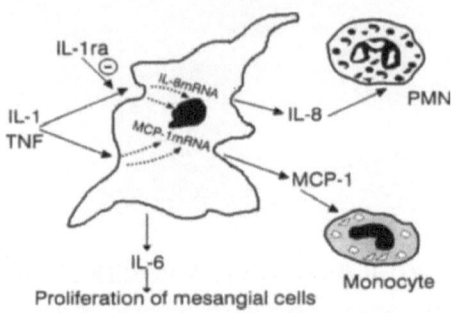

Proliferation of mesangial cells

Figure 1. Cytokine activation of mesangial cells induces IL-8, MCP-1 and IL-6 gene expression and secretion. The IL-1 receptor antagonist (IL-1ra) inhibits the generation of both chemokines as well as the cytokine IL-6.

cells with the anti-inflammatory steroid dexamethasone results in a partial inhibition of both the expression of mRNA and release of IL-8 peptide induced by either IL-1 or TNFα (7) (see Figure 1).

Furthermore, we have demonstrated that pretreatment of mesangial cells with the novel human recombinant IL-1 receptor antagonist (IL-1ra) induces a dose-dependent inhibition of expression of MCP-1 and IL-8 mRNA transcripts as well as the release of both chemotactic peptides in response to IL-1, while the receptor antagonist had no significant effect on TNFα induced chemokine generation (11, 12) (see Figure 1). Thus, we have proposed that the mesangial cell has a direct role in the initiation and propagation of inflammatory events within the glomerulus via the generation of these chemokines.

NITRIC OXIDE AS A SIGNAL TRANSDUCTION MECHANISM IN MESANGIAL CELLS

The molecular mechanisms by which IL-1 activates mesangial cells remain unclear. In order to understand the molecular mechanisms by which IL-1 regulates the expression and release of the chemokine IL-8 by mesangial cells and in particular which second messenger molecules may be involved, we have explored the role of L-arginine/nitric oxide pathway in this process (13).

Nitric oxide (NO) which accounts for the biological properties of endothelium-derived relaxing factor (EDRF), is a labile, highly reactive compound which mediates vasodilatation (13). The acute effects of a number of agonists on the renal vasculature is thought to be due to the induction of NO which has profound effects on renal blood flow and glomerular filtration rate (14).

Nitric oxide is synthesised in endothelial cells from the terminal nitrogen atom contained in the guanidino group of the amino acid L-arginine. The enzyme that initiates the conversion of L-arginine to release NO from endothelial cells is nitric oxide synthase. Numerous cell types are capable of NO production including endothelial cells, macrophages, neutrophils, neurons and some tumour cells (13, 15). The NO released has a very short half life and rapidly diffuses to vascular smooth muscle, where it activates guanylate cyclase, leading to an increase in cGMP accumulation and smooth muscle cell relaxation (16) (see Figure 2).

CHARACTERISTICS OF NO SYNTHASE

Partial characterisation of NO synthase (NOs) suggests the existence of different isoforms of NO synthase (17). Two forms of the constitutive NO synthase have been identified thus far. One from the human cerebellum has been purified, and the other from human endothelial cells has recently been cloned (17, 18). Both the constitutive forms of NOs are Ca^2+/calmodulin-dependent and result in picomolar concentrations of NO release with short lasting effects (13, 17, 19). The macrophage NO synthase is cytokine-inducible, and is independent of Ca^2+/calmodulin but requires NADPH and tetrahydrobiopterin as cofactors, the NO induced in this case being in the nanomolar concentration range (20,13). Very recently, an NO synthase has been described for rat neutrophils which is dependent on Ca^2+ but not on calmodulin (21). The glucocorticoids, such as dexamethasone, hydrocortisone and cortisol, inhibit the expression but not the activity of the inducible NO synthase *in vitro* in vascular endothelial cells, macrophages and adenocarcinoma cells following activation by cytokines (22, 13). In addition, *in vivo* induction of NO synthase in liver, lung, and vascular tissue of rats after treatment with LPS is also prevented by dexamethasone and cortisol (23).

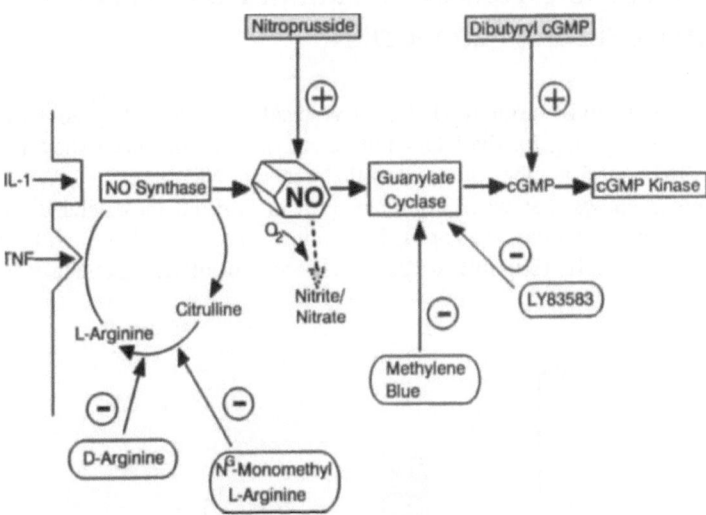

Figure 2. Biosynthetic pathway for nitric oxide (NO). The site of action of competitive inhibitors of NO and inhibitors of soluble guanylate cyclase (-) or direct activators of soluble guanylate cyclase (+).

NO AS AN EFFECTOR MOLECULE IN MESANGIAL CELLS

In coincubation experiments Shultz et al. have shown that exposure of rat glomerular mesangial cells to endothelial-derived NO resulted in increased levels of intracellular cGMP (24). Subsequent studies have demonstrated that activation of rat mesangial cells with LPS or the cytokine interferon-gamma (IFN-γ) resulted in significantly increased nitrate/nitrite levels in culture supernatants as determined by the Griess reagent method, and a concomitant stimulation of mesangial cell cGMP was also observed (25). Both the induction of nitrate/nitrite and the increase in mesangial cell cGMP were completely inhibited by pre-incubating the cells with N^G-monomethyl-L-arginine (L-NMMA) for 24 hours (25). Pfeilschifter et al. have shown that treatment of rat mesangial cells with the proinflammatory cytokines IL-1 or TNF induces a macrophage type of NO synthase (26). A significant elevation of cellular cGMP was detected in the presence of either cytokine, and in addition formation of NO was detected in the cytosol of cytokine-treated mesangial cells by activation of purified soluble guanylate cyclase (27). The induction by IL-1 of NO synthase in rat mesangial cells *in vitro* is inhibited by anti-inflammatory steroids such as dexamethasone, prednisolone, hydrocortisone, corticosterone or progesterone (28). Thus there is good evidence in mesangial cells derived from animal tissues rather than man that the L-arginine/NO pathway can be activated by pro-inflammatory agents such as LPS, IFN-γ, IL-1ß or TNFα and inhibited by the anti-inflammatory steroids. However, none of these experiments has attempted to determine the effect of NO generation on cytokine formation that occurs as a result of the initial pro-inflammatory insult e.g. LPS. In addition, to the best of our knowledge, nitric oxide production by human mesangial cells has not been described thus far.

Cultured mesangial cells provide a useful *in vitro* model system to study this signalling pathway by the use of agents which are competitive inhibitors of NO production, inhibitors of soluble guanylate cyclase or direct activators of soluble guanylate cyclase (Figure 2).

COMPETITIVE INHIBITORS OF NO PRODUCTION: EFFECT ON MESANGIAL CELL IL-8 PRODUCTION

N^G-monomethyl-L-arginine (L-NMMA) specifically prevents generation of NO from L-arginine by competing for the NO synthase enzyme and has been shown to inhibit the biological effects of a number of agonists that generate NO (13).

Figure 3 shows that when confluent growth arrested mesangial cells cultured in normal medium containing 450 μM L-arginine were stimulated with sub-maximal concentrations of IL-1α (3ng/ml) for 20 hours, 58ng/ml of IL-8 peptide was detected in

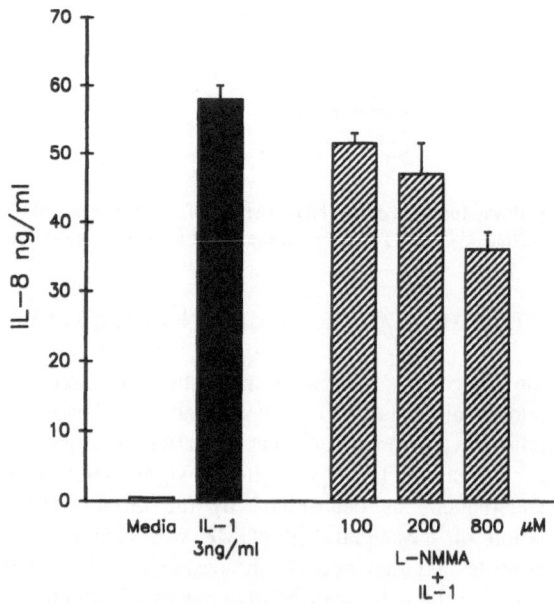

Figure 3. Effect of L-NMMA on IL-1 induced 1L-8 secretion by human mesangial cells.

culture supernatants using a double ligand ELISA method (2). Pre-treatment of cells with increasing concentrations of L-NMMA (100-800 µM) for 1 hour prior to the addition of IL-1α (3ng/ml) resulted in dose-related inhibition of IL-8 generation, a maximum inhibition (35%) of IL-8 response being observed with the highest concentration of L-NMMA of 800 µM.

One explanation for the partial inhibition observed in our system is that we need to increase the concentration of L-NMMA by 2 to 3 fold in order to inhibit the high basal levels of L-arginine present in the normal media used in our experiments. An alternative explanation is that L-NMMA is a weak inhibitor of the NO synthase in mesangial cells or is only poorly taken up into these cells, or both. Differences between the pharmacological profiles of NO synthase have been described previously; in neutrophils, L-NMMA only causes partial inhibition of synthesis following long periods of incubation with the compound (29).

In an attempt to further clarify if NO played a role in the regulation of IL-8, we next decided to investigate whether the inhibitory effect of L-NMMA on IL-8 generation could be reversed by increasing concentrations of the active precursor L-arginine but not that of the enantiomer D-arginine. For these experiments, confluent cells were cultured in either arginine free or normal media (containing 450 µM of L-arginine) for 24 hours, after which they were maintained in normal media, arginine free media or media to which either L or D-arginine (100-300 µM) was added for 1 hour. Cells cultured in normal medium resulted in good IL-8 generation following activation by IL-1 (Figure 4). However, addition of IL-1 to cells maintained in arginine free media resulted in 50% reduction of IL-1 induced IL-8 generation by mesangial cells compared with cells in normal medium. Replacement of L-arginine into the culture media resulted in partial reversal of the inhibition of IL-8 response, whereas addition of D-arginine was unable to reverse the observed inhibition of IL-8 generation. We conclude that our data suggests that inhibition of nitric oxide generation by L-NMMA, L-arginine depletion or D-arginine leads to a significant reduction of IL-1 induced IL-8 production by human mesangial cells.

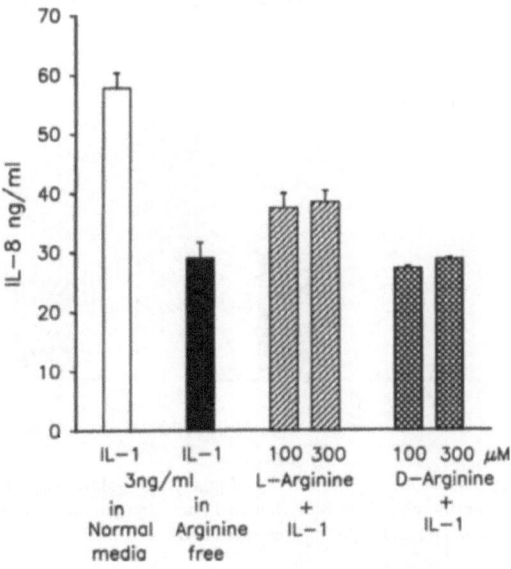

Figure 4. Effect of L-arginine (hatched bars) and D-arginine (cross hatched bars) on IL-1 induced IL-8 generation by human mesangial cell.

EFFECT OF INHIBITORS AND DIRECT ACTIVATORS OF SOLUBLE GUANYLATE CYCLASE ON IL-1 INDUCED IL-8 GENERATION OF MESANGIAL CELLS

We next turn to the question of whether modification of soluble guanylate cyclase in human mesangial cells affects the IL-8 generation.

Figure 5A shows that methylene blue (10 μM), an inhibitor of soluble guanylate cyclase (30), significantly inhibits IL-1 induced IL-8 generation when cells are cultured in normal media (arginine containing) compared with either cells treated with IL-1 alone or in the presence of dibutyryl cyclic GMP (a cell-permeable surrogate for cGMP). However, in the absence of arginine (Figure 5B), IL-1 induced IL-8 generation was significantly reduced (50%) compared with cells cultured in normal medium. In the absence of L-arginine, methylene blue treatment had no significant effect on IL-8 generation compared with IL-1 treated cells, whereas dibutyryl cyclic GMP enhanced the IL-8 generation in the absence of L-arginine.

Similarly LY 83583 (6-anilino-5,8-quinolinedione) a known inhibitor of soluble guanylate cyclase in smooth muscle cells (31, 32), induced a dose-related inhibition of IL-1 induced IL-8 generation (Figure 6). No significant toxicity from these doses of either

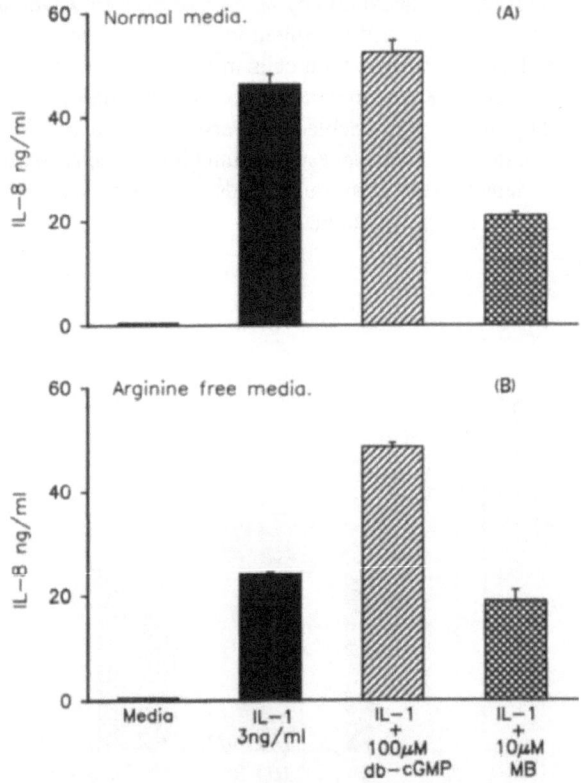

Figure 5. The effect of agents that modify the activity of guanylate cyclase on IL-1 induced IL-8 generation by mesangial cells. Inhibition by methylene blue (hatched bars) and stimulation by dibutyryl cGMP (cross hatched bars) in normal media (A) and arginine free media (B).

methylene blue (10 μM) or LY 83583 (0.1-3.0 μM) was detected by trypan blue exclusion. This data demonstrates that down regulation of guanylate cyclase by either methylene blue or LY 83583 induced significant inhibition of IL-1 induced IL-8 generation, suggesting that guanylate cyclase and cGMP pathway are important for the regulation of IL-8.

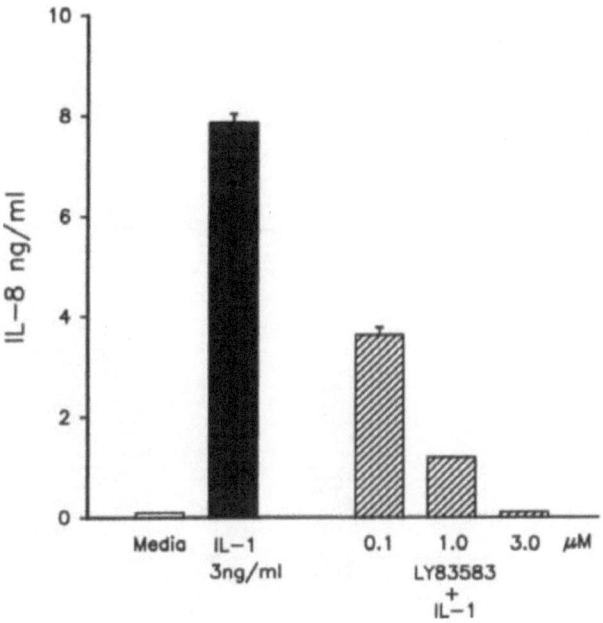

Figure 6. The effect of LY 83583 on IL-1 induced IL-8 generation by human mesangial cells.

The nitrovasodilator sodium nitroprusside (SNP) directly activates soluble guanylate cyclase by spontaneously releasing NO, which in turn activates the soluble guanylate cyclase (33). Pretreatment of arginine-depleted mesangial cells with increasing concentrations of sodium nitroprusside (1-100 μM) for 1 hour followed by the addition of IL-1 (Figure 7B) resulted in enhanced IL-8 generation at the 10 μM and 100 μM doses of SNP compared with IL-1 treated cells. In addition NO was identified in culture supernatants using a sensitive chemiluminescent method (15). Using this extremely sensitive method our data shows that sodium nitroprusside at 100μM induced release of NO from mesangial cells (Figure 7A). Following cytokine activation, mesangial cells displayed significant NO release with 10 ng/ml of IL-1α, while in contrast TNFα induced only a slight increase in NO release above basal levels (Figure 8). Thus sodium nitroprusside, a known stimulator of soluble guanylate cyclase via the generation of NO, is able to potentiate IL-1 induced IL-8 generation. In cytokine-activated mesangial cells, NO synthesis was observed only when high concentrations of IL-1 (10 ng/ml) were used as compared to the dose required to induce chemokine response (IL-1 0.3 - 3 ng/ml) in these cells. We conclude from this data that IL-1 is able to induce the generation of NO in human mesangial cells, thus confirming the earlier observation made using rat mesangial cells and supports a role for NO and cGMP as important modulators of chemokine expression in these cells.

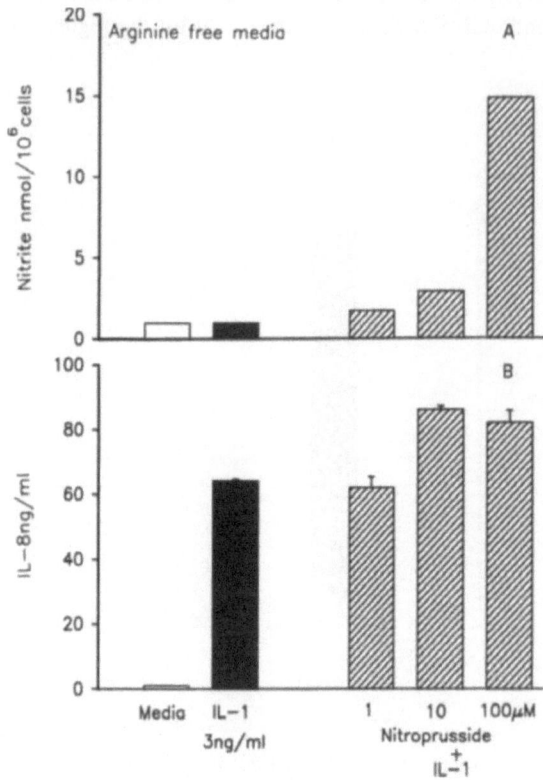

Figure 7. Effect of sodium nitroprusside on IL-1 induced NO (A) and IL-8 generation by human mesangial cells.

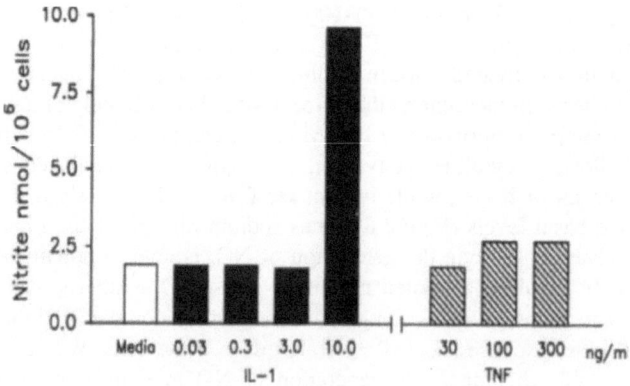

Figure 8. Nitric oxide production by IL-1 (filled bars) and TNFα (hatched bars) stimulated human mesangial cells.

DISCUSSION

Clearly the pathway leading from IL-1 binding to cells to a change in gene expression is highly complex. Changes in second messenger systems such as G-proteins, protein kinases and the activation of transcription factors are likely to be involved. In this communication we have presented evidence which supports a role for NO as a novel and alternative signal transduction mechanism for IL-1 induced expression and release of the chemokine IL-8 in human mesangial cells.

The present studies demonstrate that human mesangial cells in culture have the capacity to produce nitric oxide. We have detected significant increase in nitric oxide in culture supernatants after incubation of mesangial cells with IL-1. Similar increases in nitrite generation were observed when mesangial cells were treated with sodium nitroprusside, a known activator of NO. In addition L-NMMA and L-arginine depletion resulted in a significant reduction in IL-1 induced IL-8 production, L-NMMA is known to specifically prevent NO production from L-arginine (34). Furthermore, both of the guanylate cyclase inhibitors used in our study, methylene blue and LY 83583, resulted in significant inhibition of IL-1 induced IL-8 generation, while dibutyryl cGMP and sodium nitroprusside enhanced IL-8 generation in arginine-depleted cells. We conclude that IL-1 induces the generation of NO resulting in activation of guanylate cyclase and an increase in cGMP, which then upregulates the secretion of IL-8 in mesangial cells. This interaction suggests a novel and alternative signal transduction mechanism which may be a universal system for the regulation of chemokine generation.

ACKNOWLEDGMENTS

This work was supported by The Wellcome Trust, UK. and the National Kidney Research Fund, UK.

REFERENCES

1. Couser, W. G. 1990. Mediation of immune glomerular injury. *J. Am. Soc. Nephrol.* 1: 13-29.
2. Chemotactic cytokines: biology of the inflammatory peptide supergene family. 1991, Advances in Experimental Medicine and Biology. Eds. J Westwick, I.J.D Lindley and S.L Kunkel, Plenum, New York.
3. Westwick, J., S.W. Li, and R.D. Camp. 1989. Novel neutrophil-stimulating peptides. *Immunol. Today.* 10: 146-147.
4. Leonard, E.J., and T. Yoshimura. 1990. Human monocyte chemoattractant protein-1. *Immunol. Today.* 11: 97-101.
5. Mene, P., M.S. Simonson, and M.J. Dunn. 1989. Physiology of the mesangial cell. *Physiol Rev.* 69: 1347-1424.
6. Sedor, J.R., Y. Nakazato, and M. Konieczkowski. 1992. Interleukin-1 and the mesangial cell. *Kidney International.* 41: 595-599.
7. Brown, Z., R.M. Strieter, S.W. Chensue, M. Ceska, I. Lindley, G.H. Neild, S.L. Kunkel, and J. Westwick. 1991. Cytokine activated human mesangial cells generate the neutrophil chemoattractant-interleukin 8. *Kidney Int.* 40: 86-90.
8. Kunser, D.J., E.L. Luebbers, R.J. Nowinski, M. Konieczkowski, C.H. King, and J.R. Sedor, 1991. Cytokine and LPS-induced synthesis of interleukin-8 from human mesangial cells. *Kidney Int.* 39: 1240-1248.
9. Zoja, C., J.M. Wang, S. Bettoni, M. Sironi, D. Renzi, F. Chiaffarino, H.E. Abboud, J. Van Damme, A. Mantovani, G. Remuzzi, and A. Rambaldi. 1991. Interleukinß and TNFα induce gene expression and production of leukocyte chemotactic factors, colony-stimulating factors and interleukin-6 in human mesangial cells. *Am. J. Pathol.* 138: 991-1003.
10. Abbott, F., J.J. Ryan, M. Ceska, K. Matsushima, C.E. Sarraf, and A.J. Rees. 1991. Interleukin-1ß stimulates human mesangial cells to synthesize and release interleukin-6 and -8. *Kidney Int.* 40:597-605.

11. Brown, Z., L. Fairbanks, R.M. Strieter, G.H. Neild, S.L. Kunkel, and J. Westwick. 1991. Human mesangial cell-derived IL-8 and IL-6: modulation by an interleukin 1 receptor antagonist. in Chemotactic cytokines: biology of the inflammatory peptide supergene family. Advances in Experimental Medicine and Biology, 305: 119-126. Eds. J Westwick, I.J.D Lindley and S.L.Kunkel, Plenum, New York.

12. Brown, Z., R.M. Strieter, G.H. Neild, R.C. Thompson, S.L. Kunkel, and J. Westwick. 1992. IL-1 receptor antagonist inhibits monocyte chemotactic peptide 1 generation by human mesangial cells. *Kidney Int.* 42: 95-101.

13. Moncada, S., R.M.J. Palmer, and E.A. Higgs. 1991. Nitric oxide: physiology, pathophysiology and parmacology. *Pharma. Rev.* 43: 109-142.

14. Cairns, H.S., M.E. Rogerson, J. Westwick, and G.H. Neild. 1991. Regional heterogeneity of endothelium-dependent vasodilatation in the rabbit kidney. *J. Physiol.* 436: 421-429.

15. Palmer, R.M.J., D.S. Ashton, and S. Moncada. 1988. Vascular endothelial cells synthesise nitric oxide from L-arginine. *Nature.* 333: 664-666.

16. Rapoport, R.M., and F. Murad. 1988. Role of cyclic GMP in endothelium-dependent relaxation of vascular smooth muscle. In Relaxing and Contracting Factors. P.M. Vanhoutte, editor. Humana press, Clifton, NJ. 219-239.

17. Nathan, C. 1992. Nitric oxide as a secretory product of mammalian cells. *FASEB. J.* 6: 3051-3064.

18. Schmidt, H.H.H.W., and F. Murad. 1991. Purification and characterization of a human NO synthase. *Biochem. Biophys. Res. Commun.* 181: 1372-1377.

19. Bredt, D.S. and S.H. Snyder. 1990. Isolation of nitric oxide synthase, a calmodulin-requiring enzyme. *Proc. Natl. Acad. Sci. USA* 87: 682-685.

20. Stuehr, D., N. Kwon, and C. Nathan. 1990. FAD and GSH participate in macrophage synthesis of nitric oxide. *Biochem. Biophys. Res. Commun.* 168: 558-565.

21. Yui, Y., R. Hattori, K. Kosuga, H. Eizawa, K. Hiki, S. Ohkawa, K. Ohnishi, S. Terao, and C. Kawai. 1991. Calmodulin-independent nitric oxide synthase from rat polymorphonuclear neutrophils. *J. Biol. Chem.* 266: 3369-3371.

22. Radomski, M.W., R.M.J. Palmer, and S. Moncade. 1990. Glucocorticoids inhibit the expression of an inducible, but not the constitutive, nitric oxide synthase in vascular endothelial cells. *Proc. Natl. Acad. Sci. USA* 87: 10043-10047.

23. Knowles, R.G., M. Salter, S.L. Brooks, and S. Moncada. 1990. Anti-inflammatory glucocorticoids inhibit the induction by endotoxin of nitric oxide synthase in the lung, liver and aorta of the rat. *Biochem. Biophys. Res. Commun.* 172: 1042-1048.

24. Shultz, P.J., J.R. Sedor, and H.E. Abboud. 1990. Effects of endothelium-derived relaxing factor and nitric oxide on rat mesangial cells. *Am. J. Physiol.* 258 (Renal Fluid Electrolyte Physiol.27): F162-F167.

25. Shultz, P.J., M.A. Tayeh, M.A. Marletta, and L. Raij. 1991. Synthesis and action of nitric oxide in rat glomerular mesangial cells. *Am. J. Physiol.* 261 (Renal Fluid Electrolyte Physiol. 30): F600-606.

26. Pfeilschifter, J., P. Rob, A. Mulsch, J. Fandrey, K. Vosbeck, and R. Busse. 1992. Interleukin Iß and tumour necrosis factor induce a macrophage-type of nitric oxide synthase in rat renal mesangial cells. *Eur.J. Biochem.* 203: 251-255.

27. Pfeilschifter, J., and H. Schwarzenbach. 1990. Interleukin 1 and tumor necrosis factor stimulate cGMP formation in rat renal mesangial cells. *Febs Lett.* 273: 185.

28. Pfeilschifter, J. 1991. Anti-inflammatory steroids inhibit cytokine induction of nitric oxide synthase in rat renal mesangial cells. *Eur. J. Pharmacol.* 195: 179-180.

29. McCall, T.B., N.K. Boughton-Smith, R.M.J. Palmer, B.J.R. Whittle, and S. Moncada. 1989. Synthesis of nitric oxide from L-arginine by neutrophils. Release and interaction with superoxide anion. *Biochem. J.* 261: 293-296.

30. Martin, W., G.M. Villani, D. Jothianandan, and R.F. Furchgott. 1985. Selective blockade of endothelium-dependent and glyceryl trinitrate-induced relaxation by hemoglobin and by methylene blue in the rabbit aorta. *J Pharm. Exp. Therap.* 232: 708-716.

31. Schmidt, M.J., B.D. Sawyer, L.L. Truex, W.S. Marshall, and J.H. Fleisch. 1985. LY83583: An agent that lowers intracellular levels of cyclic guanosine 3',5'-monophosphate. *J Pharm. Exp. Therap.* 232: 764-769.

32. Mulsch, A., R. Busse, S. Liebau, and U. Forstermann. 1988. LY83583 interferes with the release of endothelium-derived relaxing factor and inhibits soluble guanylate cyclase. *J. Pharm. Exp. Therap.* 247: 283-288.

33. Katsuki, S., W. Arnold, C. Mittal, and F. Murad. 1977. Stimulation of guanylate cyclase by sodium nitroprusside, nitroglycerin and nitric oxide in various tissue preparations and comparison to the effects of sodium azide and hydroxylamine. *J. Cyclic Nucl. Res.* 3: 23-55.

34. Rees, D.D., R.M.J. Palmer, R. Schulz, H.H. Hodson, and S. Moncada. 1990. Characterization of three inhibitors of endothelial nitric oxide synthase *in vitro* and *in vivo*. *Br. J. Pharmacol.* 101: 746-752.

... ...
... ...
... ...

SOME ASPECTS OF NAP-1/IL-8 PATHOPHYSIOLOGY II: CHEMOKINE SECRETION BY EXOCRINE GLANDS

Antal Rot, Anne P. Jones and Louise M.C. Webb

Sandoz Forschungsinstitut
Brunner Strasse 59
Vienna
Austria

INTRODUCTION

In the past several years neutrophil attractant/activation protein-1 (interleukin-8, NAP-1/IL-8) had been characterized as potent neutrophil and lymphocyte agonist (1) produced by a wide variety of cells in different tissues upon stimulation either by major inflammatory cytokines interleukin-1 (IL-1) and tumor necrosis factor (TNF) or by microorganisms and their products (1,2). When injected into humans and experimental animals NAP-1/IL-8 was shown to induce, depending on the route of injection, either leukocytosis and neutrophil accumulation in the lungs (3) or leukocyte emigration into the peripheral injection sites (4-7). In addition, it was linked by a multitude of studies to the pathogenesis of human and experimental inflammatory diseases including adult respiratory distress syndrome (3), chronic inflammatory bowel disease (8), rheumatoid arthritis (9,10), gout (11), asbestosis (12), empyema (13), etc. As a result, generally accepted view emerged that the in vivo production of NAP-1/IL-8 is always associated with pathological conditions, mainly, various forms of inflammation and malignancy. In sharp contrast to this view, here we provide evidence that NAP-1/IL-8 is produced and secreted in substantial quantities by several exocrine glands under either physiological, special physiological or "physiological stress" conditions in normal healthy humans. The glands secreting NAP-1/IL-8 include eccrine sweat glands, mammary glands, lacrimal glands, major salivary glands and minor salivary glands of the oral mucosa. We speculate on the possible role this chemokine plays as a homeostatic mediator responsible for the mucosal host defenses and leukocyte extravasation into the normal tissues.

NAP-1/IL-8 SECRETION BY THE ECCRINE SWEAT GLANDS

The physiological process of sweating, facilitated secretion by the eccrine sweat glands, in humans is induced in a variety of conditions with potentially hazardous outcome for the individual; these include thermal, physical, emotional and mental stress (14,15).

The Chemokines, Edited by I.J.D. Lindley
et al., Plenum Press, New York 1993

Therefore, it is possible that the product of sweating, eccrine sweat, in addition to its well understood role in thermoregulation (14), also contributes to host defense. Previously, sweat was shown to contain molecules involved in host defense such as immunoglobulins of IgA subclass (16) and one of the major inflammatory and host-defense cytokines, interleukin-1 (17,18). Recently we investigated the leukocyte chemotactic activity of normal human eccrine sweat (19). Sweating was induced by jogging exercise performed outdoors by healthy volunteers and was collected into custom made polypropylene pouches (Figure 1) during a 1-8 km run at an approximate speed of 10-15 km/h. Sweat was aspirated from the pouch and immediately filtered through a 0.22 μm millipore filter. Using this procedure 2-100 ml of sweat were collected from individual donors. When sweat was tested in standard neutrophil and monocyte chemotaxis assays (20) using 48-well Boyden-type chemotactic chambers and 10 μm-thick polycarbonate membrane we found its chemotactic activity for both neutrophils (19) and monocytes. The mean migration index of sweat from seven different donors was 3.9 and 2.4 for neutrophils and monocytes, respectively. Using a solid phase double ligand ELISA technique (Sandoz Forschungsinstitut, Vienna, Austria, 9) we found NAP-1/IL-8 in sweat samples from five out of six donors. Its mean concentration level was 178 pg/ml of sweat (SEM=78 pg/ml).

Figure 1. Setup used for collecting sweat from normal healthy volunteers:

When the eccrine sweat was applied on C4 reversed-phase HPLC column and eluted by a gradient of acetonitrile, the neutrophil chemotactic activity separated into several peaks (19). Their number varied in the sweat of different donors between three and seven (19). One of these peaks was identified by ELISA as NAP-1/IL-8 (19) in the sweat of several different donors. Conversely, there was only one peak of monocyte chemotactic activity present in each sweat sample; it eluted in the identical position in all tested sweat samples of five different individuals. This peak had no chemotactic activity for neutrophils, clearly indicating that it is not of bacterial origin: formyl peptides are chemotactic for both monocytes and neutrophils (20). Using the same argumentation, neither of the sweat-derived neutrophil chemotactic peaks could be formyl peptides, since all of them lacked chemotactic activity for monocytes. This notion has to be emphasized, because different bacteria can be found on the surface of the human skin and potentially could have been the source of chemotactic activity present in sweat. Also, subsequently the sweat-derived peak of monocyte chemotactic activity was identified by ELISA as monocyte chemotactic peptide-1 (MCP-1, A. Rot and E. Leonard, unpublished).

In addition, we had to demonstrate that chemokines which appeared in sweat were secreted by the sweat glands rather than simply washed off the stratum corneum in the process of collecting sweat. To control for the leukocyte chemotactic activity possibly present in the stratum corneum of the skin, epidermal scrapings were obtained using a modification of a previously described method (21). The scrapings were homogenized manually using a mortar and pestle for 30 min. at room temperature. The homogenate was gradually resuspended in buffer and sedimented before filtering the supernatant through 0.22 μm filter. The filtrate was applied on HPLC column and the resulting fractions tested for chemotactic activity. Epidermal scrapings-derived filtrate contained one broad peak of chemotactic activity for both monocytes and neutrophils (19), suggesting that its nature was different from the chemotactic peaks found in sweat. This notion was further suggested by fact that in contrast to NAP-1/IL-8 and MCP-1 the stratum corneum-derived chemoattractant when applied on the heparin-Sepharose column did not bind to heparin.

Immunohistochemistry provided further evidence indicating that NAP-1/IL-8 is secreted by the sweat glands rather than stratum corneum derived (19). Immunohistochemical staining of NAP-1/IL-8 was performed on the pieces of normal human skin obtained from patients undergoing elective abdominal surgery after their informed consent. The tissue was shock-frozen, sections 5-8 μm thin were prepared on cryostat put onto siliconized glass slides, fixed for 20 min. in acetone, and then rehydrated in PBS. Sections were stained with mouse monoclonal anti-NAP-1/IL-8 (Sandoz Forschungsinstitut). The antibody binding was visualized using an APAAP detection kit (Dako, Glostrup, Denmark) according to manufacturer's instructions. The slides were counter-stained with hemalaun. Intense staining of the epithelium of the secretory and ductal portions of the coil of the sweat gland, ductal epithelial cells as well as intraluminal contents was observed (Figure 2). In contrast to previous observations by other investigators (22) using our monoclonal antibody, we could observe no staining of the epidermal layer of normal skin.

Theoretically the secreted in sweat NAP-1/IL-8 could either be produced by the sweat glands' epithelial cells or just ultrafiltered from plasma after being produced in distant location and carried to the sweat glands via blood (23). Several other molecules were shown to be secreted in sweat from blood (24). To distinguish between these two

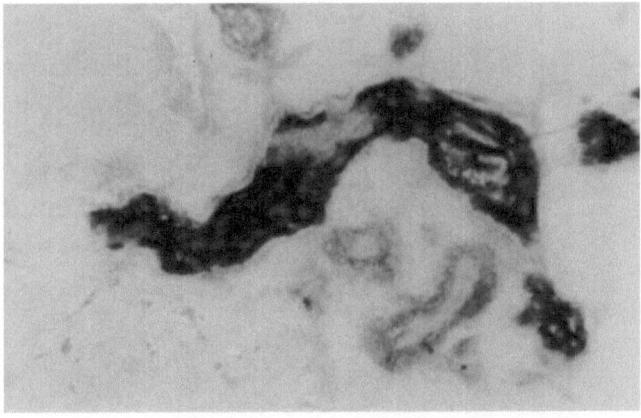

Figure 2. NAP-1/IL-8 immunoreactivity in the epithelial cells of the eccrine sweat glands.

possibilities if situ hybridization study on samples of normal human skin was performed (19). The tissue distribution of NAP-1/IL-8 mRNA was studied using digoxigenin labeled NAP-1/IL-8 cDNA (generous gift of Dr. Howard Young, NCI-FCRDC, Frederick, Maryland). Using this method abundant NAP-1/IL-8 mRNA was detected in the cytoplasm of epithelial cells of eccrine sweat glands but not other dermal or epidermal cells (19), indicating that the NAP-1/IL-8 is produced in situ by the epithelial cells of the eccrine sweat glands .

In summary, we demonstrated (19) that normal human eccrine sweat contains multiple chemoattractants including NAP-1/IL-8 and MCP-1. NAP-1/IL-8 is produced and secreted by the epithelial cells of the eccrine sweat glands.

NAP-1/IL-8 SECRETION BY OTHER EXOCRINE GLANDS

Further on, we explored the possibility that NAP-1/IL-8, in addition to sweat glands, is also produced by other exocrine glands. These studies included the immunohistological staining by monoclonal anti-NAP-1/IL-8 antibody (Sandoz Forschungsinstitut) of tissue obtained from normal non-lactating mammary glands and also tissues representing several pathological conditions of the mammary gland including fibrocystic disease and malignancies. Immunohistochemistry was also performed on tissue obtained from parotid and submandibulary glands and oral mucosa (minor salivary glands). In addition, the NAP-1/IL-8 contents of normal human colostrum of different days after delivery and tears induced in normal human volunteers by the exposure to vapor of a mild irritant (freshly cut onion) was measured using ELISA.

Using immunohistochemistry on tissue samples from normal mammary glands we could detect NAP-1/IL-8 immunoreactivity in the epithelial cells of terminal ductules and small mammary ducts (Figure 3). It has to be noted that mammary tissue samples rated here as "normal" were obtained as biopsies from patients due to the "breast lump" suspect of malignancy but found histologically devoid, on the basis of available tissue blocks, of malignant, fibrocystic, inflammatory or other disease. Thus, in spite of normal histomorphological appearance of the tissue used for immunohistochemistry we cannot conclusively exclude the presence and effect of the obscured disease or hormonal state which led to the perception of "lump" and ultimately resulted in the biopsy.

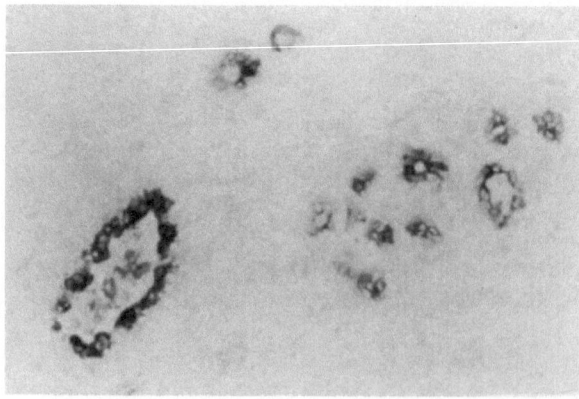

Figure 3. NAP-1/IL-8 immunoreactivity in the epithelial cells of the normal mammary gland.

The immunohistochemistry detected abundant NAP-1/IL-8 immunoreactivity also in the breast tissue biopsies diagnosed as fibrocystic disease (both proliferative and degenerative forms of it). Figures 4A and B represent adenotic and cystic type of lesions, respectively, showing NAP-1/IL-8 immunoreactivity in the epithelial cells lining ductal structures and cysts. Also, in ten out of ten studied cases of infiltrating duct carcinomas we observed NAP-1/IL-8 immunoreactivity in the epithelial cells (Figure 4C), though in different cases of various degree of expression. Various tumors have been shown to produce chemokines, mainly MCP-1; it was suggested that MCP-1 and other chemokines are responsible for the recruitment of leukocytes, mainly monocytes, into these tumors (25,26). It is possible that produced by mammary carcinoma cells NAP-1/IL-8 would cause, analogously to MCP-1, the recruitment of leukocytes into the tumor. However, in the case of NAP-1/IL-8 these cells could be either neutrophils or lymphocytes. Mammary carcinomas are among the malignancies well known to harbor variable number of lymphocytes; studies attempting to correlate the tumor prognosis with the number of tumor-infiltrating lymphocytes yielded contradictory results leaving this question without an answer. Also, tumor cell-derived NAP-1/IL-8, due to its recently described angiogenic effect (27), could contribute to the neovascularization of tumors.

Further, we studied if NAP-1/IL-8 is secreted by the mammary glands. Using ELISA we measured the levels of NAP-1/IL-8 in the human colostrum samples collected from ten mothers on different days after delivery. Figure 5 summarizes the results. We found measurable levels of NAP-1/IL-8 in all studied samples with especially high concentrations on the first couple days after the delivery.

Similarly, ELISA was used to detect NAP-1/IL-8 in tears which were induced in three normal healthy volunteers by exposing their faces to the vapors of freshly cut onion. Mean value of 4.7 ng /ml was measured.

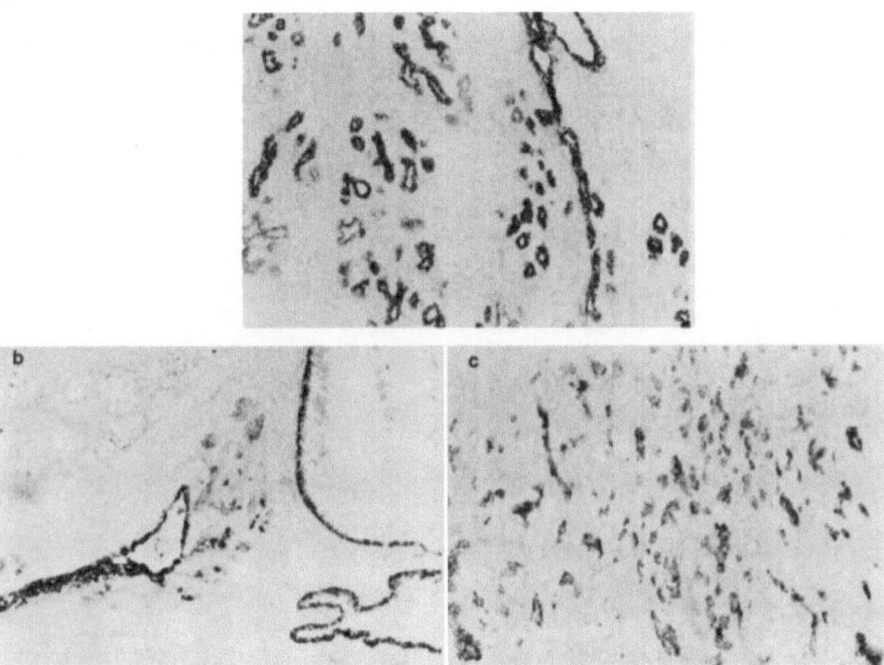

Figure 4. NAP-1/IL-8 immunoreactivity in (A) the tissue from an adenotic lesion of fibrocystic disease of the mammary gland, (B) the tissue from a cystic lesion of fibrocystic disease of the mammary gland. Note the absence of immunoreactivity in the area of apocrine metaplasia and (C) the parenchymal component of an infiltrating ductal breast carcinoma.

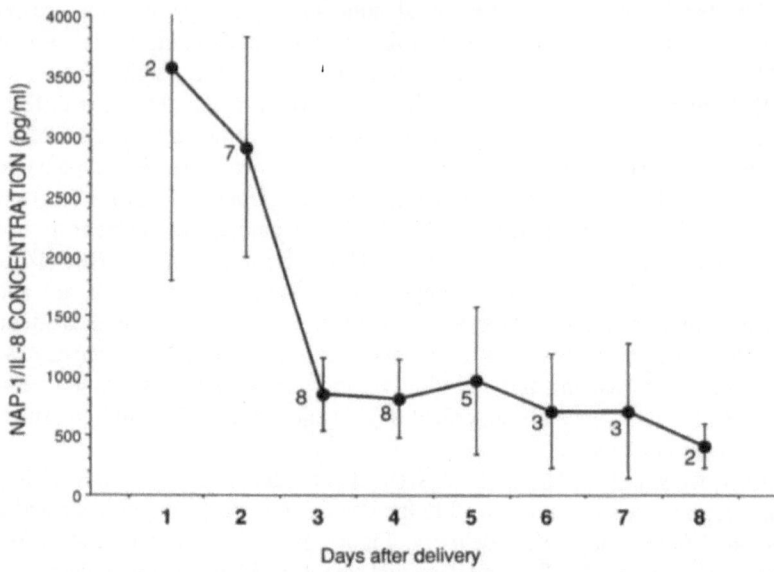

Figure 5. NAP-1/IL-8 concentration in the colostrum at different days after the delivery. Mean concentration value for each day is shown. Error bars show the SEM. Number next to each point indicates the number of observation on the particular day. The colostrum of ten volunteer mothers was studied.

Finally, immunohistochemistry detected NAP-1/IL-8 immunoreactivity in the epithelial cells of major (Figure 6) and minor salivary glands.

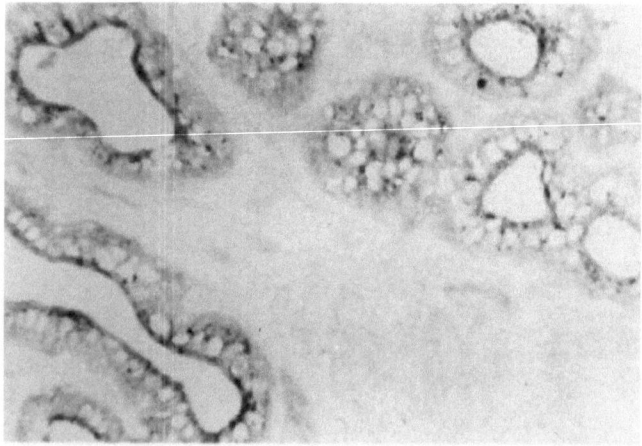

Figure 6. NAP-1/IL-8 immunoreactivity in the large salivary gland.

POSSIBLE PHYSIOLOGICAL AND PATHOPHYSIOLOGICAL ROLE OF NAP-1/IL-8 SECRETED BY EXOCRINE GLANDS

Here we provide multiple evidence indicating that NAP-1/IL-8 is secreted by several exocrine glands of normal healthy humans. NAP-1/IL-8 which is produced by these glands in different amounts and under different circumstances, and probably, plays a different physiological or pathophysiological role in each of these locations. The unifying pattern of action in different locations and the keystone of our speculative discussion on the possible role of NAP-1/IL-8 produced under physiological conditions is the ability of NAP-1/IL-8 to bind to the endothelial cells of postcapillary venules and induce leukocyte adhesion and emigration (28,29). It was suggested that this mechanism functions during the leukocyte recruitment into the lesions of different pathologies, e.g. inflammatory and tumor sites. Here we suggest that the induction of lymphocyte emigration into the normal tissues is also, in part, dependent on presence of chemokines. It is a well known phenomenon that in normal healthy individuals lymphocytes leave the circulation and accumulate in the mucosal organs and skin where they fulfill their role in immune surveillance and host defense (30). Skin and mucosal organs, in turn, by the virtue of their ability to harbor lymphocytes are considered tertiary lymphoid organs. It was suggested previously that the firm adhesion of lymphocytes to the endothelium in the process of their extravasation, both into normal and inflammatory tissues, requires the activation of lymphocyte integrins (30). Chemokines, similarly to other chemoattractans, were shown to induce such very rapid activation of leukocyte integrins (31). While chemokines were known to be present in the lesions of different pathologies, till now they were thought to be absent (at least in their active form) from the normal tissues and therefore not considered as potential inducers of lymphocyte accumulation into the tertiary lymphoid organ. Here we demonstrate that NAP-1/IL-8, the prototype chemokine, is produced and secreted in substantial amounts by several exocrine glands belonging to the tertiary lymphoid organs of normal healthy humans. Hence NAP-1/IL-8, we suggest, should be considered as the source of chemotactic activity responsible for lymphocyte accumulation into skin, mammary glands, mucosal organs, etc. For example, in skin it has been observed previously that lymphocytes accumulate in the epidermis around the acrosyringium, area of epidermis traversed by the eccrine sweat ducts (32). Also, alternative skin sources of lymphocyte chemotactic activity, including NAP-1/IL-8, have been suggested previously. Under pathological conditions NAP-1/IL-8 was shown to be produced also by keratinocytes (33) and, in addition, under the physiological conditions molecule with NAP-1/IL-8 antigenic properties was detected in epidermis (22); however, the epithelial cells of the eccrine sweat glands and not the keratinocytes are the only site in normal human skin where both the NAP-1/IL-8 immunoreactivity and mRNA could be detected (19,33).

The outcome of the NAP-1/IL-8 production by different exocrine glands seems to be very different. When secreted in sweat, NAP-1/IL-8 induces lymphocyte accumulation in the epidermis but leukocytes do not appear in the sweat itself, whereas when secreted by the mammary gland NAP-1/IL-8 also causes, we suggest, the previously described accumulation of both neutrophils and lymphocytes in the mammary tissue and also in the colostrum (34-37). Such difference is puzzling and may be explained by the different concentrations of the secreted chemokine in skin and mammary gland. So far we have no clues as to the mechanism of the induction of NAP-1/IL-8 in different exocrine glands. The secretory function of exocrine glands is regulated by different hormones and peripheral nervous system (14,24); therefore, on one hand, it seems feasible to postulate that NAP-1/IL-8 production in these glands can also be induced by these mechanisms. On the other hand, IL-1, one of the cytokines capable of inducing NAP-1/IL-8 in variety of tissues had been show to be present both in sweat and colostrum (17,18,38).

Several presented here examples show that the ability of the secretory epithelial cells to produce NAP-1/IL-8 is also retained in different pathological conditions affecting the exocrine glands. NAP-1/IL-8 produced under the pathological conditions can contribute to the pathoethiology of the disease. This can manifest e.g. as leukocyte recruitment into the malignant tumors with all the complexity of consequences(25) or as exacerbation of inflammatory reaction in several skin diseases accompanied by the disturbed free flow of sweat, e.g. include psoriasis and atopic dermatitis (39).

In conclusion, the production of NAP-1/IL-8 by several exocrine glands not only adds another interesting anatomical location to the already long list of sites were this chemokine is produced (1), but also pinpoints NAP-1/IL-8 as a molecule involved in physiological phenomena of leukocyte extravasation into the normal healthy tissues.

ACKNOWLEDGMENT

We are grateful to Kamillo Thierer for skillful technical assistance and Dr. József Timár for providing clinical samples and the Lindleys for their patience and careful reading/typing of this manuscript. Anne Jones and Louise Webb were supported by the EC fellowship COMETT II by the UETP DANUBE, Vienna, Austria.

REFERENCES

1. Oppenheim, J.J., C.O.C. Zachariae, N. Mukaida, and K. Matsushima. 1991. Properties of the novel proinflammatory supergene "intercrine" cytokine family. *Annu. Rev. Immunol.* 9: 617-48.
2. Damme, Van J., B. Decock, R. Conings, J.-P. Lenaerts, G. Opdenakker, and A. Billiau. 1989. The chemotactic activity for granulocytes produced by virally infected fibroblasts is identical to monocyte-derived interleukin 8. *Eur. J. Immunol.* 19: 1189-1194.
3. Rot, A. 1991. Some aspects of NAP-1 pathophysiology: lung damage caused by a blood-borne cytokine. *Chemotactic Cytokines.* 127-135. Edited by J. Westwick. et. al. Plenum Press, New York.
4. Foster, S.J., D.M. Aked, J.-M. Schröder, and E. Christophers. 1989. Acute inflammatory effects of a monocyte-derived neutrophil-activating peptide in rabbit skin. *Immunology* 67: 181-183.
5. Rot, A. 1991. Chemotactic potency of recombinant human neutrophil attractant/activation protein-1 (interleukin-8) for polymorphonuclear leukocytes of different species. *Cytokine* 3: 21-27.
6. Leonard, E.J., T. Yoshimura, S. Tanaka, and M. Raffeld. 1991. Neutrophil recruitment by intradermally injected neutrophil attractant/activation protein-1. *J. Invest Dermatol* 96: 690-694.
7. Swensson, O., C. Schubert, E. Christophers, and J.-M. Schröder. 1991. Inflammatory properties of neutrophil-activating protein-1/interleukin 8 (NAP-1/IL-8) in human skin: a light- and electronmicroscopic study. *J. Invest Dermatol* 96: 682-689.
8. Mahida, Y.R., M. Ceska, F. Effenberger, L. Kurlac, I. Lindley, C.J. Hawkey. 1992. Enhanced synthesis of NAP-1/IL-8 in active ulcerative colitis. *Clinical Science* 82: 273-5.
9. Peichl, P., M. Ceska, F. Effenberger, G. Haberhauer, H. Broell, I.J.D. Lindley. 1991. Presence of NAP-1/IL-8 in synovial fluids indicates a possible pathogenic role in rheumatoid arthritis. *Scand J. Immunol.* 34: 333-9.
10. Brennan, F.M., C.O.C. Zachariae, D. Chantry, et. al. 1990. Detection of interleukin 8 biological activity in synovial fluids from patients with rheumatoid arthritis and production of interleukin 8 mRNA by isolated synovial cells. *Eur. J. Immunol.* 20: 2141-4.
11. Terkeltaub, R., C. Zachariae, D. Santoro, J. Martin, P. Peveri, K. Matsushima. 1991. Monocyte-derived neutrophil chemotactic factor/interleukin 8 is a potential mediator of crystal-induced inflammation. *Arthritis Rheum.* 34: 894-903.
12. Boylan, A.M., C. Rüegg, K.J. Kim, et. al. 1992. Evidence of a role for mesothelial cell-derived interleukin 8 in the pathogenesis of asbestos-induced pleurisy in rabbits. *J. Clin. Invest.* 89: 1257-67.
13. Broaddus, V.C., C.A. Hèbert, R.V. Vitangcol, J.M. Hoeffel, M.S. Bernstein, and A.M. Boylan. 1992. Interleukin-8 is a major neutrophil chemotactic factor in pleural liquid of patients with empyema. *Am Rev. Respir. Dis.* 146: 825-830.
14. Quinton, P.M. 1983. Sweating and its disorders. *Ann. Rev. Med.* 34: 429-52.
15. Köhler, T., and U. Troester. 1991. Changes in the palmar sweat index during mental arithmetic. *Biological Psychology* 32: 143-154.

16. Okada, T., H. Konishi, M. Ito, H. Nagura, and J. Asai. 1988. Identification of Secretory Immunoglobulin A in human sweat and sweat glands. *J. Invest. Dermatol.* 90: 648-651.

17. Reitamo, S., H.S.I. Anttila, L. Didierjean, and J.-H. Saurat. 1990. Immunohistochemical identification of interleukin Iα and β in human eccrine sweat-gland apparatus. *British Journal of Dermatology* 122: 315-323.

18. Didierjean, L., D. Gruaz, Y. Frobert, J. Grassi, J.-M. Dayer, and J.-H. Saurat. 1990. Biologically active interleukin 1 in human eccrine sweat: site-dependent variations in α/β ratios and stress-induced increased excretion. *Cytokine* 2: 438-446.

19. Jones, A.P., L.M.C. Webb, E.J. Leonard, A.O. Anderson, and A. Rot. NAP-1/IL-8 secretion by the eccrine sweat glands. Submitted.

20 Rot, A., L.E. Henderson, T.D. Copeland, and E.J. Leonard. 1987. A series of six ligands for the human formyl peptide receptor: tetrapeptides with high chemotactic potency and efficacy. *Proc. Natl. Acad. Sci. USA* 84: 7967-7971.

21. Camp, R., N. Fincham, J. Ross, C. Bird, and A. Gearing. 1990. Potent inflammatory properties in human skin of interleukin-1 alpha-like material isolated from normal skin. *J. Invest. Dermatol.* 94: 735-741.

22. Sticherling, M., E. Bornscheuer, J.-M. Schröder, and E. Christophers. 1991. Localization of neutrophil-activating peptide-1/interleukin-8-immunoreactivity in normal and psoriatic skin. *J. Invest. Dermatol.* 96: 26-30.

23. Sylvester, I., T. Yoshimura, M. Sticherling, J.-M. Schröder, M. Ceska, P. Peichl, and E.J. Leonard. 1992. Neutrophil attractant protein-1-immunoglobulin G immune complexes and free anti-NAP-1 antibody in normal human serum. *J. Clin. Invest.* 90: 471-481.

24. Sato, K. 1983. The physiology and pharmacology of the eccrine sweat gland. In *Biochemistry and Physiology of the Skin*. Edited by Lowell A. Goldsmith. Oxford University Press.

25. Mantovani, A., B. Bottazzi, F. Colotta, S. Sozzani, and L. Ruco. 1992. The origin and function of tumor-associated macrophages. *Immunology Today* 13: 265-270.

26. Abruzzo, L.V., A.J. Thornton, M. Liebert, H.B. Grossman, H. Evanoff, J. Westwick, R.M. Strieter, and S.L. Kunkel. 1992. Cytokine-induced gene expression of interleukin-8 in human transitional cell carcinomas and renal cell carcinomas. *Am. J. Pathol.* 140: 365-373.

27. Koch, A.E., P.J. Polverini, S.L. Kunkel, L.A. Harlow, L.A. DiPietro, V.M. Elner, S.G. Elner, and R.M. Strieter. 1992. Interleukin-8 as a macrophage-derived mediator of angiogenesis. *Science.* 258: 1798-1801.

28. Rot, A. 1992. Binding of neutrophil attractant/activation protein-1 (interleukin 8) to resident dermal cells. *Cytokine* 4: 347-352.

29. Rot, A. 1992. Endothelial cell binding of NAP-1/IL-8: role in neutrophil emigration. *Immunology Today* 13: 291-294.

30. Picker, I.J., and E.C. Butcher. 1992. Physiological and molecular mechanisms of lymphocyte homing. *Annu Rev. Immunol.* 10: 561-591.

31. Detmers, P.A., S.K. Lo, E. Olsen-Egbert, et. al. 1990. *J. Exp. Med.* 171: 1155-1162.

32. Foster, C.A., H. Yokozeki, K. Rappersberger, F. Koning, B. Volc-Platzer, A. Rieger, J.E. Coligan, K. Wolff, and G. Stingl. 1990. Human epidermal T cells predominately belong to the lineage expressing α/β T cell receptor. *J. Exp. Med.* 171: 997-1013.

33. Gillitzer, R., R. Berger, V. Mielke, C. Müller, K. Wolff, and G. Stingl. 1991. Upper keratinocytes of psoriatic skin lesions express high levels of NAP-1/IL-8 mRNA in situ. *J. Invest. Dermatol.* 97: 73-79.

34. Riedel-Caspari, G., and F.-W. Schmidt. 1991. The influence of colostral leukocytes on the immune system of the neonatal calf. 1. Effects on lymphocyte responses. *Dtsch. tierärztl. Wschr.* 98: 102-107.

35. Chernishov, V.P., and I.I. Slukvin. 1990. Mucosal immunity of the mammary gland and immunology of mother/newborn interrelation. *Archivum Immunologiae et Therapiae Experimentalis* 38: 145.

36. Reardon, C., L. Lefrancois, A. Farr, R. Kubo, R. O'Brien, and W. Born. 1990. Expression of γ/δ T cell receptors on lymphocytes from the lactating mammary gland. *J. Exp. Med.* 172: 1263-1266.

37. Bertotto, A., R. Gerli, G. Fabietti, S. Crupi, C. Arcangeli, F. Scalise, and R. Vaccaro. 1990. Human breast milk T lymphocytes display the phenotype and functional characteristics of memory T cells. *Eur. J. Immunol.* 20: 1877-1880.

38. Munoz, C., S. Endres, J. van der Meer, L. Schlesinger, M. Arevalo, and C. Dinarello. 1990. Interleukin-1β in human colostrum. *Res. Immunol.* 141: 505-513.

39. Hu, C.-H. 1991. Sweat-related dermatoses: old concept and new scenario. *Dermatologica* 182: 73-76.

MOLECULAR MECHANISM OF INTERLEUKIN-8 GENE EXPRESSION

Shu-ichi Okamoto[1], Naofumi Mukaida[1], Kazuo Yasumoto[1,2],
Hyogo Horiguchi[1,3], and Kouji Matsushima[1]

Departments of [1]Pharmacology and [2]Surgery
Cancer Research Institute
Kanazawa University Kanazawa
920, Japan

[3]Department of Public Health
Faculty of Medicine
Toyama Medical and Pharmaceutical University
Toyama, 930-01, Japan

INTRODUCTION

Leukocyte infiltration into the lesion is a hallmark of inflammatory reactions to tissue injuries caused by various conditions including trauma, invasion of foreign particles, ischemia-reperfusion syndrome, malignant diseases, and autoimmune diseases. The types of infiltrating leukocytes depend on the kind, degree, and timing of tissue injuries. Although several leukocyte chemotactic factors including C5a, leukotriene B_4 have long been identified, these factors have chemotactic activities for any kind of leukocytes, suggesting the existence of cell type-specific leukocyte chemotactic factors.

Over the past several years, numerous members of a new family of heparin-binding polypeptide cytokines with a molecular weight of 8-10 kDa have been identified (1). These cytokines have 4 cysteine residues at well-conserved positions and are presumed to possess two disulfide bonds, resulting in the formation of a similar structure. Furthermore, it has become evident that some of these cytokines have chemotactic activity for specific type(s) of leukocytes, suggesting that these factors are bona fide chemoattractants involved in leukocyte recruitment observed in the inflammatory response. Interleukin-8 (IL-8), formerly called monocyte-derived neutrophil chemotactic factor or neutrophil activating peptide-1, has been shown to be a member of this newly discovered family of chemotactic cytokines (1).

IL-8 was originally purified from lipopolysaccharide (LPS)-stimulated human peripheral blood mononuclear cells (PBMC) as a neutrophil chemotactic factor (2). Subsequent studies revealed that IL-8 is chemotactic for T lymphocytes and basophils as well as neutrophils (3,4). IL-8 also activates neutrophils to release lysosomal enzymes including myeloperoxidase, a-mannosidase and b-glucuronidase (1). Moreover, IL-8 increases shedding of LECAM and expression of Mac-1 of neutrophils, thus inducing the adhesion of neutrophils to endothelial cells and transendothelial migration of neutrophils (5). These results suggest that the overproduction of IL-8 will lead to infiltration of leukocytes, particularly neutrophils, which is often observed in the inflammatory lesions. Thus, IL-8 gene expression should be under the tight control to prevent aberrant production of IL-8 protein.

Availability of antibodies and cDNA for IL-8 enabled us to identify additional types of IL-8 producing cells and their stimulants (Table 1). Most cell types produce little, if any, IL-8 constitutively. However, a wide variety of cell have been shown to express IL-8 mRNA accompanied with production of large amount of protein when stimulated with various types of mitogens, lectins, viruses, tumor promotors, and pro-inflammatory cytokines including IL-1 and tumor necrosis factor a (TNFa) (1). These facts support the notion that IL-8 gene expression is strictly controlled. Hence, it is important to clarify the gene activation mechanism of IL-8 which is presumed to be involved in the leukocyte infiltration observed in various inflammatory processes. Thus, elucidation of IL-8 gene activation may pave a new way to control the leukocyte infiltration, thus leading to the alleviation of inflammation.

In this article, we will discuss the molecular analysis of the IL-8 gene expression induced by various stimuli, as well as IL-8 gene repression by an immunosuppressant, FK-506, and glucocorticoids.

Table 1. Cell sources and inducers of IL-8.

Cell	Inducers
Monocytes/ macrophages	LPS, IL-1, TNF, phorbol ester, ConA uric acid crystals, plant polysaccharide
NK cells	IL-2 + anti-CD16
T lymphocytes	phorbol ester + calcium ionophore staphylococcus enterotoxin A
neutrophils	zymosan, phorbol ester
fibroblasts	IL-1, TNF, phorbol ester viruses (measles, rubella)
keratinocytes	LPS, IL-1, TNF + IFN-gamma
melanocytes/ melanoma cells	IL-1, TNF, phorbol ester
endothelial cells	LPS, phorbol ester, IL-1, TNF
hepatoma cells	IL-1, TNF
glioblastoma/ astrocytoma cells	IL-1, TNF
lung epithelial cells	IL-1, TNF
synovial cells	IL-1, TNF
gastric cancer cells	IL-1, TNF, TNF + IFN-gamma

MECHANISM OF IL-8 GENE ACTIVATION BY PROINFLAMMATORY CYTOKINES

Various kinds of cells express IL-8 mRNA rapidly and massively when stimulated with either IL-1, TNFa or phorbol ester. This increase of IL-8 mRNA is observed within 1 hr, reaches a maximal level within 3 hr after stimulation, and decreases thereafter. Simultaneous treatment of cells with a protein synthesis inhibitor, cycloheximide, did not inhibit, but rather enhanced mRNA expression induced by IL-1, TNFa, or phorbol ester, suggesting that IL-8 mRNA induction does not require *de novo* protein synthesis (6). Alternatively, it is probable that cycloheximide may activate IL-8 transcription as reported in the case of an IL-8-related chicken gene, 9E3 (7).

Nuclear run-off assays demonstrated that this mRNA induction is at least partly ascribed to the activation of transcription in several types of cells including human endothelial cells (8), a human fibrosarcoma cell line, 8387 (9), and a human astrocytoma cell line, U373 (10). Thus, we assumed that IL-8 gene activation provides a good model for studying the mechanisms of regulation of gene activation induced by IL-1, TNFa, or phorbol ester.

In order to clarify the molecular mechanisms of IL-8 gene transcription, we cloned and determined the entire sequence of genomic IL-8 including 1.5 kb 5'-upstream region (11). The analysis revealed that IL-8 genomic DNA consists of 4 exons and 3 introns with a single "TATA" and "CAT"-like structure (Fig. 1.A). S_1 nuclease protection assay revealed that the IL-8 gene has a single transcription start point (our unpublished data) in contrast to the IL-6 gene which has multiple transcription start points (12). The 5'-flanking region of the IL-8 gene shows no overall sequence similarity with other cytokines and acute phase reactant genes, whose expression is also induced by IL-1 or TNF. However, the 5'-flanking region contains several potential binding sites for known transcription factors including glucocorticoid receptor, hepatocyte nuclear factor-1, interferon regulatory factor-1, AP-1, AP-3, NF-IL6, octamer motif binding protein, and NF-kB (Fig. 1.B).

To determine the minimal enhancer region which is involved in conferring the responsiveness to IL-1, TNF, or phorbol ester, sequentially deleted 5'-flanking regions were inserted into chloramphenicol acetyl transferase (CAT) expression vectors and were transfected into a human fibrosarcoma cell line, 8387, in which IL-1, TNF, or phorbol ester

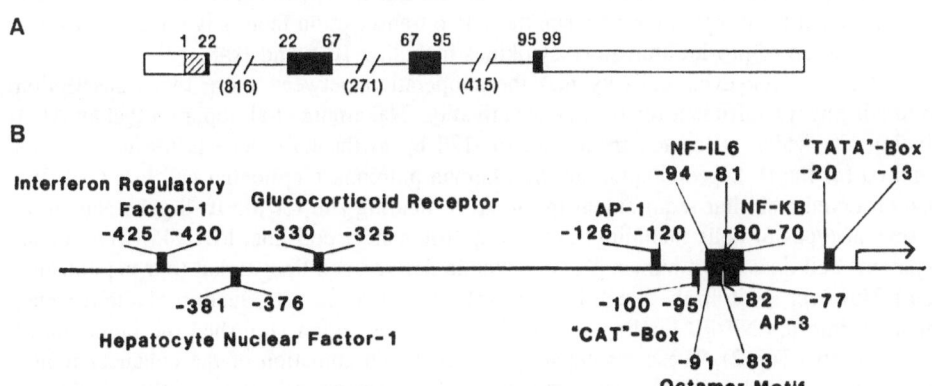

Figure 1.A. Genomic structure of human IL-8 gene. The coding and non-coding regions are indicated by closed and open boxes, respectively.
B. Potential binding sites for known transcription factors.

induces transcription of the IL-8 gene (13). The analysis revealed that the sequence from -94 to -71 bp, which probably consists of NF-kB and NF-IL6-like factor binding sites, is minimally sufficient for the responsiveness to all three stimuli.

Gel retardation assays further demonstrated that the two distinct nuclear factors bind to this region in different manners: NF-IL6-like factor binds to the region between -94 and -81 bp constitutively while IL-1, TNF, or phorbol ester induced the binding of an NF-kB-like factor to the region from -80 to -71bp (13). Thus, the cooperation between these two distinct factors seems to be essential for IL-8 gene activation induced by these stimuli. A similar mechanism seems to work in IL-1- or TNF-induced IL-8 gene activation in U373 cells and a human glioblastoma cell line, T98G (our unpublished data). The cooperation between NF-kB- and NF-IL6-like factors is also required for IL-8 gene activation induced by X-protein of hepatitis B virus as revealed by CAT assays (14).

Over the past several years, it has become evident that NF-kB consists of subunits which are the products of the rel oncogene family (15). NF-kB is classically presumed to be a heterodimer consisting of two subunits, p50 and p65. This heterodimer exists in cytosol in association with Ik-B in a resting state (16). When the cells are activated with mitogens or cytokines, NF-kB is postulated to be released from the complex after the phosphorylation of Ik-B. The dissociated p50-p65 complex translocates into the nucleus and induces gene activation after NF-kB binds to the corresponding cis-element. We previously reported that IL-1/TNF/phorbol ester all activated a common serine kinase which is distinct from protein kinase C, protein kinase A, or casein kinase (17). X-protein of hepatitis B virus has serine/threonine kinase activity by itself (18). Based on these results, we postulated that all these stimuli directly or indirectly induced phosphorylation of Ik-B, resulting in the dissociation of NF-kB heterodimer from Ik-B. Dissociated NF-kB complex translocates into the nucleus where NF-kB binds to the kB binding site on the target genes. NF-kB complex induces the gene activation in cooperation with NF-IL6 which binds to the corresponding cis-elements. This mechanism seems to work also in the IL-6 gene which contains NF-IL6 as well as an NF-kB binding element (Kishimoto et al., personal communication).

This hypothesis has been strengthened by several additional results. In collaboration with Dr. Kishimoto in Osaka University, we observed that co-transfection with p50 and p65 of NF-kB and NF-IL6 expression vectors resulted in the synergistic induction of CAT activity when either IL-6-enhancer- or IL-8-enhancer-driven CAT expression vector was used as a reporter gene (our unpublished data). Recently, LeClair et al. observed that the p50 subunit of NF-kB directly associates with NF-IL-6, and that the rel domain and leucine-zipper motif, respectively, are important for this interaction (19), further supporting the notion that the cooperation between these two transcription factors is important for the gene activation of pro-inflammatory cytokines including IL-6 and IL-8.

It seems, however, unlikely that the cooperation between these two transcription factors is always sufficient for IL-8 gene activation. Nakamura et al. reported that an AP-1 binding site which is located from -126 to -120 bp in the IL-8 gene is presumed to be required for the IL-8 gene expression in a human pulmonary epithelial cell line (20). We also observed a similar requirement for an AP-1 binding site for the IL-8 gene activation in several types of cells including a human gastric cancer cell line, MKN45 (21), human hepatoma cell lines and a human T cell leukemia-derived cell line, Jurkat (our unpublished data). However, even in these cell lines, we observed that the introduction of site-directed mutation into the NF-kB binding site of IL-8 enhancer region abolished the induction of CAT activity (Fig. 2). These results suggest that the combination of the enhancer region minimally essential for IL-8 gene activation differs among cell types, although the kB binding site is always important. The requirement for the different sets of transcription factors may be due to the different distribution of each transcription factor in each cell type although the precise mechanism remains to be investigated.

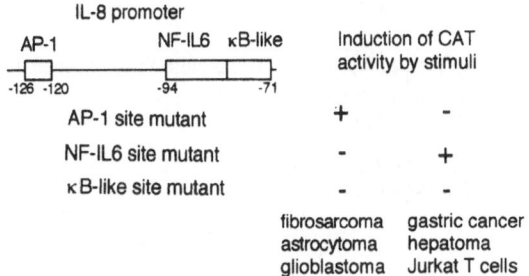

Figure 2. Determination of the enhancer activity of the 5'-flanking region of the IL-8 gene in various types of cells.

One intriguing question has been raised concerning the identity of the transcription factors that binds to kB-binding site in the IL-8 gene. As shown in Fig. 3, the nucleotide sequence of the kB-binding site in the IL-8 gene does not correspond completely with that in immunoglobulin kappa chain gene to which classical p50-p65 NF-kB complex has been shown to bind. The kB-binding site in the Igk gene contains three repeat of guanidine which is essential for NF-kB complex to recognize target genes (22). However, neither sense nor anti-sense strand of kB-binding site in the IL-8 gene contains three repeat of guanidine, raising the possibility that transcription factor(s) distinct from p50-p65 complex actually bind to this site in the IL-8 gene. Antibodies against p50 and p65 failed to interfere with complex formation between the nuclear extracts from Jurkat cells and kB-binding site in the IL-8 gene (our unpublished data), indicating that neither p50 nor p65 was involved in the complex formation. Further, antibodies against c-rel and p50-B/p49 did not have any effect on the complex formation observed in the gel retardation assay, suggesting that these rel-related factors were not responsible for the formation.

The number of rel-related transcription factors has increased rapidly over the past two years (15). These transcription factors have been demonstrated to form homo- or hetero-dimers with each other. These dimers showed different affinity towards various types of kB-binding sites in various kinds of genes. Thus, it is probable that the combination of un-identified or some identified rel-related transcription factor(s) is involved in the IL-8 gene expression. However, biochemical characterization of the transcription factor(s) that bind to kB-binding site in the IL-8 will be required to show which transcription factors are actually involved in the IL-8 gene activation.

TNF AND IFN-GAMMA SYNERGISTICALLY INDUCES IL-8 GENE ACTIVATION

Lymphocyte infiltration is often observed in the adjacent region of tumors by patho-logical examination (23). We postulated that tumor cells themselves produced IL-8, which in turn induced migration of lymphocytes into the lesion. We examined whether human gastric cancer cell lines could produce IL-8 in response to various cytokines. Among 9 cell lines that we examined, eight cell lines produced significant amount of IL-8 in the presence of IL-1 or TNF as revealed by enzyme-linked immunosorbent assay (21). Moreover, in two cell lines, MKN45 and KATO, IFN-gamma synergistically enhanced TNF-induced IL-8 production whereas IFN-gamma alone failed to induce IL-8 production. This synergistic action of IFN-gamma could not be ascribed to the effects of IFN-gamma on TNF receptor expression since IFN-gamma did not have any effects on the number and affinity of TNF receptors at all as revealed by Scatchard analysis.

```
Consensus          GGGRNNYYCC

Ig κ light chain   GGGACTTTCC          R; purines, Y; pyrimidines, N; any nucleotides
MHC class I        GGGGATTCCC
β-interferon       GGGAAATTCC
IL-6               GGGATTTTCC

IL-8               TGGAATTTCC
                   AGGAAATTCC
        sense
      anti-sense
```

Figure 3. Comparison of the sequence of the kB-binding site of the IL-8 gene with those of other genes.

Northern blotting analysis indicated that TNFa and IFN-gamma synergistically induced IL-8 mRNA expression. Moreover, IFN-gamma synergized with TNFa to induce CAT activity when IL-8-enhancer-driven CAT expression vector was employed as a reporter gene, indicating that synergism occurred at the transcriptional level. Based on further analysis using various CAT expression vectors containing either deleted or mutated 5'-flanking region (Fig. 2), we concluded that the minimal essential *cis*-elements responsible for this synergism are the AP-1 and kB-binding sites, and that both elements are indispensable for IL-8 gene activation in this cell line (21). We also observed that NF-IL6 binding site was not necessary for IL-8 gene in this cell line, as mentioned above.

In order to characterize the transcription factors which bound to either AP-1 or kB-binding site in the IL-8 gene, we performed gel retardation assays using each *cis*-element as a probe. Formation of AP-1 complex was enhanced only when cells were stimulated simultaneously with TNFa and IFN-gamma, whereas treatment of cells with either cytokine alone failed to enhance the complex formation compared with unstimulated cells (21). Moreover, IFN-gamma alone did not have any effect on kB complex formation compared with unstimulated cells. TNFa, however, induced kB complex formation which was significantly enhanced by the addition of IFN-gamma (Fig. 4) (21).

Synergism between TNF and IFN-gamma has been described on several biological responses including TNF-induced cytotoxicity against tumor cell lines (24) and expression of HLA class I antigens (25). In the former case, up-regulation of TNF receptors by IFN-gamma have been postulated to contribute to the enhancement of TNF cytotoxicity although the increase of the number of TNF receptors by IFN-gamma dose not always correlate with enhanced cytotoxicity by TNF and IFN-gamma (24). In the case of synergistic induction of HLA class I antigen expression on vascular endothelial cells, IFN-gamma did not affect the affinity and number of TNF receptors (25), consistent with our results. Taken collectively, it is likely that the signal through TNF and IFN-gamma receptors converge to activate the transcriptional factors to activate synergistically several sets of genes. At present, it remains to be investigated how this convergence of intracellular signals modify transcriptional factor(s) leading to gene activation.

SUPEROXIDE ANION-INDUCED IL-8 PRODUCTION

Cadmium, a heavy metal used in a wide variety of industries, has been recognized as causing toxic effects in the human body, including kidney injury, osteomalacia, anemia, and pulmonary emphysema (26). Pathological examinations of animals exposed to cadmium revealed the presence of infiltration of leukocytes in the lesion (26). These findings suggest that cadmium toxicity is associated with inflammatory process in which neutrophils are predominantly involved. Consequently, it is reasonable to speculate that cadmium might be an inducer of proinflammatory cytokines, in particular, IL-8, leading to inflammatory organ damage.

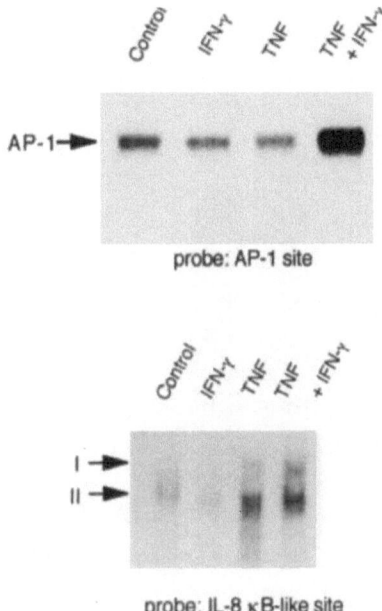

Figure 4. Gel retardation assays using either AP-1 or kB-binding site in the IL-8 gene as a probe. Nuclear proteins were extracted from either unstimulated or stimulated MKN45 cells.

Cadmium at the concentration of $10^{-4}M$ induced human PBMC to produce significant amount of IL-8 without causing cell damage. Northern blotting analysis revealed that cadmium also induced IL-8 mRNA in human PBMC, suggesting that the effect of cadmium on IL-8 production is at pretranslational level. Human PBMC produced IL-8 in the presence of other heavy metals including lead, mercury, cobalt, zinc, and chromium.

Since cadmium has been reported to induce the generation of superoxide anion radicals that are presumed to trigger the inflammatory processes, we explored the possible involvement of superoxide anion in cadmium-induced IL-8 gene activation. We confirmed that cadmium induced superoxide anion generation rapidly in human PBMC. The addition of anti-oxidant, N-acetyl-cysteine (NAC) completely blocked cadmium-induced IL-8 production in human PBMC (Fig. 4), suggesting that superoxide anions were involved in this process. The involvement of superoxide anions in IL-8 gene activation was also suggested by the result that human PBMC produced IL-8 by the stimulation with paraquat, which is a potent stimulus for generation of superoxide anions. These results indicate that superoxide anions generated intracellularly mediate IL-8 gene activation.

Recently, several independent groups reported that superoxide anions were involved in the activation of NF-kB complex (27,28). Baeuerle et al. claimed that modification of IkB by superoxide anions plays a more vital role in dissociation of the p50-p65 complex of NF-kB from IkB than the phosphorylation of IkB (15). They observed that middle surface antigen from hepatitis B virus activated the transcription of several kB-controlled reporter genes and that the transcription was inhibited by NAC (29), supporting their hypothesis on the involvement of superoxide anions in NF-kB-mediated gene activation.

At present, we have no evidence whether superoxide anions are actually involved in the IL-8 gene activation induced by other stimulants including IL-1, TNF, LPS, or phorbol ester nor which *cis*-element(s) and transcription factor(s) are responsible for superoxide anions-mediated IL-8 gene transcription. Elucidation of these points will shed new lights on the role of superoxide anions as regulators of gene expression.

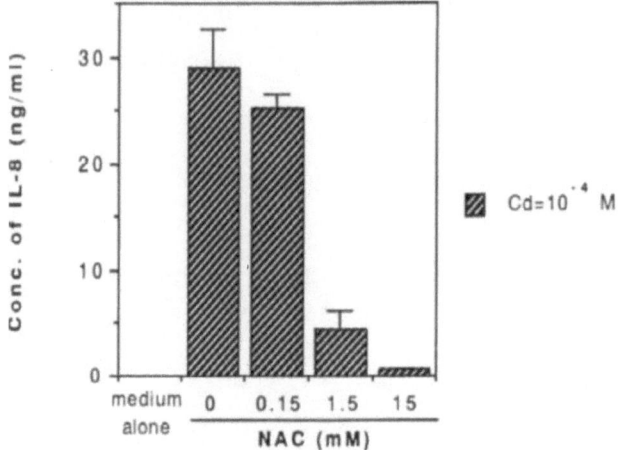

Figure 5. Effects of NAC on cadmium-induced IL-8 production by human PBMC.

MECHANISM OF IL-8 GENE REPRESSION

Considering the potential role of IL-8 in the pathogenesis of various types of inflammatory processes, it is necessary to elucidate the mechanism by which IL-8 gene transcription is suppressed. Several cytokines and agents are reported to suppress IL-8 production. Among these, IL-4 (30), transforming growth factor-b, 1,25-$(OH)_2$-vitamin D_3 (31), a-5'-lipoxygenase inhibitors, and glucocorticoids (9,12) have been demonstrated to suppress IL-8 mRNA expression in human peripheral blood mononuclear cells, a human fibrosarcoma cell line 8387, and a human glioblastoma cell line as revealed by Northern blotting analysis. Nuclear run-off assays indicate that the suppression is at least partly at the transcriptional level in 8387 cells (9).

Several independent groups reported that AP-1 binding site is indispensable for collagen gene expression and that the repression of collagen gene by glucocorticoids was ascribed to the inhibition of activity of AP-1 complex (32,33). However, in 8387 cells, the suppression of IL-8 gene transcription seems to be mediated through the interaction between glucocorticoid receptor and glucocorticoid receptor binding element in the 5'-flanking region of the IL-8 gene based on the studies of the transfection of CAT expression vector inserted with 5'-deletion mutant of the IL-8 gene (9). In a human glioblastoma cell line, T98G, besides the glucocorticoid sensitive element, the kB-binding site seems to be involved in IL-8 gene repression by glucocorticoid, based on similar studies (our unpublished data). In both cell lines, the AP-1 binding site does not seem to be responsible for IL-8 gene repression by glucocorticoid. However, It remains to be investigated whether a similar mechanisms also works in other types of cells.

In a human T cell leukemia-derived cell line, Jurkat, we observed that simultaneous stimulation with phorbol ester and calcium ionomycin induced IL-8 mRNA accompanied with IL-8 protein production. Both IL-8 protein production and mRNA expression were observed to be inhibited by an immunosuppressant, FK506. FK506, as well as cyclosporin A, inhibits the production of various cytokines including IL-2 and IL-3 at the transcriptional level, leading to the immunosuppression (34). Both are presumed to inhibit the activity of a phosphatase in cytosol, calcineurin, which is presumed to be involved in the translocation of NF-AT, an essential transcription factor for IL-2 gene transcription (34). The 5'-flanking region of IL-8 gene lacks an NF-AT binding site. The studies employing various types of CAT expression vectors revealed that both AP-1 and kB-binding sites were indispensable for IL-8 gene activation in Jurkat cells and that FK506 acted at the kB-binding site but not AP-1 binding site to inhibit IL-8 gene expression (Fig.6).

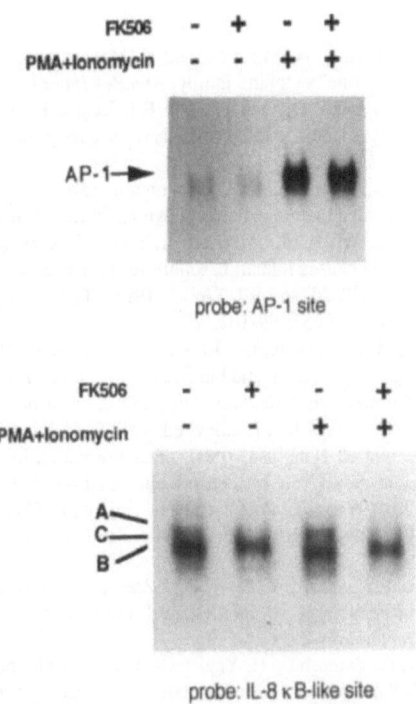

Figure 6. Gel retardation assays using either AP-1 or kB-binding site as a probe. Nuclear proteins were extracted from either stimulated or unstimulated Jurkat cells in the presence or the absence of FK506.

These results suggest that kB-binding site is involved in the IL-8 gene repression by FK506 as well as glucocorticoid. As mentioned earlier, the transcription factor(s) which bind to the kB-binding site in the IL-8 gene are at least immunochemically distinct from known rel-related transcription factors including p50, p65, p50B/p49, and c-rel. Thus, molecular characterization of the kB-binding factor(s) will be required for better understanding on the regulatory mechanism of IL-8 production.

CONCLUDING REMARKS

The accumulating evidence indicated that the kB-binding site in the IL-8 gene plays a pivotal role in IL-8 gene transcription as well as repression. Depending on the cell types, either NF-IL6- or AP-1 binding site is indispensable *cis*-element for IL-8 gene expression. The biochemical characterization of NF-kB-like factor(s) and the interaction of NF-kB-like factor(s) with additional indispensable factors will lead to the development of a new way of regulation of IL-8 production as well as better understanding of mechanism of IL-8 production at the molecular biological level.

ACKNOWLEDGEMENT

We would like to express gratitude to Dr. Bent W. Nielsen (Kanazawa University) for his critical review of this manuscript.

REFERENCES

1. Oppenheim, J.J., C.O.C. Zachariae, N. Mukaida, and K. Matsushima. 1991. Properties of the novel inflammatory supergene "intercrine" cytokine family. *Ann.Rev.Immunol.* 9: 617-648.
2. Yoshimura, T., K. Matsushima, J.J. Oppenheim, and E.J. Leonard. 1987. Purification of a human monocyte-derived neutrophil chemotactic factor that shares sequence homology with other host defense cytokines. *Proc.Natl.Acad.Sci.U.S.A.* 84: 9233-9237.
3. Larsen, C.G., A.O. Andersen, E. Appella, J.J. Oppenheim, and K. Matsushima. 1989. Neutrophil activating protein (NAP-1) is also chemotactic for T lymphocytes. *Science* 243: 1464-1466.
4. White, M.V., T. Yoshimura, W. Hook, M. Kaliner, and E.J. Leonard. 1989. Neutrophil attractant/activation protein (NAP-1) causes human basophil histamine release. *Immunol.Lett.* 22: 151-154.
5. Huber, A.R., S.L. Kunkel, R.F. Todd, and S.J. Weiss. 1991. Transendothelial neutrophil migration by endogenous interleukin-8. *Science* 254: 99-102.
6. Mukaida, N., A. Hishinuma, C.O.C. Zachariae, J.J. Oppenheim, and K. Matsushima. 1992. Regulation of the human interleukin 8 gene expression and binding of several other members of the intercrine family to receptors for IL 8. In *Chemotactic Cytokines, Advances in experimental medicine and biology, Vol. 305.* (Westwick, J., S. Kunkel, and I.J.D. Lindley, eds.) Plenum Press, New York (in press).
7. Sugano, S., M.Y. Soeckle, and H. Hanafusa. 1987. Transformation by Rous sarcoma virus induces a novel gene with homology to amitogenic platelet protein. *Cell* 49: 321-328.
8. Sica A., K. Matsushima, J.M. Wang, E. Polentarutti, E. Dejana, F. Colotta, and A. Mantovani. 1990. IL-1 transcriptionally activates the neutrophil chemotactic factor/IL-8 gene in endothelial cell. *Immunol.* 69: 548-553.
9. Mukaida, N., G.L. Gusella, T. Kasahara, Y. Ko, C.O.C. Zachariae, T. Kawai, and K. Matsushima. 1992. Molecular analysis of the inhibition of interleukin-8 production by dexamethasone in a human fibrosarcoma cell line. *Immunology* 75: 674-679.
10. Kasahara, T., N. Mukaida, K. Yamashita, H. Yagisawa, T. Akahoshi, and K. Matsushima. 1991. IL-1 and TNF-α induction of IL-8 and monocyte chemotactic and activating factor (MCAF) mRNA expression in a human astrocytoma cell line. *Immunology* 74: 60-67.
11. Mukaida, N., M. Shiroo, and K. Matsushima. 1989. Genomic structure of the human monocyte-derived neutrophil chemotactic factor (MDNCF)/interleukin 8. *J.Immunol.* 143: 1366-1371.
12. Yasukawa, K., T. Hirano, Y. Watanabe, K. Muratani, T. Matsuda, S. Nakai, and T. Kishimoto. 1987. Structure and expression of human B cell stimulatory factor-2 (BSF-2/IL-6) gene. *EMBO J.* 6: 2939-2945.
13. Mukaida, N., M. Yann, and K. Matsushima. 1990. Cooperative interaction of nuclear factor kB and cis-regulatory enhancer binding protein-like factor binding elements in activating the interleukin-8 gene by pro-inflammatory cytokines. *J.Biol.Chem.* 265: 21128-21133.
14. Mahe, Y., N. Mukaida, K. Kuno, M. Akiyama, N. Ikeda, K. Matsushima, and S. Murakami. 1991. Hepatitis B virus X protein transactivates human interleukin-8 gene through acting on nuclear factor kB and CCAAT/enhancer-binding protein-like *cis*-elements. *J.Biol.Chem.* 266: 13759-13763.
15. Schmitz, M.L., T. Henkel, and P.A. Baeuerle. 1991. Proteins controlling the nuclear uptake of NF-kB, rel and dorsal. *Trends Cell Biol.* 1: 130-137.
16. Ghosh, S. and D. Baltimore. 1990. Activation in vitro of NF-kB by phosphorylation of its inhibitor IkB. *Nature* 344: 678-682.
17. Shiroo, M. and K. Matsushima. 1989. Enhanced phosphorylation of 65 and 75 kDa proteins by tumor necrosis factor and interleukin-1 in human peripheral blood mononuclear cells. *Cytokine* 74: 13-20.

18. Wu, J.Y., Z.-Y. Zhou, A. Judd, C.A. Cartwright, and W.S. Robinson. 1990. The hepatitis B virus-encoded transcriptional *trans*-activator hbx appears to be a novel protein serine/threonin kinase. *Cell* 63: 687-695.

19. Leclair, K. P., M.A. Blanar, and P.A. Sharp. 1992. The p50 subunit of NF-kB associates with the NF-IL6 transcription factor. *Proc.Natl.Acad.Sci.U.S.A.* (in press).

20. Nakamura, H., K. Yoshimura, H.A. Jaffe, and R.G. Crystal. 1991. Interleukin-8 gene expression in human bronchial epithelial cells. *J.Biol.Chem.* 266: 19611-19617.

21. Yasumoto, K., S. Okamoto, N. Mukaida, S. Murakami, M. Mai, and K. Matsushima. 1992. Tumor necrosis factor alpha and interferon gamma synergistically induce interleukin 8 production in a human gastric cancer cell line through acting concurrently on AP-1 and NF-kB like binding site of the interleukin 8 gene. *J.Biol.Chem.* (in press).

22. Kawakami, K., C. Scheidereit, and R.G. Roeder. 1988. Identification and purification of a human immunoglobulin-enhancer-binding protein (NF-kB) that activates transcription from a human immunodeficiency virus type 1 promoter *in vitro*. *Proc.Natl.Acad.Sci.U.S.A.* 85: 4700-4704.

23. Minamoto, T., M. Mai, K. Watanabe, A. Ooi, T. Kitamura, Y. Takahashi, H. Ueda, T. Ogino, and I. Nakanishi. 1990. Medullary carcinoma with lymphocytic infiltration of the stomach. *Cancer* 66: 945-952.

24. Aggarwal, B.A., T.E. Eessalu, and P.E. Hass. 1985. Characterization of receptors for human tumor necrosis factor and their regulation by gamma-interferon. *Nature* 318: 665-667.

25. Johnson, D.R. and J.S. Pober. 1990. Tumor necrosis factor and immune interferon synergistically increase transcription of HLA class I heavy- and light-chain genes in vascular endothelium. *Proc. NAt. Acad. Sci. U.S.A.* 87: 5183-5187.

26. Frieberg, L., T. Kjellstrom, and G.F. Nordberg. 1986. Chapter 7. Cadmium. *In Handbook on the Toxicology of Metals*, 2nd ed., (Friberg, L., Nordberg, G.F., and Vouk, V. eds.) Elsevier Science Publishers B.V., Amsterdam.

27. Staal, F.J.T., M. Roeder, L.A. Herzenberg, and L.A. Herzenberg. 1990. Intracellular thiols regulate activation of nuclear factor kB and transcription of human immunodeficiency virus. *Proc.Natl.Acad.Sci.U.S.A.* 87: 9943-9947.

28. Schreck, R., P. Rieber, and P.A. Baeuerle. 1991. Reactive oxygen intermediates as apparently widely used messengers in the activation of the NF-kB transcription factor and HIV-1. *EMBO J.24* 10: 2247-2258.

29. Meyer, M., W.H. Caselmann, V. Schlueter, R. Schreck, P.H. Hofschneider, and P.A. Baeuerle, 1992. Hepatitis B virus transactivator MHBs': activation of NF-kB, selective inhibtion by antioxidants and integral membrane localization. *EMBO J.* 11: 2991-3001.

30. Standiford, T.J., Strieter, R.M., Chensue, S.W., Westwick, J., Kasahara, K., and Kunkel, S.L. 1990. IL-4 inhibits the expression of IL-8 from stimulated human monocytes. *J.Immunol.* 145: 1435-1439.

31. Larsen C.G., M. Kristensen, K. Paludan, B. Deluran, M.K. Thomsen, C. Zachariae, Kragballe, K. Matsushima, and K. Thestrup-Pedersen. 1991. 1,25(OH)$_2$-D$_3$ is a potent regulator of interleukin-1 induced interleukin-8 expression and production. *Biochem.Biophys.Res.Commun.* 176: 1020-1026.

32. Jonat, C., H.J. Rahmsdorf, K.-K. Park, A.C.B. Cato, S. Gebel, H. Ponta, and P. Herrlich. 1990. Antitumor promotion and anti inflammation: down-modulation of AP-1 (fos/jun) activity by glucocorticoid hormone. *Cell* 62: 1189-1204.

33. Yang-Yen, H.-F., J.-C. Chambard, Y.-L. Sun, T. Smeal, T.J. Schmidt, J. Drouin, and M. Karin. 1990. Transcriptional interference between c-jun and the glucocorticoid receptor: mutual inhibition of DNA binding due to direct protein-protein interaction. *Cell* 62: 1205-1215.

34. Schreiber, S.L. and G.R. Crabtree. 1992. The mechanism of action of cyclosporin A and FK506. *Immunol.Today.* 13: 136-142.

BASOPHIL ACTIVATION BY MEMBERS OF THE CHEMOKINE SUPERFAMILY

Clemens A. Dahinden, Martin Krieger, Thomas Brunner,
and Stephan C. Bischoff

Institute of Clinical Immunology
Inselspital
University of Bern
CH-3010 Bern, Switzerland

INTRODUCTION

The human chemokine superfamily represents a new class of structurally homologous cell-derived peptides with multiple functions within the immune system (1-7). According to the position of the first two cysteins in the primary sequence, the members of the PF-4 superfamily can be divided into two branches: the "C-X-C branch" (α chemokines) including interleukin 8 (IL-8)/neutrophil-activating peptide 1 (NAP-1), platelet basic protein (PBP) and the N-terminally processed forms connective tissue activating protein III (CTAP-III) and NAP-2, platelet factor 4 (PF-4), ENA-78 and the *gro* peptides α, β, γ; and the "C-C branch" (β chemokines) comprising monocyte chemotactic protein 1 (MCP-1), MCP-2/HC14, MCP-3, RANTES, MIP-1α, MIP-1β and I-309. Some of the these peptides are well characterised with regards to their biological activities, but only little information is available for others who's existence is sometimes based solely on gene expression (6,8,9). Recent studies showed that particular chemokines are strongly up-regulated in inflammatory disorders such as atherosclerosis, rheumatoid arthritis, and wound healing emphasising their clinical potential (6,10-12). Furthermore, several in vitro findings indicate that the division of this superfamily into two branches based on structural characteristics reflects also some fundamental differences in the range of biological activities of the α and the β chemokines. α chemokines such as IL-8/NAP-1, NAP-2, ENA-78 or MGSA exert their proinflammatory activity mainly through their neutrophil-activating capacity (1-4), whereas β chemokines preferentially attract different mononuclear cell types (4-7). This observation suggests that α chemokines play a central role in acute inflammatory events characterised by a polynuclear infiltrate, whereas β chemokines seem to be important in regulating chronic inflammatory processes in which monocytes and lymphocytes predominate.

Considerable progress has been made in the recent years regarding our understanding

The Chemokines, Edited by I.J.D. Lindley
et al., Plenum Press, New York 1993

of the pathogenesis of immediate hypersensitivity reaction induced by an allergen in sensitised individuals. The allergic reaction is composed of an early phase induced by IgE-mediated mast cell activation, which leads to plasma exsudation and edema, and of a late phase characterised by a cellular infiltration composed of eosinophils, basophils and lymphocytes of the Th2 phenotype (13,14). The late phase is thought to be responsible for clinically more relevant chronic allergic inflammation, since it causes tissue destruction and is correlated with the major symptoms in allergic patients (13). In recent years, some mechanisms explaining at least in part the particular pathology of allergic inflammation have been described. For example, several hematopoietic growth factors profoundly modulate the function of human eosinophils and basophils, the two major effector cells of allergic inflammation. IL-3, IL-5, and GM-CSF strongly enhance the IgE-dependent and IgE-independent release reaction in basophils, and the same set of growth factors attracts eosinophils and enhances their cytotoxicity (15-19). Confirming these *in vitro* findings, this set of cytokines is found *in vivo* at the site of allergic inflammation, and is produced by Th2 lymphocytes (14,20,21). Furthermore, at least for basophils, some IgE-independent agonists such as C5a, C3a and platelet-activating factor (PAF) could be identified providing a first explanation for the mechanism of basophil activation in inflammatory events occurring in the absence of a specific allergen (15,16,22). However, little is known about the factors regulating their migration to the site of inflammation, and only recently some cell-derived agonists triggering mediator release by basophils have been described (23-27).

In recent studies, we and others showed that IL-8/NAP-1 and MCP-1 are potent triggers for basophil degranulation (23-27). Here, we will summarize the effects of chemokines on basophil function. Furthermore, we will compare their different profiles of activity on basophils and on other effector cell types such as eosinophils, neutrophils and monocytes, and finally, we will give evidence for the presence of specific chemokine receptors on basophils.

MATERIAL AND METHODS

Cell stimuli: Recombinant human (rhu) IL-8/NAP-1 was obtained from the Sandoz Research Institute, Vienna, Austria (28). RANTES, rhu MIP-1α, and rhu MIP-1β were produced in E.coli as described previously (29). Rhu MCP-1 was from Pepro Tech Inc., Rocky Hill, NJ, USA (25). Rhu IL-3 was provided by Sandoz Ltd., Basel, Switzerland. All chemokines were stored in small aliquots at $10^{-5}M$ in Hepes buffer containing 1 mg/ml bovine serum albumin. The purified mAb 29C6 directed against the high-affinity IgE receptor a-chain was obtained from Hoffmann-La Roche, Nutley, NJ, USA.

Cell isolation: Leukocytes were obtained from blood of unselected healthy volunteers after informed consent, and were isolated and fractionated by discontinuous density centrifugation (15,30). For most histamine release experiments, mononuclear cell preparations prepared by Ficoll Hypaque density centrifugation and containing 1-8% basophils were used (15). Highly purified basophils were prepared by fractionation of leukocytes on discontinuous Percoll gradients (30) and a subsequent negative selection of basophils with immunomagnetic beads coated with mAb against CD3, CD4, CD8, CD14, CD16 and CD19 yielding basophil preparations of 80-95% purity (contaminated exclusively with small lymphocytes). All cell preparations were finally washed 3 times in HA buffer (20mM Hepes, 125 mM NaCl, 5mM KCl, 0.5mM glucose, 0.025% BSA) and resuspended in HACM buffer (HA buffer supplemented with 1mM $MgCl_2$ and 1 mM $CaCl_2$).

Basophil mediator release assay: Experiments were performed in a shaking water bath (37°C). After a warming-up period of 10 min, cells were incubated in buffer or with

IL-3 (10 ng/ml) for 10 min and then challenged with chemokines or other triggering agents at the concentrations indicated. The release reaction was stopped 20 min after addition of the triggering agent by placing the tubes in ice-cold water. Histamine and leukotrienes were measured in the cell supernatants (15,30) and expressed as % histamine release of total cellular hisatmine content and as pg leukotrienes per ng total cellular histamine content, respectively.

Measurement of in vitro basophil chemotaxis: Basophil chemotaxis was measured in a modified boyden chamber according to a previously described method (29,31). Data were expressed as the mean number of migrated adherent cells (in % of total input). The mean chemotactic efficacy (number of basophils migrated at the optimal concentration of the chemoattractant) and the mean migration index (number of basophils migrated at the optimal concentration of the chemoattractant divided by the number of eosinophils which migrated to buffer control) were determined.

Changes of intracellular calcium concentration [Ca²⁺]ᵢ: Purified cells were loaded with 0.3mmol fura-2/AM per 10^6 cells in HACM buffer for 30 min at 37°C, centrifuged and resuspended in prewarmed HACM buffer (37°C) at a concentration of 0.5-2x10^6 cells/ml. Fura-2 fluorescence changes of the cell suspensions in response to cell agonists were continuously monitored at 0.25 sec intervals and analysed as described previously (10, 16, 29). Each measurement was standardised by adding ionomycin (5 mM) leading to 100% fura-2 saturation and subsequent quenching of the fluorescence with $MgCl_2$ (1 mM). Fura-2 saturation in resting cells before stimulation was between 50 and 60 %.

RESULTS

Figure 1 shows the capacity of α-chemokines (IL-8/NAP-1, NAP-2, CTAP-III, PF-4, *gro*-α, IP-10) and of β chemokines (MCP-1, RANTES, MIP-1α, MIP-1β) to trigger basophils for mediator release. In these experiments, mononuclear cell preparations containing 1-6% basophils depleted of neutrophils and eosinophils were exposed to chemokines at 100 nM. Only MCP-1 was capable of inducing basophil degranulation directly. In the presence of IL-3, however, also IL-8 induced significant histamine release, albeit to a lesser extent than MCP-1. NAP-2 induced a marginal release reaction at 100 nM whereas all the other chemokines failed to induce significant release. However, at 1000 nM, some histamine release could be detected also in response to NAP-2, CTAP-III and PF-4, possibly due to a cationic charge effect rather than a receptor-dependent mechanism (32). Without IL-3 priming of basophils, none of the chemokines induced the production of *de novo* synthesised sulfidoleukotrienes up to 100 nM, but in the presence of IL-3, again only MCP-1 and IL-8 caused leukotriene generation. The results were different when purified basophils (>85%) were used. Figure 2 shows that purification of the basophils uncovers a clear histamine releasing capacity of RANTES and MIP-1α in IL-3 primed cells, being in a similar range than that of IL-8, whereas MIP-1β and the α chemokines apart from IL-8 consistently failed to induce basophil mediator release under these conditions.

In order to study the mechanism of basophil activation, changes of cytosolic calcium [Ca²⁺]ᵢ were measured in highly purified basophils stimulated with chemokines. Figure 3 shows that all chemokines capable of triggering for basophil mediator release (MCP-1, RANTES, MIP-1α, and IL-8) also induced a rapid and transient rise in [Ca²⁺]ᵢ. Moreover, a typical [Ca²⁺]ᵢ rise was observed after exposure to high concentrations (≥ 100 nM) of NAP-2 (32). After stimulation, the cells became unresponsive to a second challenge with the same agonist at the same concentration, but responded normally to other basophil agonists such as fMLP, C5a, C3a or PAF (data not shown). MIP-1β and the α chemokines

101

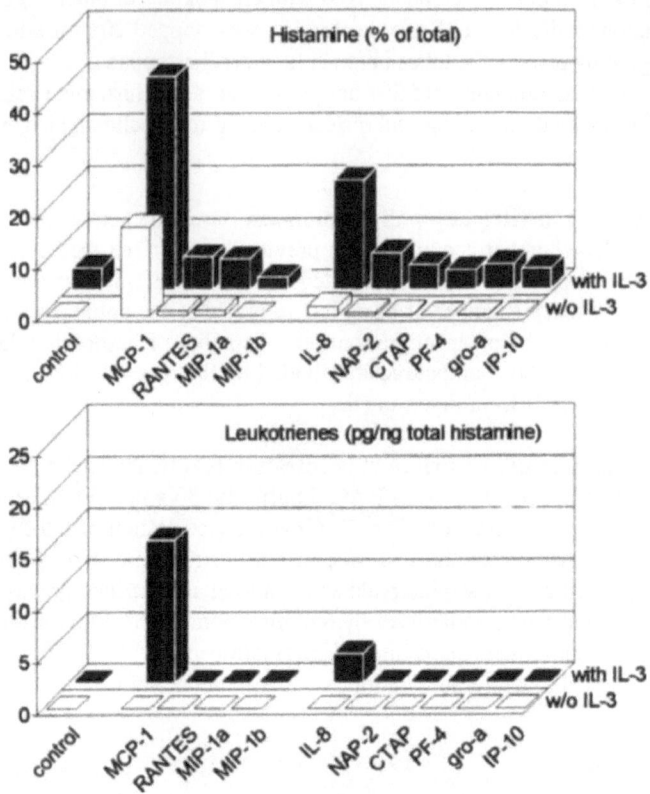

Figure 1. Basophil mediator release by chemokines. Human mononuclear cell preparations containing 1-6% basophils were preincubated for 10 min in buffer (white columns) or with IL-3 (10 ng/ml, black columns), and then exposed for 20 min to buffer (control) or to the chemokines (100 nM). Histamine release (upper panel) and leukotriene generation (lower panel) were measured in the supernatants. Mean values from 4 experiments performed in duplicates with blood from different donors are shown.

CTAP-III, PF-4 and IP-10, however, failed to induce $[Ca^{2+}]_i$ changes (data not shown). After sequential challenge with any chemokine, basophils remained fully responsive to IL-8 and to MCP-1 (25,32). Cross-desensitisation phenomena were observed between RANTES and MIP-1α (own unpublished observation) as wells as between IL-8 and NAP-2 (32), respectively. If the basophils were sequentially exposed to RANTES and then to MIP-1α, no $[Ca^{2+}]_i$ changes in response to MIP-1α occurred. However, if the sequence was reversed, RANTES still induced a slightly reduced rise in $[Ca^{2+}]_i$. A similar finding was made with regard to IL-8 and NAP-2. Exposure to NAP-2 did not affect the $[Ca^{2+}]_i$ response towards IL-8, but if basophils have been challenged with IL-8 first, and then with NAP-2, the NAP-2 response was completely abrogated (32). Interestingly, the $[Ca^{2+}]_i$ changes were completely independent on pretreatment with IL-3, which is required for the release response (32). IL-3 did not induce a rise in $[Ca^{2+}]_i$, neither modified the kinetic nor the extent of the $[Ca^{2+}]_i$ changes induced by IL-8 or other chemokines.

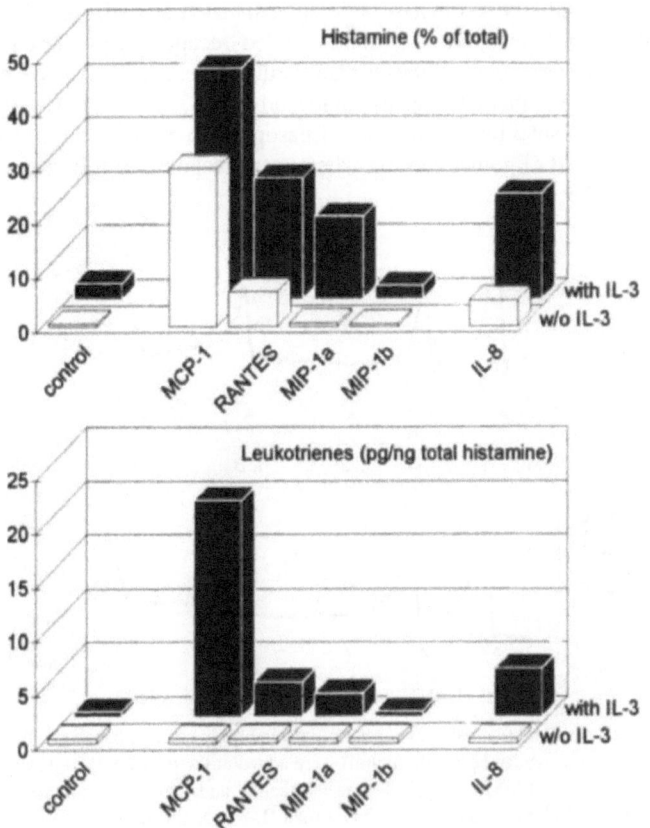

Figure 2. Histamine release (upper panel) and leukotriene generation (lower panel) by purified human basophils in response to chemokines. Basophils (80-95% purity) were contaminated exclusively with small lymphocytes. Otherwise experimental conditions as desribed in Fig 1.

The release data and in particular the $[Ca^{2+}]_i$ changes as well as the desensitisation effects strongly suggest that chemokines act through specific receptors on basophils. Binding studies with ^{125}I-IL-8 could be performed with pure basophil preparations from blood of a chronic myeloid leukaemia patient (32). These data demonstrated specific and saturable IL-8 binding and are consistent with a one-binding site model. According to these experiments, CML basophils bear approximately 3,500 receptors per cell with a Kd of 0.15 nM. Binding studies in normal human basophils are difficult to perform due to the small number of cell which can be obtained from healthy donors (max. 3-6 Mio total). However, some experiments were also performed with normal basophils at near receptor saturating conditions showing that normal basophils bear even more IL-8 receptors (approximately 10,000 per cell). Parallel experiments with neutrophils revealed 40,000 receptors per cell which is in agreement with former estimates (33,34). Apart from IL-8, only NAP-2 could displace ^{125}I-IL-8-binding, but the concentrations required were 10 to 100 times higher than for unlabeled IL-8. IL-8 binding was not affected by CTAP-III or PF-4 even at 1000 nM (32).

Since IL-8, similarly to other peptide agonists such as fMLP or C5a (35), interacts with a Pertussis toxin sensitive G-protein-coupled receptor, we examined whether the basophil response to other chemokines is also Pertussis-toxin sensitive. We could show that the calcium rise and the mediator release induced by IL-8, MCP-1, RANTES, and MIP-1α could be nearly abolished by preexposure of basophils to 5 nM Pertussis toxin, but not to the non-ribosylating B oligomer control, whereas the response to IgE receptor-crosslinking was not affected (25, 32).

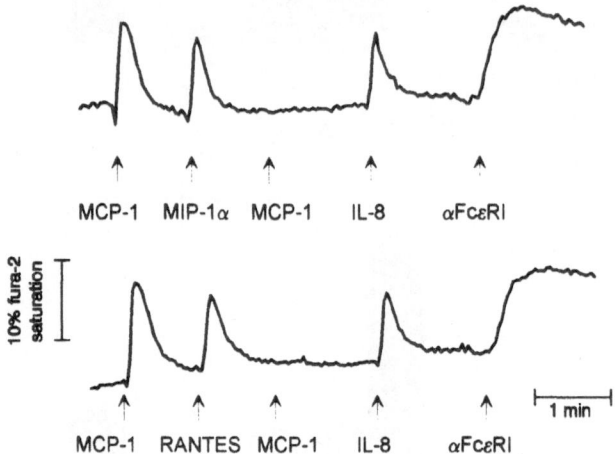

Figure 3. Changes of cytosolic free calcium concentrations in highly purified basophils. Peptide agonists (100 nM) and anti-IgE receptor Ab (aFceRI, 100 ng/ml) were added at the time points indicated by arrows. MIP-1b and the a-chemokines CTAP-III and PF-4 were ineffective under the same conditions, whereas NAP-2 induced a rise in intracellular calcium, albeit weaker than IL-8 (not shown).

It is known from previous studies that IL-8/NAP-1 is chemotactic for neutrophils, whereas the β chemokines MCP-1, RANTES and MIP-1α induce monocyte migration. Furthermore, RANTES and MIP-1α attract certain T lymphocyte types. Therefore, we were interested whether chemokines can also induce basophil migration. We found that RANTES is a particular efficacious chemotactic agent for basophils, followed by MCP-1, MIP-1α and IL-8 (Table 1). The optimal concentration of RANTES was between 10 and 30 nM, and the mean chemotactic efficacy was 45 %, the mean migration index 8.8, thus being similar efficacious as C5a (own unpublished data). MIP-1β failed to induce any chemotaxis in basophils. A similar chemotaxis response towards RANTES and MIP-1α was observed in eosinophils, except that MCP-1 was totally ineffective (29). MCP-1 also failed to induce mediator release or [Ca²⁺]ᵢ changes in eosinophils, suggesting that eosinophils lack MCP-1 receptors. In monocytes, however, MCP-1 is the most efficacious chemoattractant, followed by RANTES and MIP-1α (6). Table 2 summarises the different profiles of activities of RANTES, MIP-1α and MCP-1 in these different cell types.

DISCUSSION

In our previous studies, we showed that basophil function can be strongly modulated by the cytokines IL-3, IL-5, GM-CSF and nerve growth factor (15-17,30). The same set of hematopoietic growth factors, IL-3, IL-5 and GM-CSF, as well as the proinflammatory cytokine TNF modulate eosinophil function (18,19,36). Our recent studies (25,29,30) indicate that the newly defined chemokine family represents a new class of cytokines, from which several members profoundly regulate basophil and eosinophil function. IL-8, NAP-2, and in particular the β chemokines MCP-1, RANTES and MIP-1α are potent basophil agonists inducing a rapid change in cytosolic free calcium, a significant mediator release and a pronounced chemotaxis response. These results suggest that basophils are involved in inflammatory events other than immediate hypersensitivities in which chemokines are upregulated (6,10,11).

Table 1. Effect of chemokines on different basophil functions. Intracellular calcium concentration (Ca^{2+}), histamine release and leukotriene generation was measured as in Fig. 1+2. The chemotaxis assay is described in 'Materials and Methods'. Experiments were performed with purified basophils with (+IL-3) or without IL-3-priming (- IL-3) exposed to 100 nM of the peptides. The calcium response was not affected by IL-3 priming. Results are expressed as +++ = very strong effects, ++ = strong effects, + = moderate effects or (+) = (marginal effects); n.d. = not determined.

	Ca^{2+}	Histamine		Leukotrienes		Chemo-taxis
	-IL-3	-IL-3	+IL-3	-IL-3	+IL-3	-IL-3
IL-8	+++	+	++	--	+	(+)
NAP-2	(+)	--	--	--	--	n.d.
CTAP-III	--	--	--	--	--	n.d.
PF-4	--	--	--	--	--	n.d.
MIP-2α	n.d.	--	--	--	--	n.d.
IP-10	--	--	--	--	--	n.d.
MCP-1	+++	+++	++++	--	+++	++
RANTES	+++	+	++	--	+	+++
MIP-1α	+++	--	++	--	+	+
MIP-1β	--	--	--	--	--	--

Among the α chemokines, only IL-8 and, to a much lesser extent, NAP-2 at 100 nM are capable of inducing significant mediator release in IL-3-primed basophils by a receptor dependent mechanism. Although all the IL-8 related peptides such as CTAP-III and PF-4 may cause some release at high concentrations ($\geq$ 1000 nM), the mechanism of action is different: IL-8 and NAP-2 bind to selective basophil receptors, induce a transient $[Ca^{2+}]_i$ change and show cross-desensitisation, whereas the marginal release induced by CTAP-III and PF-4 may be due to the cationic charge of the molecules. Indeed, we could show that highly charged poly-D-lysine and histone VS cause some histamine release without inducing a rise in $[Ca^{2+}]_i$ (32).

The β chemokines seem to be even more important as basophil agonists. MCP-1 is the most efficacious trigger for basophil mediator release among all chemokines examined so far, and is the only one capable of inducing a pronounced exocytosis in unprimed basophils (25). In the presence of hematopoietic growth factors such as IL-3, also RANTES and MIP-1α, but not MIP-1β, exert basophil mediator releasing capacity. RANTES and

MIP-1α are the first basophil triggers, which consistently induce mediator release in highly purified basophil preparations only. This observation suggests that other cell types repress basophil activation through chemokines, for example by competing for the ligands through receptors or by absorption. Indeed, a recent study showed that red blood cells can bind IL-8 and MCP-1, and that IL-8 bound to erythrocytes is incapable of interacting with neutrophils (37). Other cell types such as monocytes and lymphocytes may also interact with the chemokine-induced release reaction in basophils (27). In vivo, the cellular composition of an exudate may be a further element controlling the response of individual cell types to b chemokines.

In comparison to the well-established basophil triggers such as allergen, anti-IgE Ab or the complement cleavage product C5a, the amount of mediators released by chemokines, with the exception of MCP-1, is not very pronounced. This finding was particular surprising since all these chemokines activate basophils at rather low concentrations (0.1-100nM), and rises the question whether RANTES, MIP-1α or IL-8 preferentially activate for other basophil functions. Indeed, we could show that in particular RANTES is a potent basophil chemoattractant, and that MCP-1, MIP-1α and IL-8 also induce chemotaxis, albeit with a lower efficacy. Table 2 illustrates that each β chemokine preferentially activates separate cell functions, and that the profile of activity differs between the target cells (basophils, eosinophils, monocytes). For example, MCP-1 is the most efficacious trigger for mediator release in basophils (25) and a strong chemoattractant in monocytes (1), but has no effect in eosinophils (29). In contrast, RANTES preferentially attracts basophils and eosinophils, but has only weak effects on monocytes (29, and own unpublished data). MIP-1α, however, is the most potent β chemokine on a molar basis, but exerts only a weak efficacy for histamine release or basophil migration, suggesting that MIP-1α controls other cellular functions not examined in this study. Interestingly, in the mouse system, MIP-1a activates peritoneal macrophages for cytokine production (38), a function which has not yet been examined in human cells. The fact that β chemokines selectively affect distinct cell types (4-7), and, within a given cell type, stimulate preferentially particular cell functions, as shown here, suggests that this new class of cytokines are involved in the fine tuning of inflammatory events.

To evaluate the mechanism of basophil activation by chemokines we examined the $[Ca^{2+}]_i$ changes induced by IL-8, NAP-2, CTAP-III, PF-4, MCP-1, MIP-1α, MIP-1β and

Table 2. Profile of activities of RANTES, MIP-1α and MCP-1 in different target cells. Effects on chemotaxis (CT), mediator release (MR) and other effector functions (EF) such as cytotocity in response to three b chemokines were compared in human basophils, eosinophils (29) and monocytes (6,7). The magnitude of the "+" symbol indicates the level of efficacy of each peptide to affect a particular function in the different cell types.

	Basophils		Eosinophils		Monocytes	
	CT	MR	CT	MR	CT	EF
RANTES	**✛**	+	**✛**	+	+	+
MIP-1α	+	+	+	+	+	**✛**
MCP-1	+	**✛**	--	--	**✛**	+

RANTES. All chemokines, which induced mediator release or chemotaxis in basophils, triggered a rapid and transient change in $[Ca^{2+}]_i$ with a similar kinetic to those induced by other peptide agonists such as C5a or fMLP. The similarity between the kinetics of the release reaction and the $[Ca^{2+}]_i$ changes in response to the chemokines RANTES, MIP-1α, and MCP-1 and to IL-8, C5a and fMLP, which act through G-protein-coupled receptors (32,35) suggest that also the β chemokines activate basophils directly through G-protein-coupled receptors. This hypothesis is strongly supported by the observation that the signals induced by all chemokines as well as those induced by C5a and fMLP are Pertussis toxin sensitive (25,29, and own unpublished data). Therefore, it can be assumed that not only IL-8, but also MCP-1, RANTES and MIP-1α activate basophils directly through different receptors of the rodopsin family ("seven-transmembrane spanning receptors").

Interestingly, the differences between the β-chemokines regarding the cellular function they preferentially affect, are not paralleled by differences in the $[Ca^{2+}]_i$ changes. The fact that the calcium responses towards RANTES, MIP-1α and MCP-1 are similar, indicates that $[Ca^{2+}]_i$ changes reflect a rather promiscuous cellular response evoked by many agonists and leading to the activation of different effector functions depending on the particular stimulus and yet unknown "ramifications" in the G-protein-initiated signal-transducing pathway.

To assess the specificy of the chemokine receptors we performed cross-desensitisation experiments by sequentially challenging basophils with different combinations of chemokines and other chemotactic agonists. The fact that IL-8, RANTES, MIP-1α and MCP-1 were unable to affect the $[Ca^{2+}]_i$ changes induced by other peptide agonists (C5a, fMLP) or by IgE receptor-crosslinking indicates that basophils, similarly to eosinophils (29), express novel yet to be defined receptors specific for chemokines. Since IL-8 and MCP-1 do not influence the response towards RANTES and MIP-1α, respectively, and, vice versa, neither RANTES nor MIP-1α affects the IL-8 or the MCP-1 response, basophils obviously have specific binding sites for MCP-1 and IL-8. The conclusion that basophils bear a specific MCP-1 receptor is further supported by experiments performed with eosinophils, which are activated by MIP-1α and RANTES, but do not respond to MCP-1 (29). The weak response towards NAP-2 seems to be mediated through the IL-8 receptor, since IL-8 blocks the $[Ca^{2+}]_i$ change in response to NAP-2, and NAP-2 competes the binding of labelled IL-8 on basophils (32). Data in Figure 3 suggest that RANTES and MIP-1a activate basophils through receptors distinct from the MCP-1 receptor and the IL-8 receptor. It is not yet clear whether two receptors specific for RANTES and MIP-1α, respectively, exist on basophils, or whether RANTES and MIP-1α act through a common receptor. Thus, at least 3 specific chemokine receptors exist on mature human basophils.

Platelet release products have been repeatedly proposed as mediators of allergic inflammation. In particular, it was recently reported that a mixture of CTAP-III and NAP-2 obtained from supernatants of stimulated mononuclear cells or CTAP-III from platelets induce histamine release at concentrations between 5 and 20 mg/ml (39). These results are in agreement with our findings demonstrating that histamine release, albeit to low degree, can be induced with similar concentrations of a variety of cationic peptides (32). However, it is clear that, at lower concentrations, only IL-8 and to a lesser extent, NAP-2 induce histamine release by binding to the IL-8 receptor on basophils (32). More interesting is the recent finding that platelets release RANTES when stimulated with thrombin (40). Since RANTES, MCP-1 and other members of the chemokine superfamily are also produced by monocytes (6), and since all these peptides have a similar molecular mass (8-10,000 daltons) it is tempting to speculate that the histamine-releasing activities found in supernatants of stimulated monocytes or platelets are composed of multiple effects exerted by different chemokines, which may be difficult to separate.

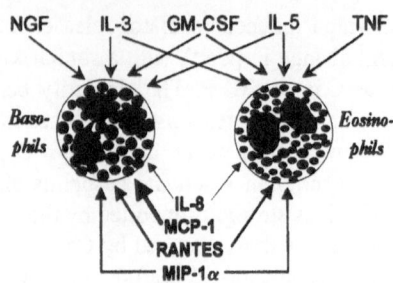

Figure 4. Cytokines affecting basophil and/or eosinophil function. Above the cells: hematopoietic growth factors modulating basophil mediator release. Below: chemokines triggering for mediator release (for details see text).

For the members of the "C-X-C branch" (IL-8 superfamily) much progress has been made regarding their biological functions and characterisation of specific receptors (1-4,33,34). The biological function of the human β-chemokines is now beginning to be unraveled. Recent studies showed that members of this subfamily are potent chemoattractants for monocytes and for eosinophils, but in contrast to a chemokines, not for neutrophils (4-7,29,40). Furthermore, RANTES and MIP-proteins are capable of attracting distinct lymphocyte types (6), and MIP-1α but not MIP-1β is a potent inhibitor of stem cell proliferation (41). This spectrum of activities suggests that β-chemokines play a role in chronic inflammation of different etiologies, whereas α-chemokines are rather involved in acute neutrophil-mediated inflammation (6). Several chronic inflammatory processes such as allergic inflammation are characterised by the infiltration with basophils, eosinophils and particular mononuclear cells (13,14). β-chemokines are so far the only peptide-agonists, which attract exactly this set of leukocyte types and may therefore be important candidates responsible for the initiation, perpetuation and regulation of such inflammatory events. For example, MCP-1 may promote inflammatory processes in which monocytes and possibly basophils play the predominant effector role, i.e. arteriosclerosis (11), whereas RANTES may partially explain the concerted emigration and activation of memory T helper cells, eosinophils and basophils, as found in the allergic late phase reaction (13,14,21) and other diseases associated with an eosinophil and basophil infiltration, e.g. certain types of intestinal inflammation (42).

ACKNOWLEDGEMENT

Parts of this study were performed in collaboration with A. Rot, Sandoz Research Institute, Vienna, Austria, Vinzenz von Tscharner and Marco Baggiolini, Theodor-Kocher-Institute, Bern, Switzerland. This work was supported in part by the Swiss National Science Foundation, grant 31-32470.91, and by the Deutsche Forschungsgemeinschaft, grant Bi-424/1-2.

REFERENCES

1. Baggiolini, M., A. Walz, and S.L. Kunkel. 1989. Neutrophil-activating peptide-1/interleukin 8, a novel cytokine that activates neutrophils. *J.Clin.Invest.* 84: 1045-1049.
2. Schröder, J.M., N.L.M. Persoon, and E. Christophers. 1990. Lipopolysaccharide-stimulated human monocytes secrete, apart from neutrophil-activating peptide 1/interleukin 8, a second neutrophil-activating protein. NH$_2$-terminal amino acid sequence identity with melanoma growth stimulatory activity. *J.Exp.Med.* 171: 1091-1100.

3. Walz, A., R. Burgener, B. Car, M. Baggiolini, S.L. Kunkel, and R.M. Strieter. 1991. Structure and neutrophil-activating properties of a novel inflammatory peptide (ENA-78) with homology to interleukin 8. *J.Exp.Med.* 174: 1355-1362.

4. Matsushima, K., and J.J. Oppenheim. 1989. Interleukin 8 and MCAF: novel inflammatory cytokines inducible by IL-1 and TNF. *Cytokine* 1: 2-13.

5. Leonard, E.J., and T. Yoshimura. 1990. Human monocyte chemoattractant protein-1 (MCP-1). *Immunol. Today* 11: 97-101.

6. Schall, T.J. 1991. Biology of the RANTES/SIS cytokine family. *Cytokine* 3: 165-183.

7. Rollins, B.J., A. Walz, and M. Baggiolini. 1991. Recombinant human MCP-1/JE induces chemotaxis, calcium flux, and the respiratory burst in human monocytes. *Blood* 78: 1112-1116.

8. Van Damme, J., P. Proost, J.P. Lenaerts, and G. Opdenakker. 1992. Structural and functional idetification of two human, tumor-derived monocyte chemotactic proteins (MCP-2 and MCP-3) belonging to the chemokine family. *J.Exp.Med.* 176: 59-65.

9. Orlovsky, A., M.S. Berger, and M.B. Prystowsky. 1991. Novel expression pattern of a new member of the MIP-1 family of cytokine-like genes. *Cell Regulation* 2: 403-412.

10. Fahey, T.J., B. Sherry, K.J. Tracey, S. van Deventer, W.G. Jones, J.P. Minei, S. Morgello, G.T. Shires, and A. Cerami. 1992. Cytokine production in a model of wound healing: the appearance of MIP-1, MIP-2, cachectin/TNF and IL-1. *Cytokine* 2: 92-99.

11. S. Ylä-Herttuala, B.A. Lipton, M.E. Rosenfeld, T. Särkioja, T. Yoshimura, E.J. Leonhard, J.L. Witztum, and D. Steinberg. 1991. Expression of monocyte chemoattractant protein 1 in macrophage-rich areas of human and rabbit atherosclerotic lesions. *Proc.Natl.Acad.Sci. USA* 88: 5252-5256.

12. Miñano, F.J., M. Vizcaino, and R.D. Myers. 1992. Action of cyclosporine on fever induced in rats by intrahypothalamic injection of macrophage inflammatory protein 1 (MIP-1). *Neuropharmacology* 31: 193-199.

13. Charlesworth, E.N., A.F. Hood, N.A. Soter, A. Kagey-Sobotka, P.S. Norman, and L.M. Lichtenstein. 1989. Cutaneous late-phase response to allergen. Mediator release and inflammatory cell infiltration. *J.Clin.Invest.* 83: 1519-1526.

14. Frew, A.J., and A.B. Kay. 1991. UCHL1[+] (CD45RO[+]) "memory" T cells predominate in the CD4[+] cellular infiltrate associated with allergen-induced late-phase skin reactions in atopic subjects. *Clin.Exp.Immunol.* 84: 270-274.

15. Kurimoto, Y., A.L. de Weck, and C.A. Dahinden. 1989. Interleukin 3-dependent mediator release in basophils triggered by C5a. *J.Exp.Med.* 170: 467-479.

16. Bischoff, S.C., A.L. de Weck, and C.A. Dahinden. 1990. Interleukin 3 and granulocyte/ macrophage-colony-stimulating factor render human basophils responsive to low concentrations of the complement component C3a. *Proc.Natl.Acad.Sci. USA* 87: 6813-6817.

17. Bischoff, S.C., T. Brunner, A.L. de Weck, and C.A. Dahinden. 1990. Interleukin 5 modifies histamine release and leukotriene generation by human basophils in response to diverse agonists. *J.Exp.Med.* 172: 1577-1582.

18. Takafuji, S., S.C. Bischoff, A.L. de Weck, and C.A. Dahinden. 1991. IL-3 and IL-5 prime normal human eosinophils to produce leukotriene C4 in response to soluble agonists. *J.Immunol.* 147: 3855-3861.

19. Lopez, A.F., C.J. Sanderson, J.R. Gamble, H.D. Champell, I.G. Young, and M.A. Vadas. 1988. Recombinant human interleukin 5 is a selective activator of human eosinophil function. *J.Exp.Med.* 167: 219-224.

20. Kay, A.B., S. Ying, V. Varney, M. Gaga, S.R. Durham, R. Moqbel, A.J. Wardley, and Q. Hamid. 1991. Messenger RNA expression of the cytokine gene cluster, interleukin 3 (IL-3), IL-4, IL-5, and granulocyte/macrophage colony-stimulating factor, in allergen-induced late-phase cutaneous reactions in atopic subjects. *J.Exp. Med.* 173: 775-778.

21. Walker, C., J.C. Virchow, P.L.B. Bruijnzeel, and K. Blaser. 1991. T cell subsets and their soluble products regulate eosinophilia in allergic and non-allergic asthma. *J.Immunol.* 146:1829-1835.

22. Brunner, T., A.L. de Weck, and C.A. Dahinden. 1991. Platelet-activating factor induces mediator release by human basophils primed with IL-3, granulocyte-macrophage colony-stimulating factor, or IL-5. *J.Immunol.* 147: 237-242.

23. Dahinden, C.A., Y. Kurimoto, A.L. de Weck, I. Lindley, B. Dewald, and M. Baggiolini. 1989. The neutrophil-activating peptide NAF/NAP-1 induces histamine and leukotriene release by interleukin 3-primed basophils. *J.Exp.Med.* 170: 1787-1792.

24. Bischoff, S.C., M. Baggiolini, A.L. de Weck, and C.A. Dahinden. 1991. Interleukin 8 - inhibitor and inducer of histamine and leukotriene release in human basophils. Biochem. *Biophys.Res.Comm.* 179: 628-633.

25. Bischoff, S.C., M. Krieger, T. Brunner, and C.A. Dahinden. 1992. Monocyte chemotactic protein 1 is a potent activator of human basophils. *J.Exp.Med.* 175: 1271-1275.

26. Kuna, P., S.R. Reddigari, D. Rucinski, J.J. Oppenheim, and A.P. Kaplan. 1992. Monocyte chemotactic and activating factor is a potent histamine-releasing factor for human basophils. *J.Exp.Med.* 175: 489-493.

27. Alam, R., M.A. Lett-Brown, P.A. Forsythe, D.J. Anderson-Walters, C. Kenamore, C. Kormos, and J.A. Grant. 1992. Monocyte chemotactic and activating factor is a potent histamine-releasing factor for basophils. *J.Clin Invest.* 89: 723-728.

28. Lindley, I., H. Aschauer, J.M. Seifert, C. Lam, W. Brunowsky, E. Kownatzki, M. Thelen, P. Peveri, B. Dewald, V. von Tscharner, A. Walz, and M Baggiolini. 1988. Synthesis and expression in *Escherichia coli* of the gene encoding monocyte-derived neutrophil-activating factor: biological equivalence between natural and recombinant neutrophil-activating factor. *Proc.Natl.Acad.Sci. USA* 85: 9199-9203.

29. Rot, A., M. Krieger, T. Brunner, S.C. Bischoff, T.J. Schall, and C.A. Dahinden. 1992. RANTES and MIP-1a induce the migration and activation of normal human eosinophil granulocytes. *J.Exp.Med.* in press.

30. Bischoff, S.C., and C. A. Dahinden. 1990. The effect of nerve growth factor upon mediator release by mature human basophils. *Blood* 79: 2662-2669.

31. Leonhard, E.J., A. Skeel, T. Yoshimura, K. Noer, S. Kutvirt, and D. Van Epps. 1990. Leukocyte specificy and binding of human neutrophil attractant/activating protein-1. *J. Immunol.* 144: 1323-1330.

32. Krieger, M., T. Brunner, S.C. Bischoff, V. von Tscharner, A. Walz, B. Moser, M. Baggiolini, and C.A. Dahinden. 1992. Activation of human basophils through the IL-8 receptor. *J. Immunol.* 149: in press.

33. Moser, B., C. Schuhmacher, V. von Tscharner, I. Clark-Lewis, and M. Baggiolini. 1991. Neutrophil-activating peptide 2 and *gro*/melanoma growth stimulatory activity interact with neutrophil-activating peptide 1/interleukin 8 receptors on human neutrophils. *J.Biol.Chem.* 266: 10666-10671.

34. Samanta, A.K., J.J. Oppenheim, and K. Matsushima. 1989. Identification and characterization of specific receptors for monocyte-derived neutrophil-chemotactic factor (MDNCF) on human neutrophils. *J.Exp.Med.* 169: 1185-1189.

35. Gerard N.P. and C. Gerard. 1991. The chemotactic receptor for human C5a anaphylatoxin. *Nature* 349: 614-617.

36. Takafuji, S., S.C. Bischoff, A.L. de Week, and C.A. Dahinden. 1992. Opposing effects of tumor necrosis factor-a and nerve growth factor upon leukotriene C4 production by human eosinophils triggered with N-formyl-methionyl-leucyl-phenylalanine. *Eur.J. Immunol.* 22: 969-974.

37. Darbonne, W.C., G.C. Rice, M.A. Mohler, T. Apple, C.A. Hébert, A.J. Valente, and J.B. Baker. 1991. Red blood cells are a sink for interleukin 8, a leukocyte chemotaxin. *J.Clin.Invest.* 88: 1362-1369.

38. Fahey, T.J., K.J. Tracey, P. Tekamp-Olson, L.S. Cousens, W.G. Jones, G.T. Shires, A. Cerami, and B. Sherry. 1992. Macrophage inflammatory protein modulates macrophage function. *J.Immunol.* 148: 2764-2769.

39. Baeza, M.L., S.R. Reddigari, D. Kornfeld, N. Ramani, E.M. Smith, P.A. Hossler, T. Fischer, C.W. Castor, P.G. Gorevic, and A.P. Kaplan. 1990. Relationship of one form of histamine-releasing factor to connective tissue-activating peptide III. *J.Clin.Invest.* 85: 1516-1521.

40. Kameyoshi, Y., A. Dörschner, A.I. Mallet, E. Christophers, and J.M. Scröder. 1992. Cytokine RANTES released by thrombin-stimulated platelets is a potent attractant for human eosinophils. *J.Exp.Med.* 176: 587-592.

41. Dunlop, D.J., E.G. Wright, S. Lorimore, G.J. Graham, T. Holyoake, D.J. Kerr, S.D. Wolpe, and I.B. Pragnell. 1992. Demonstration of stem cell inhibition and myeloprotective effects of SCI/rhMIP-1a in vivo. *Blood* 79: 2221-2225.

42. Marsh, M.N., and J. Hinde. 1985. Inflammatory components of coeliac sprue mucosa: 1. Mast cells, basophils and eosinophils. *Gastroenterology* 89: 92-101.

MONOCYTE CHEMOTACTIC PROTEINS RELATED TO HUMAN MCP-1

Jo Van Damme, Paul Proost, Jean-Pierre Lenaerts, René Conings,
Ghislain Opdenakker and Alfons Billiau

Rega Institute for Medical Research
University of Leuven
B-3000 Leuven, Belgium

INTRODUCTION

This study describes the isolation and identification of two novel monocyte chemotactic factors from human tumor cells. Since the corresponding 7.5 kD and 11 kD proteins show high structural and functional similarity with MCP-1 (1) they are designated MCP-2 and MCP-3, respectively. Based on the conservation of four cysteine residues, these molecules can be classified in the chemokine family.

PURIFICATION AND IDENTIFICATION OF HUMAN MCP-2

Human osteosarcoma cells (MG-63) were stimulated with a leukocyte-derived cytokine preparation for 5 h, washed and incubated for 48 h with medium containing 2% FCS. The monocyte chemotactic activity was measured in the agarose migration test and microchamber assay (2). It was concentrated from the conditioned medium by adsorption to controlled pore glass beads and further purified by antibody affinity chromatography (3). Separation of distinct monocyte chemotactic activities was achieved by cation-exchange FPLC (Fig. 1). A predominant peak eluted at 0.5 M NaCl in the gradient and corresponded to two major MCP-1 protein bands (10 kD and 16 kD). A second MCP-peak, was recovered at 0.8 M NaCl, whereas IL-8, eluted at 0.95 M NaCl (Fig.1.). The second MCP peak from FPLC was purified to homogeneity in parallel with MCP-1 (2). MCP-1 eluted on C8 reverse-phase HPLC at about 26% acetonitrile, whereas the second MCP peak from FPLC eluted at 30% acetonitrile (Fig. 2A). Homogeneous MCP-1 appeared to be heterogeneous upon SDS-PAGE in that multiple Mr bands (10,12,14 and 16 kD) were contained in single HPLC fractions (Fig. 2B). Identity with authentic MCP-1 was confirmed by sequencing fragments of an Asp-N endoproteinase digest. Although the second MCP activity from FPLC eluted over a wider range on HPLC, the corresponding molecule was always found to reside in a 7.5 kD protein (Fig. 2B). This protein had also a blocked NH_2-terminus, but could be identified through fragmentation of pure material with trypsin or Asp-N endoproteinase. Sequence analysis of the obtained fragments, separated by C8 reverse-phase HPLC, demonstrated that the 7.5 kD MCP has a primary structure related to, but distinct from MCP-1. The alignment of multiple sequence fragments allowed for the identification of nearly the complete primary structure of this 7.5 kD molecule (Fig. 3). This monocyte chemotactic protein, showing 62% identity with MCP-1 was therefore designated MCP-2. Although the MCP-2 sequence was not contained in the EMBL/Swiss-Prot Bank, the primary structure of this chemotactic factor fully corresponded to that of the cDNA-derived sequence of HC14. The HC14 gene encodes a protein of 99 amino acids including a signal peptide of 23 residues (4). HC14 mRNA was originally derived from human peripheral blood lymphocytes stimulated with mitogenic anti-CD2 monoclonal antibodies, but subsequent experiments revealed that HC14 mRNA is merely induced in macrophages by the T cell-derived interferon-γ. The cDNA sequence of HC14 has not yet been disclosed and the corresponding protein has not been expressed to permit biological characterization.

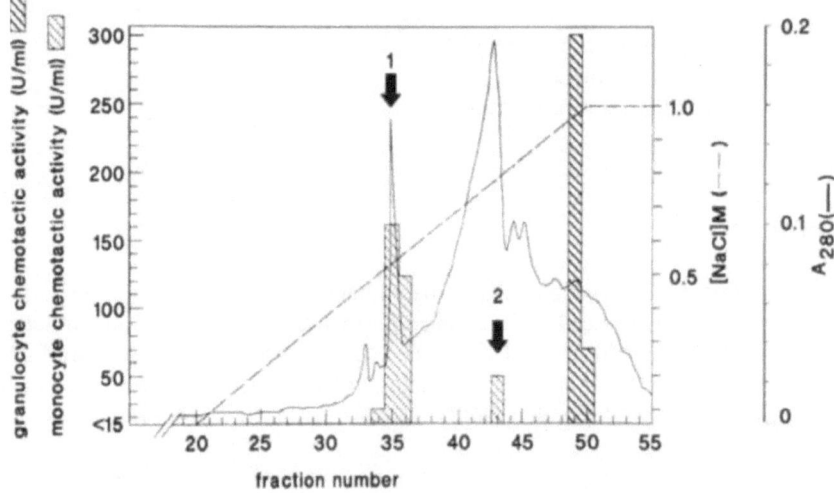

Figure 1. Separation of human MCP-1, MCP-2 and IL-8 by cation-exchange FPLC. Partially purified, concentrated MG-63 cell supernatant was loaded on a Mono S column at pH 4.0 and eluted (1 ml fractions) with a linear NaCl gradient (---). Absorbance (——) was monitored at 280 nm. Fractions were tested for monocyte and granulocyte chemotactic activity (histograms) in the agarose assay. Numbered arrows indicate MCP-1 and MCP-2, respectively.

ISOLATION AND IDENTIFICATION OF HUMAN MCP-3

To verify whether MCP-2, like the structurally related cytokines MCP-1 and IL-8, has affinity for heparin, conditioned medium of cytokine-stimulated MG-63 cells was purified on heparin-Sepharose (5). It was found that all MCP-like activity eluted in a single but broad peak at 0.75 M NaCl. Subsequent cation-exchange chromatography allowed for the separation of a major MCP-1 and a minor MCP-2 peak, as well as a weak MCP-like activity eluting in between at 0.7 M NaCl. Further purification of this third MCP-peak by HPLC showed that it eluted at 27.5% acetonitrile, a position also between the elution percentages for MCP-1 and MCP-2, respectively. Upon SDS-PAGE the corresponding protein had a Mr of 11 kDa, indicating that it was different from MCP-1 and MCP-2.

Sequence analysis of the 11 kD MCP demonstrated that also this protein was NH$_2$-terminally blocked. After digestion with trypsin, Asp-N, Lys-C and asparaginyl endoproteinase and subsequent separation of the fragments by HPLC, the 11 kD MCP could be identified as a novel protein, closely related to MCP-1 (Fig. 3). Out of 67 amino acids determined only 18 were found to differ from the corresponding residues in the sequence of MCP-1, whereas 27 residues were not identical when compared to MCP-2. Based on these differences in primary structure, this novel monocyte chemotactic protein was designated MCP-3.

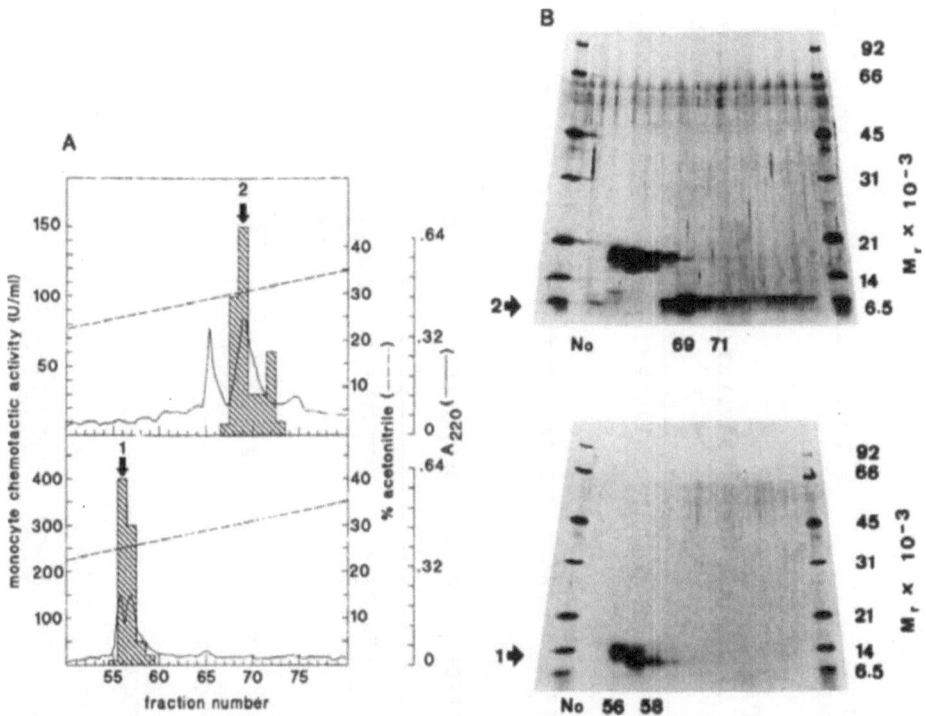

Figure 2. Purification of MCP-1 and MCP-2 to homogeneity by reverse-phase HPLC. (A) MCP-1 and MCP-2 from FPLC were loaded separately on a C-8 Aquapore RP-300 column and eluted (0.4 ml fractions) with an acetonitrile gradient (---). Absorbance was monitored at 220 nm (--). (B) SDS/PAGE of HPLC-purified MCP. Active fractions were run (20 µl/lane) on a linear gel under reducing conditions and silver stained. Numbered arrows indicate MCP-1 and MCP-2 activity (histograms) measured in the agarose assay (A) and their corresponding protein bands (B), respectively.

BIOLOGICAL CHARACTERISTICS OF MCP-2 AND MCP-3

In the agarose migration assay MCP-2 and MCP-3 have a specific activity for monocytes of about 10^4U/mg (1U corresponds to a half maximal migration distance), comparable to that of MCP-1. In contrast, at the highest dose tested, no chemotactic activity for neutrophils could be detected with all three MCP species, indicating that their specific activity on neutrophils is at least 5-10 fold lower (<10^3U/mg) than on monocytes. As a control, IL-8 was found to possess a specific activity of more than 10^5U/mg in neutrophil chemoattraction. The biological effect of the two novel chemotactic factors was confirmed using the Boyden microchamber assay (Table 1). When tested at different dilutions, MCP-1, -2 and -3 were equally chemotactic (specific activity of about 5×10^5U/mg) for monocytes in this assay, whereas they failed to be active on neutrophils.

MCP-2 tryptic fragments

```
              IPIQR(1) a)        EAVIFK(2)      WVR(1)   HLDQIFQNLKP(3)
              KIPIQR(2)  ITNIQCPK(5)    EV?ADPK(2)
                  LESYTR(2)              GKEVCADPK(2)
```

MCP-2 ASP-N fragments   DSVSIPITCCFNVINRKIPIQRL(2)   ERWVR(1)   DQIFQNLKP(1)
```
                         ESYTRITNIQ?PKEAVIFKT(2)        DPKERWVR(1)
                                                          DSMKHL(2)
```

MCP-2 sequence: DSVSIPITCCFNVINRKIPIQRLESYTRITNIQCPKEAVIFKT   GKEVCADPKERWVRDSMKHLDQIFQNLKP

MCP-1 sequence: QPDAINAPVTCCYNFTNRKISVQRLASYRRITSSKCPKEAVIFKTIVAKEICADPKQKWVQDSMDHLDKQTQTPKT

MCP-3 sequence: KSTTCCYRFINKKIPKQRLESYRRTTSSHCPREAVIFK   DKEICADPTQKWVQDFMKHLDKKTQTPKL

MCP-3 tryptic fragments   LESYR(2)   EAVIFK(7)   EI?ADP(3)
```
                           QRLESYR(4)                  WVQDFMK(4)
                              ?TSS??P(2)            ?D?EI?ADP(1)
```

MCP-3 ASP-N fragments   ESYRRTTSS??P(2)   DKEICA(3)   DFMKHL(2)
```
                                                  DPTQKWVQ(2)   DKKTQTPKL(4)
```

MCP-3 LYS-C fragments   QRLESYRRTTSSHCPREAVIF(2)   WVQDFMK(1)

MCP-3 ASN-C fragments   KSTTCCYRFIN(2)
                          KKIPKQRLESYRRTTS?H?PREAVIF

a) Numbers between brackets indicate the number of times the sequences were confirmed.

Figure 3. Amino acid sequences of MCP-2 and MCP-3 fragments.

MCP-1 and IL-8 have been reported to possess chemotactic activity in vivo in that they caused both locally and systematically an increased appearance of monocytes and neutrophils, respectively (5,6). Similarly, MCP-2 and MCP-3 were found to induce monocyte infiltration after intradermal injection in rabbit skin (7). The monocyte infiltration was characterized by clustering of cells, visible monocyte adherence to endothelia (margination), and accumulation of monocytes around the site of injection.

Table 1. Comparison of MCP-1, -2 and -3 in the microchamber migration assay

Chemo-attractant	Concentration (nM)	Chemotactic index[a]	
		monocytes	granulocytes
MCP-1	0.03	1.8	-
	0.1	2.4	0.4
	0.3	3.6	0.6
	1.0	4.4	0.5
	3.0	3.5	0.6
MCP-2	0.03	0.6	-
	0.1	1.2	0.4
	0.3	1.2	0.3
	1.0	4.1	0.7
	3.0	2.7	1.5
MCP-3	0.03	0.8	
	0.1	1.2	0.8
	0.3	0.9	0.5
	1.0	1.8	0.9
	3.0	2.4	0.8

[a]   chemotactic index corresponds to the number of migrated cells by the chemoattractant divided by the number of migrated cells in the control cultures.

RELATION OF MCP-2 AND MCP-3 WITH OTHER MEMBERS OF THE CHEMOKINE FAMILY

The cellular sources of MCP-1 are multiple and include mononuclear leukocytes (8-12), fibroblasts (2,13,14), endothelial cells (15-17), smooth muscle cells (18), epithelial cells (19), melanocytes (20), keratinocytes (21), in addition to tumor cells (2,6,22-25). Although mRNA for HC14/MCP-2 could be measured in stimulated peripheral blood lymphocytes and monocytes, until now we have been unable to purify MCP-2 (or MCP-3) from mitogen or lipopolysaccharide treated mononuclear cells. Under those conditions the relative amount of MCP-1 produced by leukocytes is significantly lower than by MG-63 tumor cells treated with IL-1ß. Taking into consideration that the amount of MCP-2 and MCP-3 produced by these tumor cells was about five times lower than of MCP-1, it is possible that the quantity of the former two factors was below the detection limit when leukocytes were used as a cell source.

Other human molecules of this subfamily such as MIP-1 (26), RANTES (27), HC21/G-26/Act-2/pat744/hH400 (28-32), LD78/pat464/hSIS/GOS19 (31-34) and I309 (29) have been reported to be secreted by leukocytes or tumor cells. Although the structural similarity of these proteins with MCP-1 is much weaker than that of MCP-2 and -3, some of these, e.g. RANTES have been reported to possess monocyte chemotactic activity (27). IFN-γ has been reported to be a major inducer of MCP-2 in monocytes (4). IL-1 and TNF-α might be major inducers of MCP-2 and -3 on tumor cells, since these cytokines significantly increase MCP-1 levels in tumor cells, as well as in fibroblasts, endothelial cells and monocytes (2,13,16,17,19). Induction of MCP-3 was observed in MG-63 cells stimulated with pure IL-1ß under serumfree conditions.

With regard to sequence similarity human MCP-1, -2 and -3 are equally related to mouse MCP/JE (14) and rat MCP (35). It can therefore not be deduced from their sequences whether these rodent MCPs are the homologues of human MCP-1 (Table 2). However, rabbit (36) and bovine (37) MCP-1 seem to possess a protein sequence that is more related to human MCP-1 than to MCP-2 and MCP-3.

Table 2. Amino acid sequence similarity of animal MCPs

Species	% Identity when compared to[a]		
	hu MCP-1	hu MCP-2	hu MCP-3
Human MCP-1	100	62	73
Human MCP-2	62	100	60
Human MCP-3	73	60	100
Mouse MCP/JE	55	53	51
Rat MCP-1	51	43	45
Rabbit MCP-1	75	51	63
Bovine MCP-1	72	51	64

[a] alignment of the 76 NH2-terminal residues (ref. 4,7,8,9,14,35-37) except for MCP-3 (67 residues)

POSSIBLE ROLE OF CHEMOTACTIC FACTORS IN TUMOR INVASION

Many tumor cells e.g. sarcoma, carcinoma, glioma and leukemic cells, produce chemotactic factors *in vitro* (1,38). These tumor cell derived chemotactic factors are indistinguishable from those produced by inflammatory cells (2,11). In addition, tumor-derived monocyte chemotactic activity is correlated with the presence of macrophages in the tumor (39). However, a tumor cell specific chemotactic factor (explaining active chemoattraction of tumor cells in invasion) has not yet been found. Therefore, chemotactic factors produced by tumor cells might passively complement tumor cell invasion. Instead of being promiscuous factors that help to attract immune cells to eliminate the tumor, they might be as well used by the tumor for its own spreading in a passive way. In this way the chemoattractant producing tumor cell resembles very much an inflammatory focus or a wound. For decennia pathologists have observed and described the presence of inflammatory cells in invading and metastasing tumors. It has been speculated that these inflammatory cells (macrophages, lymphocytes ...) play a role in tumor surveillance by the immune system (6,39). It might, however, at the same time be that especially those tumor cells, that are generous in the production of chemotactic factors for neutrophils and monocytes use for invasion a natural *aspecific* defense mechanism against microorganisms (40).

ACKNOWLEDGEMENTS

The authors thank W. Put for technical assistance and D. Brabants for editorial help. This work was supported by the National Fund for Scientific Research (N.F.W.O.), the General Savings and Retirement Fund (A.S.L.K.) and the Belgian Ministry of Science Policy. G.O. is Research Associate of the N.F.W.O.

REFERENCES

1. Leonard, E.J., and T. Yoshimura. 1990. Human monocyte chemo-attractant protein-1 (MCP-1). *Immunology Today* 11: 97.
2. Van Damme, J., B. Decock, J.-P. Lenaerts, R. Conings, R. Bertini, A. Mantovani, and A. Billiau. 1989. Identification by sequence analysis of chemotactic factors for monocytes produced by normal and transformed cells stimulated with virus, double-stranded RNA or cytokine. *Eur.J. Immunol.* 19: 2367.
3. Van Damme, J., S. Cayphas, J. Van Snick, R. Conings, W. Put, J.-P. Lenaerts, R.J. Simpson, and A. Billiau. 1987. Purification and characterization of human fibroblast-derived hybridoma growth factor identical to T-cell-derived B-cell stimulatory factor-2 (interleukin-6). *Eur.J. Biochem.* 168: 543.
4. Chang, H.C., F. Hsu, G.J. Freeman, J.D. Griffin, and E.L. Reinherz. 1989. Cloning and expression of a γ-interferon-inducible gene in monocytes: a new member of a cytokine gene family. *International Immunology* 1: 388.
5. Van Damme, J., J. Van Beeumen, G. Opdenakker, and A. Billiau. 1988. A novel, NH_2-terminal sequence-characterized human monokine possessing neutrophil chemotactic, skin-reactive, and granulocytosis-promoting activity. *J.Exp.Med.* 167: 1364.
6. Zachariae, C.O.C., A.O. Anderson, H.L. Thompson, E. Appella, A. Mantovani, J.J. Oppenheim, and K. Matsushima. 1990. Properties of monocyte chemotactic and activating factor (MCAF) purified from a human fibrosarcoma cell line. *J.Exp.Med.* 171: 2177.
7. Van Damme, J., P. Proost, J.-P. Lenaerts, and G. Opdenakker. 1992. Structural and functional identification of two human, tumor-derived monocyte chemotactic proteins (MCP-2 and MCP-3) belonging to the chemokine family. *J.Exp.Med.* 176: 59.
8. Yoshimura, T., N. Yuhki, S.K. Moore, E. Appella, M.I. Lerman, and E.J. Leonard. 1989. Human monocyte chemoattractant protein-1 (MCP-1). Full length cDNA cloning, expression in mitogen-stimulated blood mononuclear leukocytes, and sequence similarity to mouse competence gene JE. *FEBS Letters* 244: 487.
9. Furutani, Y., H. Nomura, M. Notake, Y. Oyamada, T. Fukui, M. Yamada, C.G. Larsen, J.J. Oppenheim, and K. Matsushima. 1989. Cloning and sequencing of the cDNA for human monocyte chemotactic and activating factor (MCAF). *Biochem.Biophys.Res.Commun.* 159: 249.
10. Yoshimura, T., E.A. Robinson, S. Tanaka, E. Appella, and E.J. Leonard. 1989. Purification and amino acid analysis of two human monocyte chemoattractants produced by phytohemagglutinin-stimulated human blood mononuclear leukocytes. *J.Immunol.* 142: 1956.
11. Decock, B., R. Conings, J.-P. Lenaerts, A. Billiau, and J. Van Damme. 1990. Identification of the monocyte chemotactic protein from human osteosarcoma cells and monocytes: detection of a novel N-terminally processed form. *Biochem. Biophys.Res.Commun.* 167: 904.
12. Rollins, B.J., P. Stier, T. Ernst, and G.G. Wong. 1989. The human homolog of the JE gene encodes a monocyte secretory protein. *Molecular and Cellular Biology* 9: 4687.
13. Larsen, C.G., C.O.C. Zachariae, J.J. Oppenheim, and K. Matsushima. 1989. Production of monocyte chemotactic and activating factor (MCAF) by human dermal fibroblasts in response to interleukin-1 or tumor necrosis factor. *Biochem.Biophys.Res.Commun.* 160: 1403.
14. Rollins, B.J., E.D. Morrison, and C.D. Stiles. 1988. Cloning and expression of JE, a gene inducible by platelet-derived growth factor and whose product has cytokine-like properties. *Proc. Natl.Acad.Sci. USA* 85: 3738.
15. Takehara, K., E.C. LeRoy, and G.R. Grotendorst. 1987. TGF-ß inhibition of endothelial cell proliferation: alteration of EGF binding and EGF-induced growth-regulatory (competence) gene expression. *Cell* 49: 415.
16. Strieter, R.M., R. Wiggins, S.H. Phan, B.L. Wharram, H.J. Showell, D.G. Remick, S.W. Chensue, and S.L. Kunkel. 1989. Monocyte chemotactic protein gene expression by cytokine-treated human fibroblasts and endothelial cells. *Biochem.Biophys.Res.Commun.* 162: 694.
17. Sica, A., J.M. Wang, F. Colotta, E. Dejana, A. Mantovani, J.J. Oppenheim, C.G. Larsen, C.O.C. Zachariae, and K. Matsushima. 1990. Monocyte chemotactic and activating factor gene expression induced in endothelial cells by IL-1 and tumor necrosis factor. *J.Immunol.* 144: 3034.
18. Valente, A.J., D.T. Graves, C.E. Vialle-Valentin, R. Delgado, and C.J. Schwartz. 1988. Purification of

a monocyte chemotactic factor secreted by nonhuman primate vascular cells in culture. *Biochemistry* 27: 4162.

19. Elner, S.G., R.M. Strieter, V.M. Elner, B.J. Rollins, M.A. Del Monte, and S.L. Kunkel. 1991. Monocyte chemotactic protein gene expression by cytokine-treated human retinal pigment epithelial cells. *Lab.Invest.* 64: 819.

20. Zachariae, C.O.C., K. Thestrup-Pedersen, and K. Matsushima. 1991. Expression and secretion of leukocyte chemotactic cytokines by normal human melanocytes and melanoma cells. *J.Invest. Dermatol.* 97: 593.

21. Barker, J.N.W.N., M.L. Jones, C.L. Swenson, V. Sarma, R.S. Mitra, P.A. Ward, K.J. Johnson, J.C. Fantome, V.M. Dixit, and B.J. Nickoloff. 1991. Monocyte chemotaxis and activating factor production by keratinocytes in response to IFN-γ *J.Immunol.* 146: 1192.

22. Graves, D.T., Y.L. Jiang, M.J. Williamson, and A.J. Valente. 1989. Identification of monocyte chemotactic activity produced by malignant cells. *Science* 245: 1490.

23. Yoshimura, T., E.A. Robinson, S. Tanaka, E. Appella, J.-I. Kuratsu, and E.J. Leonard. 1989. Purification and amino acid analysis of two human glioma-derived monocyte chemoattractants. *J.Exp. Med.* 169: 1449.

24. Matsushima, K., C.G. Larsen, G.C. Dubois, and J.J. Oppenheim. 1989. Purification and characterization of a novel monocyte chemotactic and activating factor produced by a human myelomonocytic cell line. *J.Exp.Med.* 169: 1485.

25. Bottazzi, B., F. Colotta, A. Sica, N. Nobili, and A. Mantovani. 1990. A chemoattractant expressed in human sarcoma cells (tumor-derived chemotactic factor, TDCF) is identical to monocyte chemoattractant protein-1/monocyte chemotactic and activating factor (MCP-1/MCAF). Int. J. *Cancer* 45: 795.

26. Davatelis, G., P. Tekamp-Olson, S.D. Wolpe, K. Hermsen, C. Luedke, C. Gallegos, D. Coit, J. Merryweather, and A. Cerami. 1988. Cloning and characterization of a cDNA for murine macrophage inflammatory protein (MIP), a novel monokine with inflammatory and chemokinetic properties. *J.Exp.Med.* 167: 1939.

27. Schall, T.J., K. Bacon, K.J. Toy, and D.V. Goeddel. 1990. Selective attraction of monocytes and T lymphocytes of the memory phenotype by cytokine RANTES. *Nature* 347: 669.

28. Chang, H.-C. and E.L. Reinherz. 1989. Isolation and characterization of a cDNA encoding a putative cytokine which is induced by stimulation via the CD2 structure on human T lymphocytes. *Eur.J.Immunol.* 19: 1045.

29. Miller, M.D., S. Hata, R. De Waal Malefyt, and M.S. Krangel. 1989. A novel polypeptide secreted by activated human T lymphocytes. *J.Immunol.* 143: 2907.

30. Lipes, M.A., M. Napolitano, K.-T. Jeang, N.T. Chang, and W.J. Leonard. 1988. Identification, cloning, and characterization of an immune activation gene. *Proc.Natl.Acad. Sci. USA* 85: 9704.

31. Zipfel, P.F., J. Balke, S.G. Irving, K. Kelly, and U. Siebenlist. 1989. Mitogenic activation of human T cells induces two closely related genes which share structural similarities with a new family of secreted factors. *J.Immunol.* 142: 1582.

32. Brown, K.D., S.M. Zurawski, T.R. Mosmann, and G. Zurawski. 1989. A family of small inducible proteins secreted by leukocytes are members of a new superfamily that includes leukocyte and fibroblast-derived inflammatory agents, growth factors, and indicators of various activation processes. *J.Immunol.* 142: 679.

33. Obaru, K., M. Fukuda, S. Maeda, and K. Shimada. 1986. A cDNA clone used to study mRNA inducible in human tonsillar lymphocytes by a tumor promoter. *J.Biochem. (Tokyo)* 99: 885.

34. Blum, S., R.E. Forsdyke, and Forsdyke, D.R. 1990. Three human homologs of a murine gene encoding an inhibitor of stem cell proliferation. (1990). DNA and Cell Biology 9:589.

35. Yoshimura, T., M. Takeya, and K. Takahashi. 1991. Molecular cloning of rat monocyte chemoattractant protein-1 (MCP-1) and its expression in rat spleen cells and tumor cell lines. *Biochem.Biophys.Res. Commun.* 174: 504.

36. Yoshimura, T. and N. Yuhki. 1991. Neutrophil attractant/activation protein-1 and monocyte chemoattractant protein-1 in rabbit. cDNA cloning and their expression in spleen cells. *J.Immunol.* 146: 3483.

37. Wempe, F., A. Henschen, and K.H. Scheit. 1991. Gene expression and cDNA cloning identified a major basic protein constituent of bovine seminal plasma as bovine monocyte-chemoattractant protein-1 (MCP-1). *DNA and Cell Biology* 10: 671.

38. Van Damme, J. 1991. Granulocyte and monocyte chemotactic factors: stimuli and producer cells. In: *Chemotactic Cytokines*, Ed. J. Westwick et al., Plenum Press, New York, pp. 1-9.

39. Mantovani, A. 1990. Tumor-associated macrophages. Current Opinion in Immunology 2: 689.

40. Opdenakker, G. and J. Van Damme. 1992. Cytokines and proteases in invasive processes: molecular similarities between inflammation and cancer. *Cytokine* 4: 251.

PLATELETS SECRETE AN EOSINOPHIL-CHEMOTACTIC CYTOKINE WHICH IS A MEMBER OF THE C-C-CHEMOKINE FAMILY

Jens-Michael Schröder, Yoshikazu Kameyoshi and Enno Christophers

Department of Dermatology
University of Kiel
W-2300 Kiel
Germany

INTRODUCTION

Eosinophilic granulocytes (Eos) represent an inflammatory cell type hypothesized to play a crucial role in the development of chronic asthma (1), after allergen challenge in sensitized animals (2) and atopic subjects (3).

Eos contain potent toxic proteins with the potential to mediate tissue damage. Immunofluorescent localization of eosinophil granule proteins has shown that eosinophils disrupt in tissue and deposit toxic granule proteins. This deposition is vastly out of proportion to the number of identifiable cells, which indicates that eosinophil involvement in diseases cannot be judged alone by the number of intact eosinophils observed in the tissue sections. Eos represent a minor part of the granuloytes in the circulation. Therefore their prevalence in tissue samples of patients with parasitic infections as well as allergic diseases raises the question of whether preferential eosinophil chemotactic factors lead to the migration of Eos into the tissue.

EOSINOPHIL ATTRACTANTS

A number of eosinophil attracting factors have been described in the past. The complement fragment C5a represents a powerful (potent and efficient) chemotaxin for human eosinophils (4). However, it also is a powerful chemotaxin for neutrophils and monocytes. Leukotriene B_4 (LTB_4) has been suggested to be an important chemotactic factor for eosinophils for a long time. Indeed, in guinea pig eosinophils it elicits strong chemotactic responses (5). In human eosinophils, however, only a low response is observable (6). Later, platelet activating factor (PAF) was identified as the most efficient chemotactic factor for human eosinophils, albeit attracting neutrophils as well (7). We recently

The Chemokines, Edited by I.J.D. Lindley
et al., Plenum Press, New York 1993

observed that eosinophils themselves are capable of producing their own attractant, which has been identified as a novel and potent eosinophil chemotactic eicosanoid in original studies (8). Very recently we could structurally characterize this chemotactic factor to be identical with 5-oxo-15-hydroxyeicosatetraenoic acid (9), which is as efficient as PAF in attracting human Eos. Apart from lipid-like Eo-attractants and C5a also other proteinaceous Eo-chemotaxins have been described. Interleukin 5, which was originally described as T-cell-replacing factor and is known to stimulate bone marrow precursor cells to differentiate into mature eosinophils, is also a specific chemoattractant for human eosinophils (10). Similarly granulocyte macrophage colony-stimulating factor (GMCSF) and interleukin 3 (IL-3) represent selective eosinophil attractants (11,12).

Another cytokine, termed lymphocyte-derived chemotactic factor (LCF), originally detected as a lymphocyte and monocyte chemotaxin which binds to CD4, has been reported to represent a powerful and selective Eo-attractant, which does not attract neutrophils (13). All these chemotactic cytokines have relative molecular masses > 15 kD.

In recent years a number of chemotactic cytokines belonging to a superfamily of structurally related polypeptides have been isolated and characterized, which have some cell selectivity representing chemotaxins for neutrophils, monocytes and lymphocytes (for review see (14-16). It has been shown that apparently members of the C-X-C branch of the "chemokine family" such as IL-8 (17,18), MGSA/gro (19,20) and NAP-2 (21) are chemotactic for neutrophils and in part also lymphocytes and basophils, but not however monocytes and eosinophils. Members of the C-C branch of the gene family represent cytokines which are chemotaxins for monocytes and lymphocyte subsets but not for neutrophils. These findings raised the question of whether this gene family also contains chemotactic cytokines for eosinophils, thus completing the spectrum of leukocyte-selective attractants.

DETECTION OF EO-CHEMOTACTIC PROTEINS IN SUPERNATANTS OF MONONUCLEAR CELL PREPARATIONS

In order to test the working hypothesis that peripheral blood mononuclear cell (PBMC) preparations secrete apart from IL-8 (17) or MCP-1 (22) also Eo-chemotactic cytokines, PBMC preparations were stimulated for 48 hrs with bacterial lipopolysaccharide (1 µg/ml) together with concanavalin A (10 µg/ml).

Supernatants of PBMC preparations were separated by preparative reversed phase HPLC and analyzed for Eo-chemotactic activity using purified human Eo-preparations (17).

As shown in Fig. 1 fractions eluting at 47 min showed a single peak of Eo-chemotactic activity. When its M_r was determined by the use of size exclusion-HPLC, biological activity came from the column in fractions corresponding to a M_r near 8 kD, exactly in the fractions where IL-8 elutes. In order to evaluate the cellular origin of the 8 kD eosinophil chemotactic protein we analyzed supernatants of purified, LPS-stimulated monocytes. As shown in Fig. 2 under conditions known to elicit maximum release of neutrophil-chemotactic IL-8, no eosinophil chemotactic activity could be detected.

Indeed, preparations of lymphocytes obtained after elutriation of mononuclear cells released Eo-chemotactic proteins when incubated for 2 days in the presence of Concanavalin A. These preparations usually contained variable numbers of platelets. Since platelets are known to be a rich source of some members of the IL-8 family, such as platelet factor 4 (23), a structurally-related attractant we termed NAP-4 (24) and platelet basic protein or its truncation products connective tissue activating peptide (CTAP) III and ß-thromboglobulin (25), we investigated whether platelets contain Eo-chemotactic proteins.

When lysates of platelets were analyzed for Eo-chemotaxins strong Eo-chemotactic activity could be detected. To evaluate whether Eo-chemotactic activity is also released from platelets upon physiologic stimulation, platelets were stimulated with thrombin. Consequently, the majority of Eo-chemotactic activity was detected in supernatants after incubation with 1 U/ml thrombin for 15 min (26).

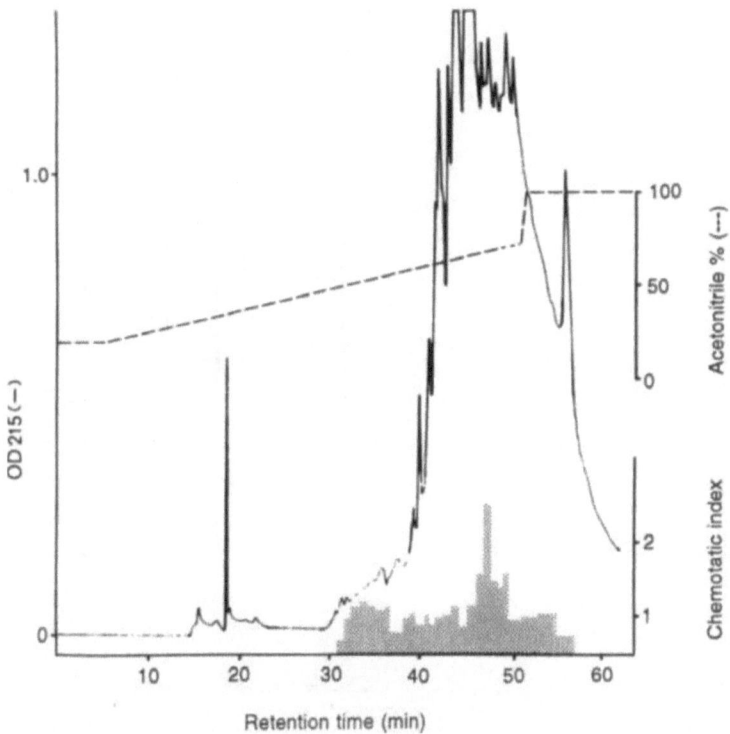

Figure 1. Preparative reversed phase (RP-8) HPLC of a supernatant of a peripheral blood mononuclear cell preparation. Peripheral blood mononuclear cell preparations obtained after Ficoll^R-centrifugation of human blood were stimulated with a mixture of LPS (1 μg/ml, Salmonella minnesota) and Concanavalin A (10 μg/ml) for 48 hrs and incubation of supernatants were separated by preparative RP-8 HPLC. Human eosinophil chemotactic activity (shaded area) was tested in aliquots of each fractions using the indirect cell counting method recently described (17). Note the presence of a single peak of activity.

PLATELETS SECRETE AN EO-CHEMOTACTIC PROTEIN

When these supernatants were separated by TSK-2000 size exclusion HPLC a major peak of Eo-chemotactic activity appeared in fractions corresponding to a M_r near 8 kD. Furthermore the elution behavior of this activity indicated that it does not come from the highly efficient Eo-chemotaxin PAF (7), which is known to be produced by platelets (27) and which elutes from the TSK-2000 HPLC column at different time. Furthermore, PAF can be easily diafiltered through a YM-5 Amicon membrane, which is used for concentrating supernatants of thrombin-stimulated platelets. For purification of Eo-chemotactic proteins supernatants were applied to a preparative RP-8 reversed phase HPLC column after acidification with trifluoroacetic acid (TFA) and proteins were eluted with an increasing gradient of acetonitrile containing 0.1 % TFA.

Fractions were tested for Eo-chemotaxis using the indirect cell counting Boyden chamber technique with purified human Eos as described (17). A single peak of biological activity was obtained, which usually coeluted with a small peak absorbing at 215 nm appearing between the two major peaks of connective tissue activating peptide III (CTAP III) (elution time: 21 min) and platelet factor 4 (PF4) (elution time: 25 min) (Fig. 3).

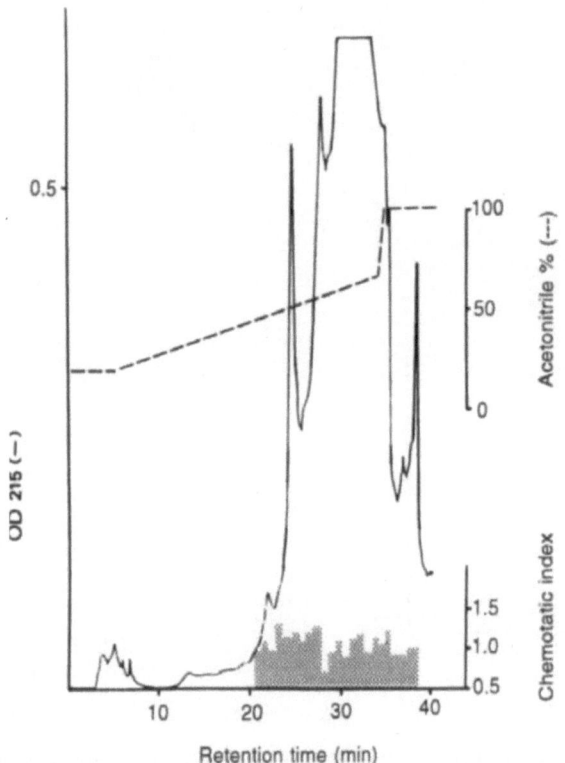

Figure 2. Preparative RP-8 HPLC of a supernatant of purified human monocytes stimulated with LPS. Human monocytes purified by counter-current elutriation were cultivated for 24 hrs in the presence of LPS (1 µg/ml) and supernatants collected after this time interval were applied to a preparative RP-8 HPLC column. Proteins were eluted with an increasing gradient of acetonitrile containing 0.1 % TFA and the effluent was analyzed as 30 µl aliquots for Eo-chemotactic activity (shaded area). Note the absence of fractions containing significant Eo-chemotactic activity. IL-8 is present in large amounts in the effluent eluting at 29 min (not shown).

Fractions off RP-8-HPLC were further purified by the use of cyanopropyl (CN) reversed phase HPLC with n-propanol as eluent. Eo-chemotactic activity was present in the effluent at 22 min with the second major peak absorbing at 215 nm (not shown). These fractions were collected and contaminating high molecular mass proteins were separated by TSK-2000 size exclusion HPLC.

Final purification was achieved by the use of narrow pore RP-18-HPLC. Eo-chemotactic activity eluted in two peaks absorbing at 215 nm tentatively termed Eo-chemotactic protein 1 (EoCP-1), the earlier eluting compound representing a broadened peak, and EoCP-2, the later eluting compound representing a sharp peak (26).

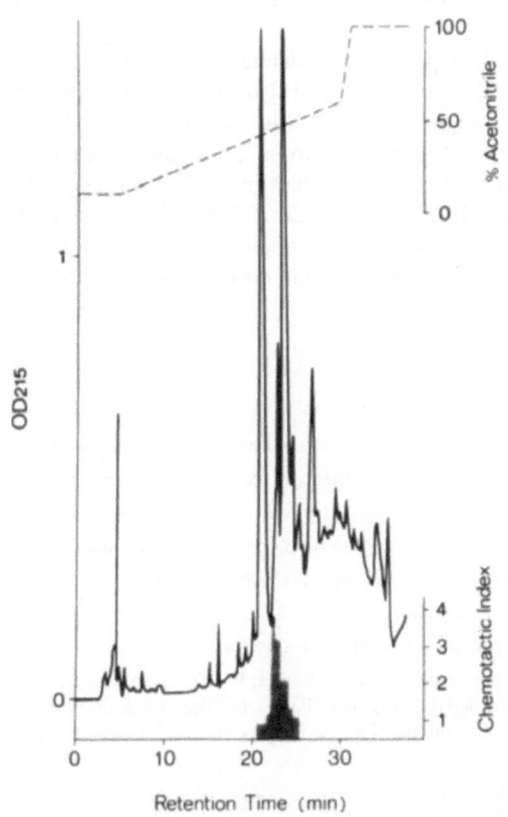

Figure 3. Preparative RP-8 HPLC of supernatants of thrombin-stimulated human platelets. Purified platelets were stimulated for 30 min with thrombin (1 U/ml) and supernatants were separated by preparative RP-8 HPLC. Proteins were eluted with a gradient of acetonitrile containing 0.1 % TFA. Eosinophil-chemotactic activity in the fractions is presented by the shaded area. Note the presence of a single peak of Eo-chemotactic activity.

Upon SDS-PAGE analysis the mobility of EoCP-2 is identical with that seen for Ser-IL-8_{72} (26), whereas EoCP-1, the quantitatively predominant Eo-chemotaxin in platelet supernatants, shows a higher mobility with a calculated M_r of 7 kD (26), when the Tris/-Tricene method without urea was used.

The molecular weight of both, EoCP-1 and EoCP-2 was determined by electrospray mass spectrometry (ESP-MS). Whereas EoCP-1 revealed a calculated molecular mass of 8,355 ± 10, for EoCP-2 the molecular mass was determined to be 7,862.8 ± 1.1 (26).

IDENTIFICATION OF EO-CHEMOTACTIC PROTEINS AS CYTOKINE RANTES

When both EoCP-1 and EoCP-2 were analyzed by gas phase amino acid sequencing in both preparations a single sequence was obtained. As shown in Figure 4 this sequence is identical for both preparations. However, in the EoCP-1-preparation the residues No. 4 and 5 could not be determined, whereas in EoCP-2 at the same position two serine residues were detected.

Both amino acid sequences are identical to that deduced from a cytokine cDNA termed RANTES (28) (Regulated and Normal T-cell expressed and secreted). The calculated molecular weight of RANTES is 7,847.03, which differs from EoCP-1 by 508 mass units and from EoCP-2 by 15.8 mass units. Since N-glycosylation sites are absent from RANTES, most likely serine residues in EoCP-1 are O-glycolysated, which leads to broadening of the RP-HPLC peak. In EoCP-2 the difference of 15.8 mass units can be accounted for by the assumption of oxidation having taken place. Since the difference corresponds to a single oxygen, oxidation on the single methionine residue No. 67 is the most likely cause.

HuMIP-1α	ADTPTAC-CFSYTSRQI-PQNFIAD-Y-FETSSQ-CSKPGVIF-LTKRSRQVCADPSEEWVQKYV--SDLELSA
HuMIP-1β	APMGSDPPTSC-CFSYTARKL-PHNFVVD-Y-YETSSL-CSQPAVVF-QTKRGKQVCADPSESWVQEYV--YDLELN
MCP-1/MCAF	QPDAINAPVTC-CYNFTNRKI-SVQRLAS-YRRITSSK-CPKEAVIF-KTIVAKEICADPKQKWVQDSM--DHLDKQTQTPKT
I-309	VDSKSMQVPFSRC-CFSFAEQEI-PLRAILC-Y-RNTSSI-CSNEGLIF-KLKRGKEACALDTVGWVQRHR--KMLRHCPSKRK
RANTES	SPYSSDTTPC-CFAYIARPL-PRAHIKE-Y-FYTSGK-CSNPAVVF-VTRKNRQVCANPEKKWVREYI--NSLEMS
EoCP-1	SPYXXDTTPX-XFAYIA
EoCP-2	SPYSSDTTPX-XFAYIARPL-PRAXXXE-Y-FYXXG

Figure 4. Amino acid sequence alignments of EoCPs. EoCP sequences were experimentally determined, the other sequences are cDNA derived (15). The single letter code for amino acids is used. X represents amino acids which could not be determined. Note the identify of EoCP sequences with that of cytokine RANTES.

Oxidation of methionine is very common, especially in the case of samples that have been exposed to the atmosphere before analyses.

Both natural RANTES cytokine forms were analyzed for Eo-chemotactic properties and showed similar potency in eliciting Eo-chemotaxis using the Boyden chamber method. (26). It is interesting to note that chemotactic efficacy, the number of Eos having migrated through the chemotaxis filter per unit time, varied depending upon the donor for both forms of RANTES (Fig. 5).

Since half-maximal chemotactic stimulation is identical for both natural forms of RANTES, derivatization at serine residue numbers 4 and 5 appears not to affect potency and efficacy of Eo-chemotactic stimulation. Since recombinant RANTES is now commercially available, this preparation was also tested for eosinophil chemotactic properties.

Recombinant RANTES induces chemotactic responses in human eosinophils with similar ED_{50} and efficacy as the natural forms (26). None of the RANTES preparations, including rRANTES showed any significant neutrophil chemotactic activity (Fig. 6) - in accordance with previous findings with rRANTES (29). In support of the observation that

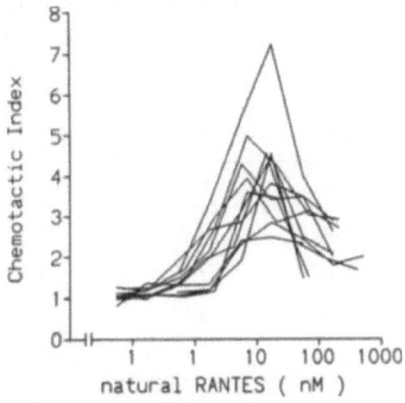

Figure 5. Eo-chemotactic activity of natural RANTES. Platelet-derived RANTES (EoCP-1) was investigated for Eo-chemotactic activity elicited in Eos of different donors. Chemotactic index represents the quotient of Eos migrating under stimulatory conditions and random migration. Note the variable efficacy (chemotactic index) of the dose response curve.

rRANTES also attracts human monocytes, both natural RANTES forms were found to be chemotactic for monocytes (our unpublished results).

When rRANTES was tested for Eo-chemokinetic activity using a checkerboard analysis in a Boyden chamber migration assay system, significant chemokinetic activity was detected at concentrations higher than 50 ng/ml, which is similar to the concentration eliciting significant Eo-chemotaxis (Table 1).

The migratory response, however, at chemokinetic stimulation is lower than that seen at chemotactic stimulation. Therefore RANTES is rather a chemotactic than chemokinetic cytokine.

Cross desensitization studies of Eo-chemotaxis revealed that pretreatment of human eosinophils with RANTES reduces chemotactic responses only to RANTES, but not to other chemotaxins such as C5a or PAF, a finding which is consistent with the hypothesis that RANTES binds to a separate receptor on human eosinophils (Y. Kameyoshi and J.- M. Schröder, unpublished observation).

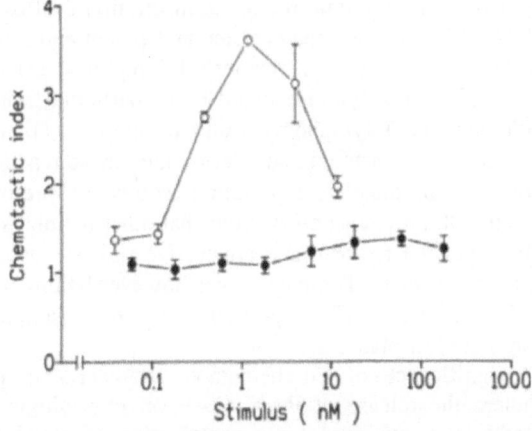

Figure 6. Comparison of chemotactic stimulation of neutrophils and Eos by RANTES. Chemotactic activity of neutrophils (-•-) and eosinophils (-o-) is shown as a function of the RANTES dose.

Table 1. Checkerboard analysis of Eo-chemotactic activity of rRANTES

rRANTES in lower chamber	rRANTES in upper chamber (ng/ml)				
(ng/ml)	0	10	25	50	100
0	1.00	0.88	0.93	0.93	1.17
10	0.83	1.01	1.03	0.96	1.06
25	1.14	1.03	1.08	1.33	1.35
50	1.96	1.60	1.60	1.76	1.35
100	2.52	2.56	2.56	2.56	1.59

Results are expressed as migration index of an experiment performed in duplicate. A representative out of three experiments is shown.

POSSIBLE ROLE OF RANTES IN EOSINOPHILIC INFLAMMATION

RANTES was originally was identified as an apparently T-lymphocyte-specific inducible gene, which was found to be expressed by cultured T-cell lines that were antigen-specific and growth factor depedent (28). RANTES mRNA expression has been found to be inducible in peripheral blood leukocyte preparations. It is therefore likely, but yet not proven, that supernatants of stimulated lymphocytes contain Eo-chemotactic RANTES. The absence of 8 kD Eo-chemotactic proteins in supernatants of stimulated human monocytes points towards a lack of RANTES mRNA expression in monocytes, which is supported by PCR investigations using RANTES-specific primers (P. Kiene, unpublished results).

Eosinophils or products of Eos are found in a variety of inflammatory reactions including late phase reactions, allergic diseases and asthma (for review see ref. 30). Release of toxic substances such as cationic Eo granule proteins and peptidoleukotrienes are believed to be important in Eo-derived inflammatory reactions. The finding of predominantly Eos, and only to a lesser extent neutrophils in such inflammatory diseases led to the hypothesis of a local production of Eo-selective or preferential (relative to neutrophils) attractants in the tissues. Eo-chemotactic cytokines such as GMCSF, IL-5, IL-3 and LCF are suggested to be important for tissue infiltration by Eos.

The cytokine RANTES has now been detected as a potent and selective Eo-chemotaxin and has already been described as a memory T-lymphocte selective (CD45 RO+) chemotactic factor. Therefore this cytokine could be of particular importance in clinical situations where both memory T-lymphocytes and eosinophils are present in affected tissues, such as atopic dermatitis, asthma and allergic late phase reactions.

The finding that human platelets represent a source of preformed RANTES is unexpected and fits well with previous observations that other members of the chemotactic cytokine family such as platelet factor 4 and connective tissue activating peptide III are stored in the α-granules of platelets. These cytokines however belong to the C-X-C branch of this family (14). Therefore rRANTES represents the first example of a C-C branch member which is also stored in platelets.

The biological significance of Eo-chemotactic RANTES in platelets is as yet speculative. Nevertheless the release of RANTES upon physiological stimulation with thrombin serves as additional evidence for a contribution of platelets to inflammatory reactions (31).

From guinea pig models it was suggested that platelets are a prerequisite component in allergic asthma, since platelet depletion reduced eosinophil infiltration into the lung following PAF-or allergen-exposure of sensitized animals (20). Similarly PAF antagonists or PGI$_2$ pretreatment produced similar inhibitory activities.

Moreover, bronchial eosinophil accumulation was reduced without a significant change in neutrophil infiltration after antigen challenge in thrombocytopenic allergic rabbits compared to control animals (32). Clinical observations, such as elevation of plasma platelet factor 4 (33) or ß-thromboglobulin (34) also suggest platelet activation associated with allergen exposure in allergic asthmatics (33,34).

It is therefore tempting to speculate that after antigen challenge in these situations RANTES might play a role for selective eosinophil infiltration, being released directly or indirectly, possibly via PAF- or antigen-stimulation in vivo.

ACKNOWLEDLGEMENT

The authors wish to gratefully acknowledge Jutta Quitzau for HPLC analyses, Albrecht Dörschner for performing amino acid sequence analyses, Tony Mallet for performing Electrospray Mass Spectrometrical analyses and Ilse Brandt for expert secreterial support. Part of this work was supported by DFG, grant Schr 305/1-3.

REFERENCES

1. Frigas, E., G.J. Gleich. 1986. The eosinophil and the pathophysiology of asthma. *J. Allergy Clin. Immunol.* 77: 527-537.
2. Lellouch-Tubiana, A., J. Lefort, M.-T. Simon, A. Pfister, and B.B. Vargaftig. 1988. Eosinophil recruitment into guinea pig lungs after PAF-acether and allergen administration. Modulation by prostacyclin, platelet depletion, and selective antagonists. *Am. Rev. Respir. Dis.* 137: 948-954.
3. Leiferman, K. M. 1989. Eosinophils in atopic dermatitis. *Allergy* 44: 20-26.
4. Kay, A.B., H.S. Shin, K.F. Austen. 1973. Selective attraction of eosinophils and synergism between eosinophil chemotactic factor of anaphylaxis (ECF-A) and a fragment cleaved from the fifth component of complement (C5a). *Immunology* 24: 969-976.
5. Czarnetzki, B.M., W. König, L.M. Lichtenstein. 1976. Eosinophil chemotactic factor (ECF). I. Release from polymorphonuclear leukocytes by the calcium ionophore A 23187. *J. Immunol.* 117: 229-234.
6. Morita, E., J.-M. Schröder, E. Christophers. 1989. Differential sensitivities of purified human eosinophils and neutrophils to defined chemotaxis. *Scand.J.Immunol.* 29: 709-716.
7. Wardlaw, A.J., R. Moqbel, O. Cromwell, A.B. Kay. 1986. Platelet-activating factor. A potent chemotactic and chemokinetic factor for human eosinophils. *J. Clin. Invest.* 78: 1701-1712.
8. Morita, E., J.-M. Schröder, E. Christophers. 1990. Identification of a novel and highly potent eosinophil chemotactic lipid in human eosinophils treated with arachidonic acid. *J. Immunol.* 144: 1893-1900.
9. Schwenk, U., E. Morita, R. Engel, J.-M. Schröder. 1992. Identification of 5-Oxo-15-hydroxy-6,8,11,13-eicosatetraenoic acid as a novel and potent human eosinophil chemotactic eicosanoid. *J. Biol. Chem.* 267: 12482-12488.
10. Wang, J.M., A. Rambaldi, A. Biondi, Z.G. Chen, C.J. Sanderson, A. Mantovani. 1989. Recombinant human interleukin 5 is a selective eosinophil chemoattractant. *Eur. J. Immunol.* 19: 701-705.
11. Warringa, R.A.J., L. Koenderman, P.T.M. Kok, J. Kreukmiet, P.L.B. Bruijnzeel. 1991. Modulation and induction of eosinophil chemotaxis by granulocyte-macrophage colony stimulating factor and interleukin3. *Blood* 77: 2694-2700.
12. Kameyoshi, Y., E. Morita, J.-M. Schröder, E. Christophers. 1992. Recombinant human interleukin-3 is a chemoattractant for human eosinophils, but not for neutrophils. Submitted for publication.
13. Rand, T.H., W.W. Cruikshank, D.M. Center, P.F. Weller. 1991. CD4-mediated stimulation of human eosinophils: lymphocyte chemoattractant factor and other CD4-binding ligands elicit eosinophil migration. *J. Exp. Med.* 173: 1521-1528.
14. Baggiolini, M., A. Walz, S.L. Kunkel. 1989. Neutrophil- activating peptide-1/interleukin 8, a novel cytokine that activates neutrophils. *J. Clin. Invest.* 84: 1045-1049.
15. Schall T.J. 1991. Biology of the RANTES/SIS cytokine family. *Cytokine* 3: 165-183.

16. Schröder, J.-M. 1992. Chemotactic cytokines in the epidermis. *Exp. Dermatol.* 1: 12-19.

17. Schröder, J.-M., U. Mrowietz, E. Morita, E. Christophers. 1987. Purification and partial biochemical characterization of a human monocyte-derived neutrophil-activating peptide that lacks Interleukin 1 activity. *J. Immunol.* 139: 3474-3483.

18. Yoshimura, T., K. Matsushima, S. Tanaka, E.A. Robinson, E. Appella, J.J. Oppenheim, E.J. Leonard. 1987b. Purification of a human monocyte-derived neutrophil chemotactic factor that shares sequence homology with other host defense cytokines. *Proc. Natl. Acad. Sci. USA* 84: 9233.

19. Schröder, J.-M., N. Persoon, E. Christophers. 1990. Lipopolysaccharide-stimulated human monocytes secrete apart from NAP-1/IL-8 a second neutrophil-activating protein: NH_2-terminal aminoacid sequence-identity with melanoma growth stimulatory activity (MGSA/gro). *J. Exp. Med.* 171: 1091-1100.

20. Moser, B., I. Clark-Lewis, R. Zwahlen, M. Baggiolini. 1990. Neutrophil-activating properties of the melanoma growth-stimulatory activity. *J. Exp. Med.* 171: 1797-1802.

21. Walz, A., B. Dewald, V. von Tscharner, M. Baggiolini. 1989. Effects of the neutrophil-activating peptide NAP-2, platelet basic protein, connective tissue-activating peptide III, and platelet factor 4 on human neutrophils. *J. Exp. Med.* 170: 1745-1750.

22. Yoshimura, T., N. Yuhki, S.K. Moore, E. Appella, M.I. Lerman, E.J. Leonard. 1989. Human monocyte chemoattractant protein-1 (MCP-1): Full-length cDNA cloning, expression in mitogen-stimulated blood mononuclear leukocytes, and sequence similarity to mouse competence gene JE. *FEBS Lett.* 244: 487-493.

23. Deuel, T.F., R.M. Senior, D. Chang, D.L. Griffin, R.L. Heinrikson, E.T. Kaiser. 1981. Platelet factor 4 is chemotactic for neutrophils and monocytes. *Proc. Natl. Acad. Sci. USA* 78: 4584-4587.

24. Schröder, J.-M., M. Sticherling, N.-L. M. Perssoon, E. Christophers. 1990. Identification of a novel platelet-derived neutrophil-chemotactic polypeptide with structural homology to platelet factor 4. *Biochem. Biophys. Res. Commun.* 172: 898-904.

25. Castor, C.W., J.W. Miller, D.A. Walz. 1983. Structural and biological characteristics of connective tissue activating peptide (CTAP-III), a major human platelet-derived growth factor. *Proc. Natl. Acad. Sci. USA* 80: 765-769.

26. Kameyoshi, Y., A. Dörschner, A.I. Mallet, E. Christophers, J.-M. Schröder. 1992. Cytokine RANTES released by thrombin-stimulated platelets is a potent attractant for human eosinophils. *J. Exp. Med.* 176: 587-592.

27. Chignard, M., J.P. Le Conedic, B.B. Vargaftig, J. Benveniste. 1980. Platelet-activating factor (PAF-acether) secretion from platelets: effect of aggregating agents. *Br. J. Haematol.* 46: 455-464.

28. Schall, T.J., J. Jongstra, B.J. Dyer, J. Jorgensen, C. Claylberger, M.M. Davis, A.M. Krensky. 1988. A human T cell-specific molecule is a member of a new gene family. *J. Immunol.* 141: 1018-1025.

29. Schall, T.J., K. Bacon, K.J. Toy, D.V. Goeddel. 1990. Selective attraction of monocytes and T lymphocytes of the memory phenotype by cytokine RANTES. *Nature (Lond.)* 347: 669-671.

30. Leiferman, K.M. 1991. A current perspective on the role of eosinophils in dermatologic diseases. *J. Am. Acad. Dermatol.* 24: 1101-1112.

31. Page, C.P. 1989. Platelets as inflammatory cells. *Immunopharmacology* 17: 51-59.

32. Coyle, A.J., C.P. Page, L. Atkinson, R. Flanagan, W.J. Metzger. 1990. The requirement for platelets in allergen-induced late asthmatic airway obstruction. Eosinophil infiltration and hieghtened airway responsiveness in allergic rabbits. *Am. Rev. Respir. Dis.* 142: 587-593.

33. Knauer, K.A., L.M. Lichtenstein, N.F. Adkinson, Jr., and I.F. Fish. 1981. Platelet activation during antigen-induced airway reactions in asthmatic subjects. *N. Engl. J. Med.* 304: 1404-1407.

34. Gresele, P., S. Grasselli, T. Todisco, and G.G. Nenci. 1985. Platelets and asthma. *Lancet*, ii: 347.

NEUTROPHIL-ACTIVATING PEPTIDE ENA-78

Alfred Walz, Robert M. Strieter[1] and Silvia Schnyder[1]

Theodor-Kocher Institute
University of Bern
CH-3000 Bern 9
Switzerland

[1]Department of Internal Medicine
The University of Michigan Medical School
Ann Arbor
Michigan, USA

INTRODUCTION

Neutrophil-activating peptide ENA-78 was recently isolated from a human type II-like epithelial cell line A549 (1). ENA-78 is a peptide of 78 amino acids and belongs to the C-X-C supergene family of the chemotactic cytokines (now called chemokines). The C-X-C family contains five other peptides with neutrophil-activating properties, interleukin-8 (IL-8) (2), three closely related members of the GRO family, GROα, GROβ and GROγ (3,4) and neutrophil-activating peptide 2 (NAP-2) (5). Two other members of the same subfamily, platelet factor 4 and interferon-γ induced peptide 10 (γ-IP10) have been shown not to be active on human neutrophils (6,7).

The six peptides with neutrophil-activating properties exhibit very similar activities on neutrophils, however, they differ markedly in their cellular origin and/or kinetics of induction, suggesting a different pathophysiological role. IL-8 and GROα appear to be produced upon stimulation by almost any cell type (for review see ref. 8), whereas GROβ and GROγ are poorly expressed in fibroblasts and other cell types. NAP-2 is produced exclusively by proteolytic cleavage from a precursor peptide contained in the alpha granules of the platelets (9). ENA-78, on the other hand, appears to be produced predominantly by epithelial cells and differs in its gene expression in human monocytes from IL-8 and GROα.

STRUCTURE OF ENA-78

ENA-78 was isolated from supernatants of interleukin-1 (IL-1) or tumor necrosis factor-α (TNFα) stimulated alveolar type II-like epithelial cells. The complete structure as determined by amino acid sequence analysis and recombinant technologies yielded a peptide of 78 amino acids (see Fig. 1) (1). ENA-78 has a molecular weight of 8,357 daltons and a calculated isoelectric point of 8.73. The peptide contains the same four characteristic cysteine residues in identical position as present in IL-8, NAP-2 and GROα. Interestingly, ENA-78 has only a 22% sequence identity to IL-8, but 53% and 52% to NAP-2 and GROα, respectively.

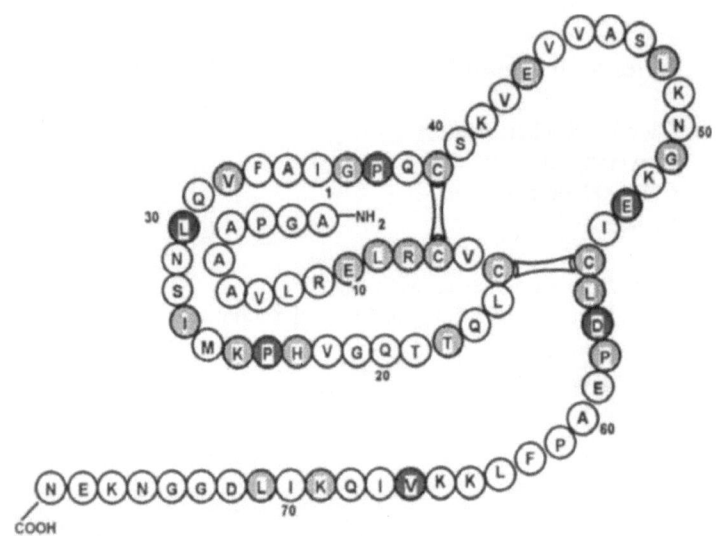

Figure 1. Structure of ENA-78. Hatched circles indicate identical amino acids in ENA-78, IL-8, GROα and NAP-2. Hatched and cross-hatched circles together indicate identical amino acids in ENA-78 and IL-8.

BIOLOGICAL ACTIVITIES OF ENA-78

ENA-78 was purified on the basis of its elastase-releasing activity on cytochalasin-B treated human neutrophils. Highly purified natural as well as recombinant ENA-78 was active at concentrations ranging between 1 and 100 nM, as was the case for NAP-2. By contrast, IL-8 was more active. ENA-78 induced a concentration-dependent chemotaxis between 0.1 and 100 nM in vitro. This response was identical to NAP-2, but somewhat reduced than for IL-8 (1). The lower potency of ENA-78 with respect to IL-8, was also apparent by the measurements of stimulus-dependent changes in cytosolic free calcium.

Amino-terminal truncation of the 77-amino acid form of IL-8 to the most abundant 72-amino acid variant, found in supernatants of stimulated mononuclear cells, can leads to a 5 to 10-fold increase in potency for neutrophil chemotactic activity (10). A similar difference in activity for the two IL-8 variants can also be observed for neutrophil adhesion to IL-1-stimulated endothelial cells (11). Furthermore, an extension of 3-5 residues at the amino-terminus of NAP-2 can result in a 10-fold drop in potency (9). To evaluate the effect of amino terminal truncations on ENA-78, we have studied the influence of various commercial proteases on the change of neutrophil-stimulating activity. As measured by elastase-release from human neutrophils, a transient increase in activity was observed when ENA-78 was exposed to human leukocyte cathepsin G and bovine chymotrypsin, suggesting the formation of amino-terminal variants of ENA-78 with elevated biological activity (Fig. 2). From this result we may assume that under physiological or pathophysiological conditions, e.g. in the pulmonary alveolus, ENA-78 released from type II-like epithelial cells can be processed by alveolar macrophages through the release of cathepsin G, to molecules that have the same potency as IL-8. Incubation of ENA-78 with trypsin caused a gradual loss of activity with no evidence for the production of an intermediate molecule with increased activity (Fig. 2).

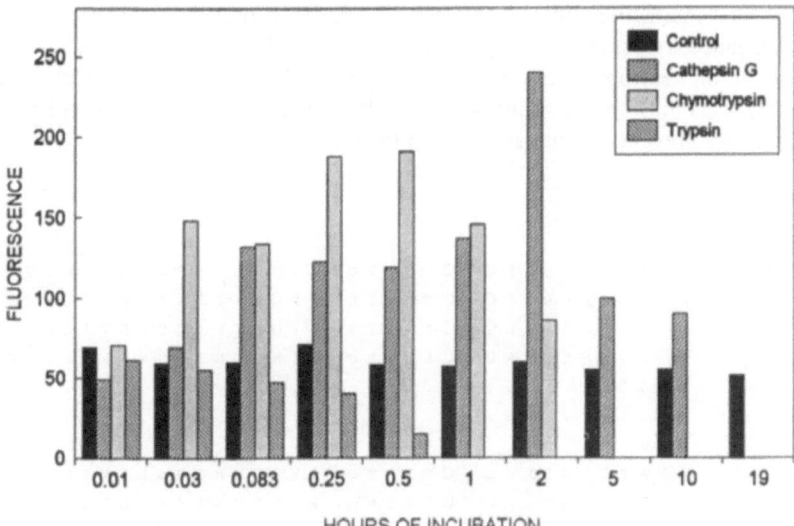

Figure 2. The effect of cathepsin G, chymotrypsin and trypsin on ENA-78 activity. ENA-78 (5 ug) was incubated in 500 ml of phosphate-buffered saline with either 0.2 mg human leukocyte cathepsin G, bovine chymotrypsin or trypsin and incubated at 37°C for times indicated. The control reaction was carried out similarly but without addition of enzymes.

In contrast to IL-8 and GROα, neither NAP-2 nor ENA-78 induced a transient rise of cytosolic free calcium in human monocytes (12). This suggests a similar receptor specificity for both NAP-2 and ENA-78.

ENA-78 RECEPTOR BINDING

Several reports have demonstrated that IL-8 binds with high affinity to human neutrophils and myeloid cell lines (13-15). Peripheral blood neutrophil receptor numbers for IL-8 range from 20,000 to 75,000 per cell with mean affinity constants (Kd) ranging from 0.18 to 4 nM. Two different cDNAs for the human IL-8 receptor have been reported (16,17). Functional studies have shown desensitization of neutrophils when they were subsequently stimulated with IL-8, NAP-2 or GROα (12), indicating an overlap in receptor usage for the three peptides. This has been confirmed by competition studies of ^{125}I-IL-8 with unlabeled IL-8, NAP-2 and GROα. Two classes of binding sites have been reported for NAP-2 and GROα, high affinity binding (Kd: 0.34 and 0.14 nM) and low affinity binding (Kd:100 and 130 nM) (18). For ENA-78 the affinity constant has not been determined. Cross-desensitization studies measuring the rise of intracellular calcium real-time in response to sequential stimulation with IL-8 and ENA-78, or IL-8 and NAP-2, or vice-versa, show essentially the same pattern for ENA-78 as observed for NAP-2 (1), predicting two binding sites for ENA-78 and a similar affinity constant as observed for NAP-2.

Amino-terminal truncations in IL-8 suggest that the ELR (Glu-Leu-Arg) element (adjacent to the first cysteine residue) is of importance for receptor binding (Fig. 3). Truncations or changes in this structure dramatically change the binding affinity for the receptor (19,20). IL-8, NAP-2, GROα, GROβ and GROγ, and ENA-78 contain this feature, however, this amino acid sequence is not found in the inactive members of the C-X-C subfamily, PF-4 or γIP-10. Short peptides containing the ELR element are inactive, which implicates that interactions of other domains of the IL-8 molecule with the receptor are required. Connective tissue activating peptide III (CTAP-III), a precursor molecule of NAP-2, is inactive on neutrophils and does not bind to the IL-8 receptor, although it contains ELR (6,21). This may be due to the additional 16 amino acids upstream from ELR, which could sterically prevent the interaction with the receptor.

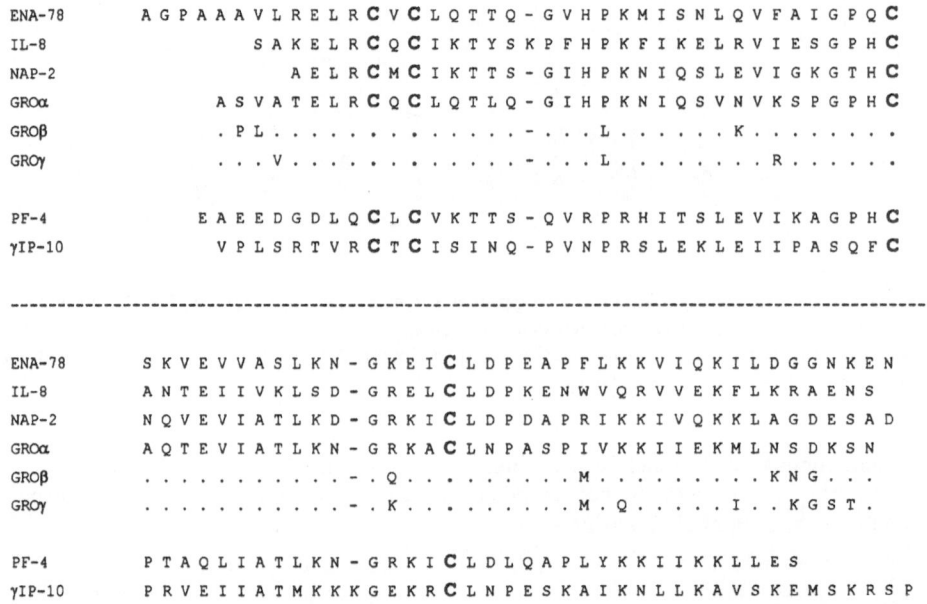

Figure 3. Amino acid sequences of ENA-78 and related peptides of the C-X-C subfamily of the chemokines. Peptides were aligned according to their four conserved cysteine residues (*bold*).

PRODUCTION OF ENA-78 BY VARIOUS CELLS

ENA-78 was originally purified from supernatants of human type II-like epithelial cell line, A549. In the absence of a stimulus, these cells release low levels of ENA-78 and GROa. In the presence of either IL-1ß or TNFα, a rapid induction of ENA-78 can be observed, whereas lipopolysaccharide had no effect on augmenting ENA-78 expression above basal levels. In addition to ENA-78 and GROα, IL-1ß and TNFα-induced A549 cells to release three other neutrophil-stimulating peptides, two of which were biochemically characterized. IL-8 was present predominantly as the 77-amino acid form, which was obtained earlier in minor quantities in monocyte cultures (22), and was the most abundant variant in cultures of endothelial cells stimulated with IL-1ß (23). Furthermore, low levels of GROγ and an unidentified activity were also present.

Using a synthetic 43-mer oligonucleotide (5'GCACTGTGGGCCTATGG CGAACACTTGCAGATTACTGATCATT 3') a number of cell types were tested for their capacity to synthesize ENA-78 mRNA. As observed for type II-like epithelial cells, unstimulated renal cortical epithelial cells contained low amounts of mRNA, which was strongly upregulated upon induction with IL-1ß (R. Strieter, unpublished results). As summarized in Table 1, lung fibroblasts, monocytes, endothelial cells and mesothelial cells can produce ENA-78 mRNA after simulation with LPS or IL-1ß whereas in keratinocytes and alveolar macrophanges no such mRNA has been detected (S. Schnyder, R. Strieter, unpublished results).

Table 1. Detection of ENA-78 mRNA by northern blotting

Cell Type	Control	LPS	IL-1β
Type II-like epithelial cells (A549)	+	+	+++
Renal cortical epithelial cells	+	+	++
Keratinocytes	-	n.d.	-
Adult lung fibroblasts (3229)	-	n.d.	+
Monocytes	-	++	+
Alveolar macrophages	-	n.d.	-
Endothelial cells	-	-	+
Mesothelial cells	-	-	+

n.d., not determined; -, no mRNA detected; +, ++, +++, relative amounts of ENA-78 mRNA present.

We have recently developed a sensitive method for the detection of ENA-78 from cellular supernatants using a sandwich enzyme-linked immunosorbent assay (ELISA) (24). The ENA-78 ELISA is based on polyclonal rabbit antibodies. Rabbit anti-human ENA-78 serum was purified using a protein A-agarose column. A portion of the purified antibody was then biotinylated and used as detection antibody. The assay is specific for ENA-78 and does not detect the closely related members of the C-X-C subfamily IL-8, GROα, NAP-2, γIP-10, nor monocyte chemoattractant protein 1 (MCP-1) or IL-1α, IL-1β, IL-1 receptor antagonist (IRAP), IL-4, IL-6, IL-7, TNFα, or transforming growth factor-β (TGFβ). The ENA-78 ELISA has a threshold sensitivity of 100 pg/ml. As shown in Table 2, various cell types were incubated in the absence and presence of LPS (100 ng/ml; 0111:B4) and IL-1β (2 ng/ml) for 24 and/or 40 hrs (monocytes) in a CO_2-incubator at 37°C. Supernatants were then analyzed for the presence of ENA-78. The results obtained by ENA-78 ELISA were in agreement with the data obtained by northern blots. In the absence or presence of LPS, alveolar type II-like, as well as renal cortical epithelial cells expressed low levels of mRNA and produced none or only low levels of ENA-78. Stimulation with IL-1β and TNFα (data not shown), however, induced high levels of mRNA and the subsequently released of ENA-78. In normal human monocytes, LPS was an efficient inducer for ENA-78 mRNA, as well as peptide production. Induction of ENA-78 mRNA was transient with maximal levels observed at about 20 hrs after stimulation. In agreement with these findings, ENA-78 peptide levels started to raise after approximately 20 hrs and were still increasing beyond 48 hrs. These observations suggest that in human monocytes, ENA-78 is regulated differently than the more prominent chemotactic cytokine IL-8. The activity of IL-8 was clearly downregulated by the time ENA-78 was released from monocytes.

Table 2. Detection of ENA-78 by ELISA

Cell type	Control	LPS	IL-1β
Type II epithelial cells (A549)	+	+	++
Renal cortal epithelial cells	-	+	+++
Normal pulmonary fibroblasts	+	n.d.	++
Monocytes	-	++	+
Endothelial cells	-	-	+
Neutrophils	-	+	n.d.

n.d., not determined; -, none or less than 100 pg/ml ENA-78; +, ++, +++, relative amounts of ENA-78 present.

PATHOPHYSIOLOGICAL RELEVANCE OF ENA-78

Leukocyte emigration from the vasculature to the tissue may be considered as one of the basic mechanisms of the inflammatory process. Newly discovered neutrophil-activating peptides have been shown to be potent chemoattractants and activators of neutrophils in vitro and in vivo. Once formed IL-8 and related peptides are very resistant to inactivation and have a slow clearance (long-lasting effect after intradermal injection) (25,26). Other chemoattractants such as fMet-Leu-Phe, C5a, LTB4 and PAF are inactivated rapidly by oxidation or hydrolysis (25). Therefore, neutrophil-activating peptides may contribute significantly to the influx of neutrophils into diseased tissue.

In fact, elevated levels of IL-8 have been detected in the skin of patients with psoriasis and plantar pustulosis (27-29) and in the synovial fluid of patients with rheumatoid arthritis (30-32). Furthermore, IL-8 levels are elevated in the bronchoalveolar lavage (BAL) fluid of patients with inflammatory lung diseases, such as adult respiratory distress syndrome (33) and idiopathic pulmonary fibrosis (34). Enhanced chemotactic activity for neutrophils has also been observed in BAL of patients with asbestosis (35).

ENA-78, which is formed at high level by pulmonary type II-like epithelial cells and fibroblasts might be expected to contribute significantly to the recruitment of neutrophils into the lung. Since a specific detection system for ENA-78 has only recently been developed, little published data is available (24). ENA-78 has been detected in bronchoalveolar lavage (BAL) of patients with idiopathic pulmonary fibrosis and cystic fibrosis (A. Walz, unpublished results), and in BAL of patients with adult respiratory distress syndrome (S. Donnelly, unpublished results). ENA-78 has been also found in the synovial fluid of patients with rheumatoid arthritis (RA) at 239 ng/ml, compared to 2.6 ng/ml in osteoarthritis (36). It is noteworthy that ENA-78 levels in the synovial fluid of RA patients are about 15-times higher than IL-8 levels (15 ng/ml). Interestingly, elevated levels of ENA-78 were also found in peripheral blood of RA patients (70 ng/ml), in comparison to normal levels of 0.12 ng/ml (37). That ENA-78 was observed in the peripheral blood of normal individuals separates this cytokine from IL-8, which is not detectable in normal sera.

The major function attributed to IL-8, and the other members of the C-X-C subfamily, so far, has been neutrophil attraction and activation. However, IL-8 appears to functions as a mediator of angiogenesis at similar concentrations as reported for other agents such as fibroblast growth factor, TNFα and angiogenin (37). It is likely that other relevant functions in physiology and pathology will be found in the future.

Acknowledgments

We thank Brigitte Walz for critical reading of the manuscript and Regula Müller & Marie Burdick for excellent technical assistance.

REFERENCES

1. Walz, A., R. Burgener, B. Car, M. Baggiolini, S.L. Kunkel, and R.M. Strieter. 1991. Structure and neutrophil-activating properties of a novel inflammatory peptide (ENA-78) with homology to interleukin 8. *J. Exp. Med.* 174: 1355-1362.
2. Baggiolini, M., A. Walz, and S.L. Kunkel. 1989. Neutrophil-activating peptide-1/interleukin 8, a novel cytokine that activates neutrophils. *J. Clin. Invest.* 84: 1045-1049.
3. Haskill, S., A. Peace, J. Morris, S.A. Sporn, A. Anisowicz, S.W. Lee, T. Smith, G. Martin, P. Ralph, and R. Sager. 1990. Identification of three related human *GRO* genes encoding cytokine functions. *Proc. Natl. Acad. Sci. USA* 87: 7732-7736.
4. Moser, B., I. Clark-Lewis, R. Zwahlen, and M. Baggiolini. 1990. Neutrophil-activating properties of the melanoma growth-stimulatory activity. *J. Exp. Med.* 171: 1797-1802.
5. Walz, A. and M. Baggiolini. 1989. A novel cleavage product of b-thromboglobulin formed in cultures of stimulated mononuclear cells activates human neutrophils. *Biochem. Biophys. Res. Commun.* 159: 969-975.
6. Walz, A., B. Dewald, V. von Tscharner, and M. Baggiolini. 1989. Effects of the neutrophil-activating peptide NAP-2, platelet basic protein, connective tissue-activating peptide III and platelet factor 4 on human neutrophils. *J. Exp. Med.* 170: 1745-1750.
7. Dewald, B., B. Moser, L. Barella, C. Schumacher, M. Baggiolini, and I. Clark-Lewis. 1992. IP-10, a gamma-interferon-inducible protein related to interleukin-8, lacks neutrophil activating properties. *Immunol. Lett.* 32: 81-84.
8. Baggiolini, M., B. Dewald, and A. Walz. 1992. Interleukin-8 and related chemotactic cytokines. *In*: "Inflammation: Basic principles and clinical correlates", Gallin, J.I., Goldstein, I.M., and Snyderman, R. eds., Raven Press, New York, pp. 247-263.
9. Walz, A. and M. Baggiolini. 1990. Generation of the neutrophil-activating peptide NAP-2 from platelet basic protein or connective tissue-activating peptide III through monocyte proteases. *J. Exp. Med.* 171: 449-454.
10. Schröder, J.-M., M. Sticherling, H.H. Henneicke, W.C. Preissner, and E. Christophers. 1990. IL-1a or tumor necrosis factor-a stimulate release of three NAP-1/IL-8-related neutrophil chemotactic proteins in human dermal fibroblasts. *J. Immunol.* 144: 2223-2232.
11. Hébert, C.A., F.W. Luscinskas, J.-M. Kiely, and et al. 1990. IL-1a or tumor necrosis factor-a stimulated release of three NAP-1/IL-8-related neutrophil chemotactic proteins in human dermal fibroblasts. *J. Immunol.* 144: 2223-2232.
12. Walz, A., F. Meloni, I. Clark-Lewis, V. von Tscharner, and M. Baggiolini. 1991. $[Ca^{2+}]_i$ changes and respiratory burst in human neutrophils and monocytes induced by NAP-1/interleukin-8, NAP-2, and *gro*/MGSA. *J. Leukocyte Biol.* 50: 279-286.
13. Besemer, J., A. Hujber, and B. Kuhn. 1989. Specific binding, internalization, and degradation of human neutrophil activating factor by human polymorphonuclear leukocytes. *J. Biol. Chem.* 264: 17409-17415.

14. Samanta, A.K., J.J. Oppenheim, and K. Matsushima. 1989. Identification and characterization of specific receptors for monocyte-derived neutrophil chemotactic factor (MDNCF) on human neutrophils. *J. Exp. Med.* 169: 1185-1189.

15. Grob, P.M., E. David, T.C. Warren, R.P. DeLeon, P.R. Farina, and C.A. Homon. 1990. Characterization of a receptor for human monocyte-derived neutrophil chemotactic factor/inter-leukin-8. *J. Biol. Chem.* 265: 8311-8316.

16. Murphy, P.M. and H.L. Tiffany. 1991. Cloning of complementary DNA encoding a functional human interleukin-8 receptor. *Science* 253: 1280-1283.

17. Holmes, W.E., J. Lee, W.-J. Kuang, G.C. Rice, and W.I. Wood. 1991. Structure and functional expression of a human interleukin-8 receptor. *Science* 253: 1278-1280.

18. Moser, B., C. Schumacher, V. von Tscharner, I. Clark-Lewis, and M. Baggiolini. 1991. Neutrophil-activating peptide 2 and *gro*/melanoma growth-stimulatory activity interact with neutrophil-activating peptide 1/interleukin 8 receptors on human neutrophils. *J. Biol. Chem.* 266: 10666-10671.

19. Clark-Lewis, I., C.M. Schumacher, M. Baggiolini, and B. Moser. 1991. Structure-activity relationships of interleukin-8 determined using chemically synthesized analogs. Critical role of NH_2-terminal residues and evidence for uncoupling of neutrophil chemotaxis, exocytosis, and receptor binding activities. *J. Biol. Chem.* 266: 23128-23134.

20. Hébert, C.A., R.V. Vitangcol, and J.B. Baker. 1991. Scanning mutagenesis of interleukin-8 identifies a cluster of residues required for receptor binding. *J. Biol. Chem.* 266: 18989-18994.

21. Leonard, E.J., T. Yoshimura, A. Rot, K. Noer, A. Walz, M. Baggiolini, D.A. Walz, E.J. Goetzl, and C.W. Castor. 1991. Chemotactic activity and receptor binding of neutrophil attractant/activation protein-1 (NAP-1) and structurally related host defense cytokines: Interaction of NAP-2 with the NAP-1 receptor. *J. Leukocyte Biol.* 49: 258-265.

22. Lindley, I., H. Aschauer, J.M. Seifert, C. Lam, W. Brunowsky, E. Kownatzki, M. Thelen, P. Peveri, B. Dewald, V. von Tscharner, A. Walz, and M. Baggiolini. 1988. Synthesis and expression in Escherichia coli of the gene encoding monocyte-derived neutrophil-activating factor: Biological equivalence between natural and recombinant neutrophil-activating factor. *Proc. Natl. Acad. Sci. U.S.A.* 85: 9199-9203.

23. Gimbrone, M.A. Jr., M.S. Obin, A.F. Brock, E.A. Luis, P.E. Hass, C.A. Hébert, Y.K. Yip, D.W. Leung, D.G. Lowe, W.J. Kohr, W.C. Darbonne, K.B. Bechtol, and J.B. Baker. 1989. Endothelial interleukin-8: A novel inhibitor of leukocyte-endothelial interactions. *Science* 246: 1601-1603.

24. Strieter, R.M., S.L. Kunkel, M.D. Burdick, P.M. Lincoln, and A. Walz. 1992. The detection of a novel neutrophil-activating peptide (ENA-78) using a sensitive ELISA. *Immunol. Invest.* 21: 589-596.

25. Colditz, I.G., R.D. Zwahlen, and M. Baggiolini, M. 1990. Neutrophil accumulation and plasma leakage induced in vivo by neutrophil-activating peptide-1. *J. Leukocyte Biol.* 48: 129-137.

26. Zwahlen, R., A. Walz, and A. Rot. 1993. In vitro and in vivo activity and pathophysiology of human interleukin-8 and related peptides. *Int. Rev. Exp. Pathol.* 34B: 27-42.

27. Sticherling, M., E. Bornscheuer, J.-M. Schröder, and E. Christophers. 1991. Localization of neutrophil-activating peptide-1/interleukin-8-immunoreactivity in normal and psoriatic skin. *J. Invest. Dermatol.* 96: 26-30.

28. Nickoloff, B.J., G.D. Karabin, J.M.W.M. Barker, C.E.M. Griffiths, V. Sarma, R.S. Mitra, J.T. Elder, S.L. Kunkel, and V.M. Dixit. 1991. Cellular localization of interleukin-8 and its inducer, tumor necrosis factor-a in psoriasis. *Am. J. Pathol.* 138: 129-140.

29. Anttila, H.S.I., S. Reitamo, P. Erkko, M. Ceska, B. Moser, and M. Baggiolini. 1992. Interleukin-8 immunoreactivity in the skin of healthy subjects and patients with palmoplantar pustulosis and psoriasis. *J. Invest. Dermatol.* 98: 96-101.

30. Brennan, F.M., C.O.C. Zachariae, D. Chantry, C.G. Larsen, M. Turner, R.N. Maini, K. Matsushima, and M. Feldmann. 1990. Detection of interleukin 8 biological activity in synovial fluids from patients with rheumatoid arthritis and production of interleukin 8 mRNA by isolated synovial cells. *Eur. J. Immunol.* 20: 2141-2144.

31. Seitz, M., B. Dewald, N. Gerber, and M. Baggiolini. 1991. Enhanced production of neutrophil-activating peptide-1/interleukin-8 in rheumatoid arthritis. *J. Clin. Invest.* 87: 463-469.

32. Koch, A.E., S.L. Kunkel, J.C. Burrows, H.L. Evanoff, G.K. Haines, R.M. Pope, and R.M. Strieter. 1991. Synovial tissue macrophage as a source of the chemotactic cytokine IL-8. *J. Immunol.* 147: 2187-2195.

33. Miller, E.J., A.B. Cohen, S. Nagao, D. Griffith, R.J. Maunder, T.R. Martin, J.P. Weiner-Kronish, M. Sticherling, E. Christophers, and M.A. Matthay. 1992. Elevated levels of NAP-1/interleukin-8 are present in the airspaces of patients with the adult respiratory distress syndrome and are associated with increased mortality. *Am. Rev. Respir. Dis.* 146: 427-432.

34. Carré, P.C., R.L. Mortenson, T.E. King Jr., P.W. Noble, C.L. Sable, and D.W.H. Riches. 1991. Increased expression of the interleukin-8 gene by alveolar macrophages in idiopathic pulmonary fibrosis. A potential mechanism for the recruitment and activation of neutrophils in lung fibrosis. *J. Clin. Invest.* 88: 1802-1810.

35. Hayes, A.A., T.J. Venaille, A.H. Rose, A.W. Musk, and B.W. Robinson. 1990. Asbestos-induced release of a human alveolar macrophage-derived neutrophil chemotactic factor. *Exp. Lung. Res.* 16: 121-130.

36. Koch, A.E., S.L. Kunkel, L.A. Harlow, D. Mazarakis, G.K. Haines, M.D. Burdick, R.M. Pope, A. Walz, and R.M. Strieter. 1992. Expression of the novel chemotactic cytokines macrophage inflammatory protein-1a and ENA-78 in arthritis. *Clin. Res.* 40: 742. (Abstract)

37. Koch, A.E., P.J. Polverini, S.L. Kunkel, L.A. Harlow, L.A. DiPietro, V.M. Elner, S.G. Elner, and R.M. Strieter. 1992. Interleukin-8 as a macrophage-derived mediator of angiogenesis. *Science* 258: 1798-1801.

THE EFFECTS OF HUMAN RECOMBINANT MIP-1α, MIP-1β, AND RANTES ON THE CHEMOTAXIS AND ADHESION OF T CELL SUBSETS

Dennis D. Taub, Andrew R. Lloyd, Ji-Ming Wang, Joost J. Oppenheim, and David J. Kelvin

Laboratory of Molecular Immunoregulation
Biological Response Modifiers Program
Division of Cancer Research
NCI-FCRDC
Frederick, MD 21702, USA

INTRODUCTION

Different members of the chemokine superfamily chemoattract and activate specific leukocyte subsets. For example, IL-8 predominantly chemoattracts neutrophils in vivo, while MCAF is selectively chemotactic for monocytes and induces the accumulation of monocytes in vivo (1). Some of these chemokines can also induce enzyme release, intracellular calcium flux, shape change and alterations in adhesiveness to an endothelial monolayer in neutrophils and monocytes (1-5). The directional migration of lymphocytes into inflamed tissues is presumeably also mediated by a variety of chemotactic molecules. The role of cytokines in the regulation of T cell adhesion and chemotaxis is presently unknown. We have explored the biological activity of the available chemokines on T lymphocyte function. In this report, we will describe our recent studies demonstrating the effects of recombinant human MIP-1α, MIP-1β, and RANTES on T cell chemotaxis and adhesion to human endothelial cells.

MATERIALS AND METHODS

Peripheral blood T cells and T cell subsets were prepared from normal donors by an extensive purification process. Briefly, peripheral blood mononuclear cells obtained by leukapheresis were passaged over Ficoll-Hypaque to remove eythrocytes, granulocytes and cellular debris. Small lymphocytes were obtained by sequential depletion of adherent cells in plastic flasks, B lymphocytes on nylon wool columns, and large granular lymphocytes by Percoll gradient centrifugation. This isolation procedure typically yielded >90% CD3+

CD16⁻ lymphocytes. For T cell subset isolation, isolated lymphocytes were washed twice in cold phosphate buffered saline (PBS) and resuspended at 2-3 x 10^7 cells/ml. Saturating concentrations of anti-CD4, anti-CD8, anti-CD45RA, or anti-CD29 were added to the T cell suspension for overnight incubation at 4^0C. Purified CD4⁺, CD8⁺, CD45RA⁺, and CD29⁺ lymphocytes were obtained through negative selection using Advanced Magnetics goat anti-mouse-specific Fc beads (Cambridge, MA) to selectively deplete the cell populations of particular T cell subsets as described previously (6). This procedure routinely yielded a >94% purified T cell subpopulation. For antibody activation of T cells or T cell subsets, the lymphocytes were adjusted to a concentration of 5 X 10^6 cells/ml and cultured on untreated tissue culture plates or plates previously coated with a 10 µg/ml suspension of anti-CD3 monoclonal antibody. The T cells were cultured for 6-8 hrs, harvested and used in assays of T cell chemotaxis. Cell migration was evaluated by using a 48-well microchemotaxis chamber technique (7). Polycarbonate filters (5 µm pore size) were used for monocyte, neutrophil, and T cell migration assays. T lymphocyte migration was quantitated by a modified technique. Briefly, T cells were suspended at 1 x 10^7 cell/ml in RPMI 1640 plus 0.5% FCS, and placed in the top wells of a 48-well microchemotaxis chamber. All polycarbonate filters utilized in these experiments were coated with collagen type IV approximately 24 hr prior to use in the assay (8). The chambers were incubated for 4 hr at 37°C in 5% CO_2 and a humidified atomosphere. The filters were fixed in methanol, stained, and the number of migrating cells in five high powered fields were counted for each well. The results represent the average number of migrating cells per high powered field (±SD).

Human umbilical vein endothelial cells (HUVEC) were prepared by collagenase treatment of umbilical cords as previously described (9). Endothelial cells at passage three or less were plated at 7.5 X 10^4 cells/well onto gelatin-coated 24 well plates (Costar) and cultured to confluence over 48 hours. Prior to the adhesion assays, HUVEC were washed once with RPMI 1640/5% FCS and then incubated in this medium for six hours at 37°C with or without the addition of 10ng/ml of rhIL-1α.

The measurement of T cell adhesion was based upon the binding of ⁵¹Cr-labelled T cells as previously described (10). Briefly, preincubated HUVEC were washed twice with medium (RPMI 1640/5% FCS) and 500,000 ⁵¹Cr-labelled T cells were added to each well in a final volume of 500 µl. The plates were incubated at 37°C for 60 minutes and then washed three times with prewarmed medium using a standardized wash procedure. The contents of each well were lysed with 300 µl 1% Triton X-100, and γ-emissions counted. Wells were inspected prior to lysis to ensure that the integrity of the HUVEC monolayer had been maintained. Total counts added and spontaneously released counts were determined for each variable in the assays. Percentage adhesion was calculated according to the following formula:

Percentage adhesion = Measured cpm/(Total cpm - spont. released cpm) x 100.

RESULTS

We initially compared the in vitro T cell chemotactic effects of a variety of recombinant human chemokines, MIP-1α, MIP-1β, IP-10, PF4, I-309, IL-8, MCAF, and GRO to that of rhRANTES which has been previously reported to selectively chemoattract CD4⁺ memory T lymphocytes (11,12). The results in Table I show that rhMIP-1α, rhMIP-1β, and rhRANTES were more active in chemoattracting stimulated than unstimulated peripheral blood T cells in a dose dependent manner (Figure 1). Various concentrations of each of these chemokines were assayed for their chemotactic activity on unstimulated and anti-CD3 antibody stimulated T lymphocytes. T cell chemotactic activity was observed at rhMIP-1α and rhMIP-1β concentrations between 1 and 25 ng/ml with maximal migration being demonstrated at 10 ng/ml. RANTES showed chemotactic activity for activated T cells over a similar dose range of 1 to 20 ng/ml. Unstimulated T cells were also chemoattracted by RANTES over a similar concentration range, but unstimulated T cells showed no reproducible chemotactic response to rhMIP-1α and rhMIP-1β over a wide dose range. The response curves were typical bell-shaped curves, similar to those observed with monocyte and neutrophil chemoattractants. By contrast, the chemokines IL-8, MCAF, I-309 and PF4 showed no significant chemotactic activity on either stimulated or unstimulated T cells over a wide dose range (Table 1). Using a checkerboard analysis, we have consistently observed that T cell migration in response to rhMIP-1α, rhMIP-1β and rhRANTES was dependent on a concentration gradient and less than 10% of the migration was due to chemokinesis.

Table 1. Chemotactic response of leukocyte subsets to various chemokines.

	Leukocyte Population		
Chemokine	**T Cell**	**Monocyte**	**PMN**
MIP-1α	2+	2+	-
MIP-1β	2+	2+	-
RANTES	3+	1+	-
IP-10	1+	1+	-
IL-8[*]	1+	-	3+
MCAF	-	3+	-
GRO	-	-	3+
PF4	-	-	ND
I-309	-	2+	-

[*] Chemotactic for Jurkat cells, a T lymphocyte cell line.

T lymphocytes consist of phenotypically identifiable subpopulations with specialized functions such as CD4+ T cells with predominantly helper activities and CD8+ T cells with suppressor and cytotoxic functions. In addition, these subsets can be further subfractionated into naive (CD45RA+) and memory (CD29+) T cell subpopulations. Memory T cells are thought to be long-lived effector cells that preferentially participate in the generation of chronic T cell mediated inflammatory responses. The previous report demonstrating that rhRANTES selectively induces CD4+ memory T cell migration (11,12) prompted us to investigate the effects of rhMIP-1α and rhMIP-1β on subsets of T lymphocytes. In Figure 2, a representative experiment shows considerable migration of anti-CD3 stimulated CD4+ T cells to rhMIP-1β and RANTES but not rhMIP-1α. Only RANTES showed T cell chemotactic activity for unstimulated T lymphocytes. Stimulated CD8+ T lymphocytes migrated in response to rhMIP-1α, and RANTES. Recombinant hMIP-1β also exhibited a low chemoattractant effect on stimulated CD8+ lymphocytes. Further subfractionation of T lymphocytes into naive (CD45RA+) and memory (CD29+) T cell subsets demonstrated that both naive and memory T cell populations migrated in response to rhMIP-1α, rhMIP-1β and rhRANTES (Figure 2b) only if activated. As expected, unstimulated CD45RA+ and CD29+ failed to respond to either rhMIP-1α or rhMIP-1β. RANTES also chemoattracted unstimulated CD29+ but not naive unstimulated T cells.

Although the previous reports have demonstrated that RANTES preferentially induces CD4[+] CD45RO[+], but not naive CD4[+] CD45RA[+] or CD8[+] T lymphocytes, we find that RANTES also has activity on previously stimulated CD45RA[+] and CD29[+] T cells. This discrepancy may be based on the use of collagen-coated polycarbonate filters in our chemotaxis assay versus uncoated nitrocellulose membranes used by Schall et al. (11). In our experience, matrix-coated filters provide a more sensitive reproducible assay of T lymphocyte chemotaxis. These results show that MIP-1α, MIP-1β, and RANTES each specialize in chemoattracting distinct subsets of T lymphocytes: MIP-1α and MIP-1β act preferentially on activated CD8[+] and CD4[+] memory T cells, respectively, whereas RANTES attracts memory cells from both subsets.

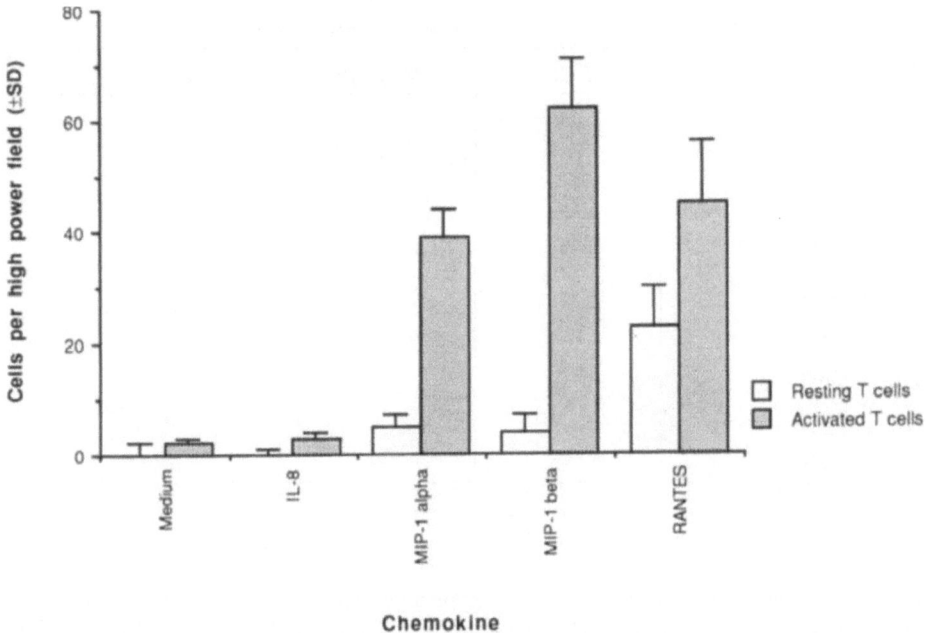

Figure 1. Human peripheral blood T lymphocyte migration in response to rhMIP-1α, rhMIP-1β, and rhRANTES as described in methods.

T cell adhesion to endothelial cells plays a key role in the development of inflammatory and immunological responses. Contact between endothelial cells and T cells is required for transmigration of T cells into tissues. Cytokines such as IL-1, TNFα, and IFNγ have been shown to promote these functions by induction of adhesion molecules such as ICAM-1 and VCAM on endothelial cells (13). IL-8 has been previously shown to regulate neutrophil adherence to an endothelial monolayer (13). We designed additional experiments to determine whether the T cell chemoattractants rhMIP-1α, rhMIP-1β and rhRANTES could influence the adhesion of T lymphocytes to human umbilical vein

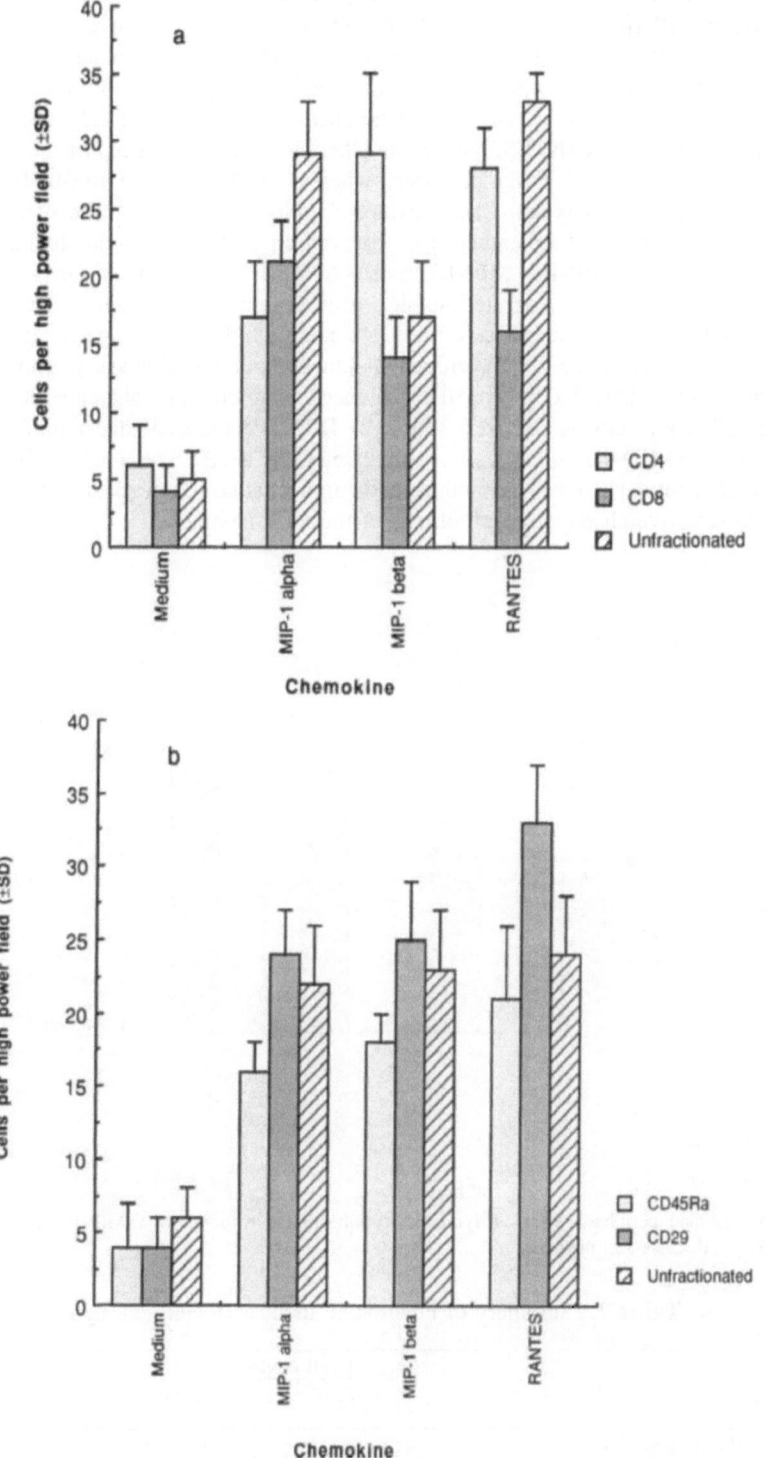

Figure 2. Phenotype of T lymphocytes migrating to rhMIP-1α, rhMIP-1β, and RANTES. a) The chemotactic activity of CD4+ and CD8+ T cell subsets. b) The chemotactic activity of CD29+ and CD45RA+ T cell subsets.

endothelium. Peripheral blood T cells were pretreated with various concentrations of rhMIP-1α, rhMIP-1β and rhRANTES at 37⁰C for 6 hr on uncoated or anti-CD3 antibody coated plates. After washing, lymphocytes were incubated with human umbilical cord endothelium pretreated with optimal concentrations of rhIL-1 for the adhesion assay. The results in Figure 3 demonstrate that all of these chemokines augment T cell adhesion to IL-1-treated endothelium. RANTES, but not the other chemokines, induced an increase in the adhesion of unstimulated T lymphocytes, whereas rhMIP-1α and rhMIP-1β induced substantial adhesion of only anti-CD3 activated T lymphocytes. No significant chemokine-induced increase in T cell adherence was observed on unstimulated endothelium. These results suggest that MIP-1α, MIP-1β, and RANTES may augment the ability of T lymphocytes to adhere and migrate into sites of antigenic challenge and inflammation.

We further subfractionated T cells into CD4+ and CD8+ subsets using negative selection techniques and tested these T cells with chemokines in our adhesion assay. Similar to the chemotaxis data, rhMIP-1α and rhMIP-1β induced a preferential adherence of CD8+ and CD4+ stimulated T cells, respectively (Table 2). RANTES induced adhesion of stimulated and unstimulated CD4+ T cell as well as stimulated CD8+ T cell. These results showing that these T cell chemotactic cytokines induce and augment adhesion of specific T cell subsets mirror the selective chemoattractant effects of these chemokines.

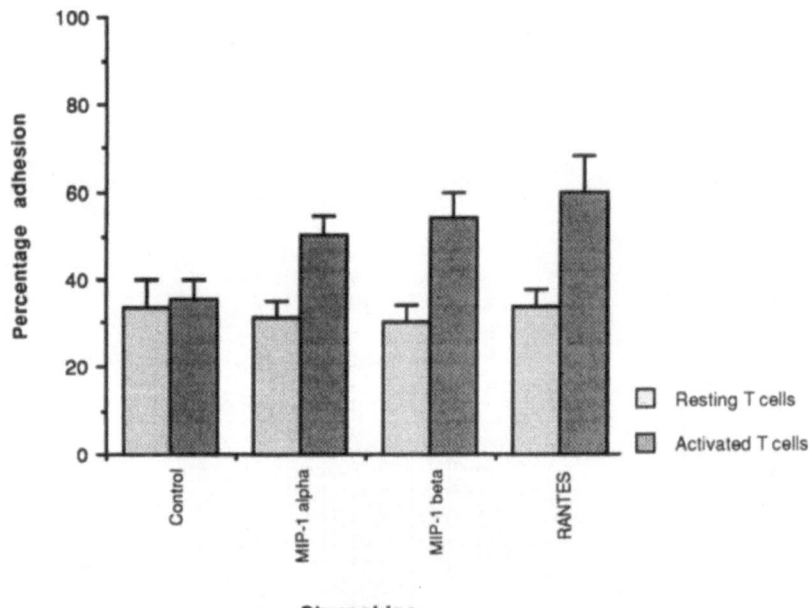

Figure 3. Human peripheral blood T lymphocyte adhesion in response to rhMIP-1α, rhMIP-1β, and rhRANTES as described in methods.

Table 2. Summary of chemokine-induced T cell adhesion.

	Unfractionated		T cell subset CD4+		CD8+	
Active/Rest	+	-	+	-	+	-
MIP-1α	3+	-	-	-	3+	-
MIP-1β	3+	-	2+	-	-	-
RANTES	3+	2+	3+	1+	2+	-

144

DISCUSSION AND CONCLUSION

The present results demonstrate that in addition to chemoattracting monocytes, rhMIP-1α, rhMIP-1β and rhRANTES are T cell chemoattractants with varying potencies that have preferential chemotactic and adhesive activity for distinct T cell subsets. MIP-1α stimulated predominantly CD8$^+$ T cell migration and adhesion, while MIP-1β preferentially induced CD4$^+$ T cell chemotaxis and adhesion. We have documented that, upon activation, T cells upregulate their expression of receptors for MIP-1α and MIP-1β resulting in an increased responsiveness to these chemokines (unpublished observations). The fact that RANTES induces unstimulated memory T cells to adhere and migrate suggests that a previous round of T cell activation may have induced constitutive expression of the RANTES receptor, permitting T cell responsiveness to this chemokine in various activation states.

Both differential and overlapping biological functions for MIP-1α and MIP-1β have been reported. MIP-1 originally represented an inflammatory activity that was subsequently determined to be due to two distinct molecules copurified from the supernatants of LPS-stimulated murine macrophages (14,15). Preparations of this mixture of natural MIP-1α and MIP-1β were capable of inducing footpad swelling associated with neutrophil and monocyte infiltration in vivo and had a low level of chemotactic activity for neutrophils in vitro (16). MIP-1α and MIP1β have been independently cloned for both mouse (17,18) and human (19,20). The homology between MIP-1α and MIP1β is 70% at the amino acid level. Natural mMIP-1α differs from natural mMIP-1β in properties such as the ability to modulate macrophage function and suppress the proliferation of less differentiated stem cells (21-23). In some of these studies, natural mMIP-1β antagonized the inductive effects of MIP-1α. In this report, we have also observed differential effects of MIP-1α and MIP-1β on T cell subsets. These results are perplexing in light of the recent data from our laboratory demonstrating that rhMIP-1α and rhMIP-1β bind with equal affinity to a receptor on monocytes and are equipotent chemoattractants for human monocytes (24). Despite these observations, it appears that both MIP-1α and MIP-1β elicit a number of opposing effects on macrophages, T cell subsets and stem cells, presumably by interaction with distinct receptors.

In a separate study (Taub et al. manuscript submitted), a stable recombinant human IP-10 (provided by Dr. K. Matsushima from Kanazawa University, Japan) was evaluated. These studies revealed that rhIP-10 was a moderately active chemoattractant for human monocytes and activated T lymphocytes of both CD4$^+$ and CD8$^+$ phenotypic subsets. In addition, rhIP-10 was a potent promoter of T cell adhesion to HUVEC.

The activation of T cells through the CD3 complex is known to result in an affinity change of adhesion molecules for their prospective ligands, resulting in a greater T cell adhesion (13). The ability of chemokines to augment the adhesion process suggests that these molecules may alter the density and/or the affinity of adhesion molecules on the T cell surface. Studies are currently in progress to elucidate the mechanism of chemokine-induced T cell adhesion.

ACKNOWLEDGEMENTS

The authors thank Drs. D. Longo for helpful discussion and criticism on the manuscript. We are grateful to Ms R. Unger for the secretarial assistance in the preparation of the manuscript.

REFERENCES

1.   Oppenheim, J.J., C.O.C. Zachariae, N. Mukaida, and K. Matsushima. 1991. Properties of the novel proinflammatory supergene "Intercrine" cytokine family, *Ann. Rev. Immunol.* 9: 617.

2. Djeu, J.Y., K. Matsushima, J.J. Oppenheim, K. Shiotsuki, and D.K. Blanchard. 1990. Functional activation of human neutrophils by recombinant monocyte-derived neutrophil chemotactic factor/IL-8, *J. Immunol.* 144: 2205.

3. Schall, T.J. 1991. Biology of the RANTES/SIS cytokine family. *Cytokine* 3: 165.

4. Carveth, H.J., J.F. Bohnsack, T.M. McIntyre, M. Baggiolini, S.M. Prescott, and G.A. Zimmerman. 1989. Neutrophil activating factor (NAF) induces polymorphonuclear leukocyte adherence to endothelial cells and to subendothelial matrix proteins, *Biophys. Res. Comm.* 162: 387.

5. Farina, P.R., R. DeLeon, A. Graham, P. Grob, E. David, R. Barton, J. Ksiazek, R. Rothlein, E. Mainolfi, and K. Matsushima. 1989. Monocyte-derived neutrophil chemotactic factor (MDNCF): a stimulator of neutrophil function, *FASEB J.* 3:A1333.

6. June, C.H., J.A. Ledbetter, M.M. Gillespie, T. Lindsten, and L.B. Thompson. 1987. T cell proliferation involving the CD28 pathway is associated with cyclosporine-resistant interleukin 2 gene expression, *Mol. Cell. Biol.* 7: 4472.

7. Falk, W.R., R.H. Goodwin Jr., and E.J. Leonard. 1980. A 48-well microchemotaxis assembly for rapid and accurate measurement of leukocyte migration, *J. Immunol. Meth.* 33: 239.

8. Pilaro, A.M., T.J. Sayers, K.L. McCormick, C.W. Reynolds, and R.H. Wiltrout. 1990. An improved *in vitro* assay to quantitate chemotaxis of rat peripheral blood large granular lymphocytes (LGL), *J. Immunol. Meth.* 135: 213.

9. Jaffe, E. A., R.L. Nachman, C.G. Becker, and C.R. Minick. 1973. Culture of human endothelial cells derived from umbilical veins, *J. Clin. Invest.* 52: 2745.

10. Shimizu, Y., S. Shaw, N. Graber, T.V. Gopal, K.J. Horgan, G. Van Seventer, and W. Newman. 1991. Acivation-independent binding of human memory T cells to adhesion molecule ELAM-1, *Nature* 349: 799.

11. Schall, T.J., K. Bacon, K.I. Toy, and D.V. Goedell. 1990. Selective attraction of monocytes and T lymphocytes of the memory phenotype by the cytokine RANTES. *Nature* 34: 669.

12. Schall, T.J. 1991. Biology of the rantes/sis cytokine family, *Cytokine* 3: 165.

13. Dustin, M.L. and T.J. Springer. 1991. Role of lymphocyte adhesion receptors in transient interactions and cell locomotion, *Ann. Rev. Immunol.* 9: 27.

14. Wolpe, S., G. Davatelis, B. Sherry, B. Beutler, D.G. Hesse, H.T. Nguyen, C.F. Moldawer, C.F. Nathan, S.F. Lowry, and A. Cerami. 1988. Macrophages Secrete a Novel Heparin-Binding Protein with Inflammatory and Neutrophil Chemokinetic Properties, *J. Exp. Med.* 167: 570.

15. Wolpe, S.D. and A. Cerami. 1990. Macrophage Inflammatory Proteins 1 and 2: Members of a Novel Superfamily of Cytokines, *FASEB J.* 3: 2565.

16. Davatelis, G., S.D. Wolpe, B. Sherry, J-M. Dayer, R. Chicheportiche, and A. Cerami. 1989. Macrophage inflammatory protein-1: a prostaglandin-independent endogenous pyrogen, *Science* 243: 1066.

17. Sherry, B., P. Tekamp-Olson, C. Gallegos, D. Bauer, G. Davatelis, S.D. Wolpe, F. Masiarz, D. Coit, and A. Cerami. 1988. Resolution of two components of Macrophage Inflammatory Protein 1, and Cloning and Characterization of One of These Components, Macrophage Inflammatory Protein 1B. *J. Exp. Med.* 168: 2251.

18. Davatelis, G., P. Tekamp-Olson, S.D. Wolpe, K. Hermsen, C. Luedke, C. Gallegos, D. Coit, J. Merryweather, and A. Cerami. 1988. Cloning and Characterization of a cDNA for Murine Macrophage Inflammatory Protein (MIP), A Novel Monokine With Inflammatory and Chemokinetic Properties. *J. Exp. Med.* 167: 1939.

19. Lipes, M.A., M. Napolitano, K-T. Jeang, N.T. Chang, and W. Leonard. 1988. Identification, Cloning, and Characterization of an Immune Activation Gene. *Proc. Natl. Acad. Sci. USA* 85: 9704.

20. Obaru, K., T. Hattori, Y. Yamamura, K. Takatsuki, H. Nomiyama, S. Maeda, and K. Shimada. 1989. A cDNA Clone Inducible in Human Tonsillar Lymphocytes by a Tumor Promoter Codes for a Novel Protein of the B-Thromboglobulin Superfamily. *Mol. Immunol.* 26: 423.

21. Fahey III,T.J., K.J. Tracey, P. Temamp-Olson, L.S. Cousens, W.G. Jones, G.T. Shires, A. Cerami, and B. Sherry. 1992. Macrophage inflammatory protein 1 modulates macrophage function, *J. Immunol.* 148: 2764.

22. Broxmeyer, H.E., B. Sherry, L. Lu, S. Cooper, K-O. Oh, P. Tekamp-Olson, B.S. Kwon, and A. Cerami. 1990. Enhancing and Suppressing Effects of Recombinant Murine Macrophage Inflammatory Proeyins on Colony Formation In Vitro by Bone Marrow Myeloid Progenitor Cells. *Blood.* 76: 1110.

23. Maze, R., B. Sherry, B.S. Kwon, A. Cerami, and H.E. Broxmeyer. 1992. Myelosuppressive effects in vivo of purified recombinant murine macrophage inflammatory protein-1. *J. Immunol.* 149: 1004.

24. Wang, J-M., B. Sherry, D.J. Kelvin, and J.J. Oppenheim. Human recombinant macrophage inflammatory protein-1 alpha and beta (MIP-1α and MIP-1β) and MCAF utilize common and unique receptors in the induction of monocyte chemotaxis, *J. Immunol.* In press.

PROMISCUITY OF LIGAND BINDING IN THE HUMAN CHEMOKINE BETA RECEPTOR FAMILY

David J. Kelvin[1], Ji-Ming Wang[1], Dan McVicar[2], and
Joost J. Oppenheim[1]

[1]Laboratory of Molecular Immunoregulation,
[2]Laboratory of Experimental Immunology
Biological Response Modifiers Program
Division of Cancer Treatment
NCI-FCRDC
Frederick, MD 21702, USA

INTRODUCTION

The chemokine beta subfamily is comprised of cytokines (chemokines) having structural similarity as well as amino acid homology and includes Monocyte Chemotactic and Activating Factor (MCAF or MCP-1), RANTES, Macrophage Inflammatory Protein-1 alpha (MIP-1α), and Macrophage Inflammatory Protein-1 beta (MIP-1β) (1,2). They are distinguished from the chemokine alpha subfamily members by the absence of an intervening amino acid between the first two conserved cysteine residues. MCAF/MCP-1, RANTES, human MIP-1α and human MIP-1β all induce monocyte chemotaxis (3,4) (Taub et al., this volume) while human MIP-1α (LD78) human MIP-1β (ACT-2) and RANTES also stimulate T-cell chemotaxis and adhesion (Taub et al. this volume). To gain insight into the nature of the receptors for this class of ligands we have studied the binding of MCAF, human MIP-1α, human MIP-1β and RANTES to ligand specific sites on monocytes and the human monocytic cell line THP-1. We have identified at least three distinct functional receptors for these four chemokines, two of which display promiscuous behavior in binding multiple chemokines.

METHODS AND MATERIALS

Cytokines: rhRANTES was purchased from Pepro Tech Inc. (lot. 8721, Rocky Hill, NJ). Radioiodinated rhRANTES preparations, with the ^{125}I located on either histidine or tyrosine residues, were kind gifts of Dr. B. Brown, DuPont NEN, MA. Both preparations had a specific activity of 5 X 10^5 cpm/ng protein. The sources of other cytokines were: rhMIP-1α from Dr. M. Tsang of the R & D Systems, Minneapolis, MN ; rhIL-8 was a gift from Dainippon Pharmaceutical Co. LTD; rhMCAF and rhGRO from Pepro Tech. ^{125}I-MCAF, ^{125}I-MIP-1α, and ^{125}I-MIP-1β were prepared as previously described (5).

Cells: The human acute monocytic leukemic cell line, THP-1, was obtained from the ATCC and maintained in RPMI 1640 medium (Whittaker, Walkersville, MD) supplemented with 10% FCS (Hyclone, Logan, UT) and penicillin/streptomycin. Human peripheral blood monocytes were purified from normal blood (National Institutes of Health Clinical Center Transfusion Medicine Department, Bethesda, MD) with an isosmotic Percoll (Pharmacia, Uppsala, Sweden) gradient as described elsewhere (5). The monocyte preparations were always >90% pure as assessed by morphological criteria.

Binding of ^{125}I-chemokines to monocytes and THP-1 cells: Binding conditions for MIP-1α, MIP-1β, and MCAF were carried as previously described (5). ^{125}I-RANTES binding was carried out as follows: In steady-state binding assays, 2 X 10^6 cells were incubated in duplicate with increasing concentrations of the ^{125}I-RANTES in a modified binding medium (RPMI-1640 containing 1mg/ml BSA and 25mM HEPES, pH 7.4) in a total volume of 200ul. The residual nonspecific binding was determined by parallel incubation of ^{125}I-RANTES in the presence of 100-fold excess of unlabeled RANTES. After incubation at 4°C for 4h, the cells were pelleted through a 10% sucrose/PBS cushion. The tips of the tubes containing cells were cut and radioactivity was determined in a gamma counter (Gamma 4000, Beckman). The residual non-specifically bound radioactivity associated with cells in the presence of unlabeled RANTES was subtracted from total bound radioactivity to yield specific binding. The data was analyzed using previously described methods (6). The displacement curves of ^{125}I-RANTES binding to monocytic cells by different cytokines were generated by incubating cells with a constant concentration of ^{125}I-RANTES for 4h at 4°C in the presence of increasing amounts of unlabeled ligands. The cells were then pelleted through a sucrose cushion and the residual radioactivity determined.

Analysis of intracellular calcium mobilization: Analysis of the changes in intracellular Ca^{+2} concentration ($[Ca^{+2}]_i$) of THP-1 cells was carried out using a Deltascan photometer (Photon Technologies International, Princeton NJ). Briefly, THP-1 cells were incubated at a density of 10^7/ml for 30 minutes at 37° C in medium containing Indo-1AM (Molecular Probes, Eugene OR). After 30 minutes, fresh medium was added diluting the cells to 5x10^6/ml, and cells were further incubated for 30 minutes at 37° C. Cells were washed once with medium after loading with Indo-1, and held at room temperature in the dark until analysis. The $[Ca^{+2}]_i$ was monitored with the Indo-1 loaded cells suspended at 37° C in DPBS with Ca^{+2} and Mg^{+2} supplemented with 5mM Glucose. Indo-1 excitation was at 358nm with detection of bound dye at 402nm (Violet) and free dye at 486 (Blue)[ref] (7).

RESULTS

In order to assess the binding characteristics of RANTES, MCAF, MIP-1α, and MIP-1β, we examined the specific binding of these ligands to peripheral blood monocytes. Table one summarizes Scatchard analyses for ^{125}I labelled MCAF, RANTES, MIP-1α, and MIP-1β on monocytes. Approximately 2,700 receptors were found for MIP-1α and MIP-1β with comparable affinities of 300-500 pM. MCAF had the most specific binding sites (18,000) with the lowest affinity of 10nM. RANTES had only 570 receptors with an affinity of 444 pM.

Table 1. Affinity and receptor numbers for chemokine beta members on peripheral blood monocytes

Chemokine	Affinity (Kd)	Specific Binding Sites
MCAF	10nM	18,000
MIP-1α	385pM	2,760
MIP-1β	469pM	2,660
RANTES	440pM	570

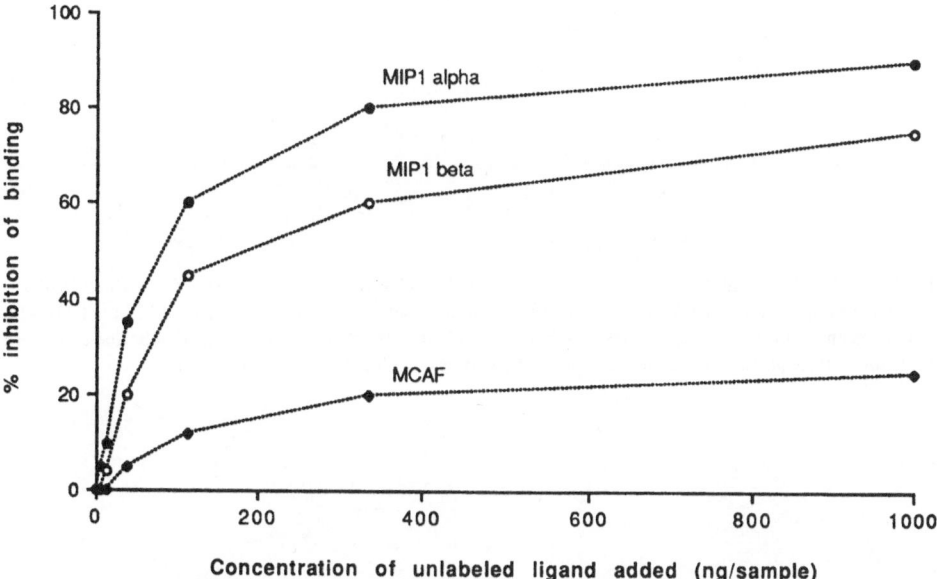

Figure 1. Displacement of ^{125}I-MCAF binding to THP-1 cells by unlabeled MCAF, MIP-1α and MIP-1β. Duplicate samples of 2 X 10[6] THP-1 cells in 200 μl binding medium were incubated with 0.1 nM ^{125}I-MCAF in the presence of increasing quantities of unlabeled cytokines. After incubation at 4°C for 4 hr, the cells were centrifuged through sucrose cushion and the radioactivity in cell pellets were measured.

To circumvent potential donor variation we utilized the human monocytic THP-1 cell line for the subsequent experiments. This cell line displayed similar numbers of receptors with similar affinities as those on monocytes, and could functionally respond to MCAF, RANTES, MIP-1α and MIP-1β with mobilization of CA^{+2} (8) (See below).

We performed competitive inhibition studies to determine the binding relationship between various chemokines. [125]I labelled MCAF could be completely inhibited by unlabeled MCAF and partially competed for by MIP-1α and MIP-1β (Fig. 1); other cytokines such as IL-8 and IP-10 could not compete for binding (data not shown). ^{125}I-MIP-

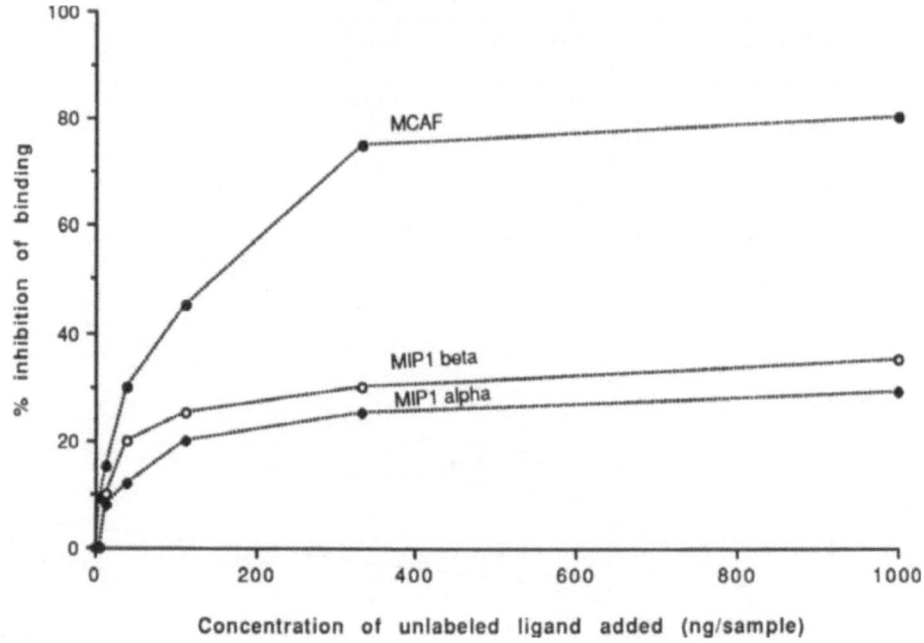

Figure 2. Displacement of ^{125}I-MIP-1α binding to THP-1 cells by unlabeled MCAF, MIP-1α and MIP-1β. Duplicate samples of 2 X 10[6] THP-1 cells in 200 μl binding medium were incubated with 0.1 nM ^{125}I-MIP-1α in the presence of increasing quantities of unlabeled cytokines. After incubation at 4°C for 4 hr, the cells were centrifuged through sucrose cushion and the radioactivity in cell pellets were measured.

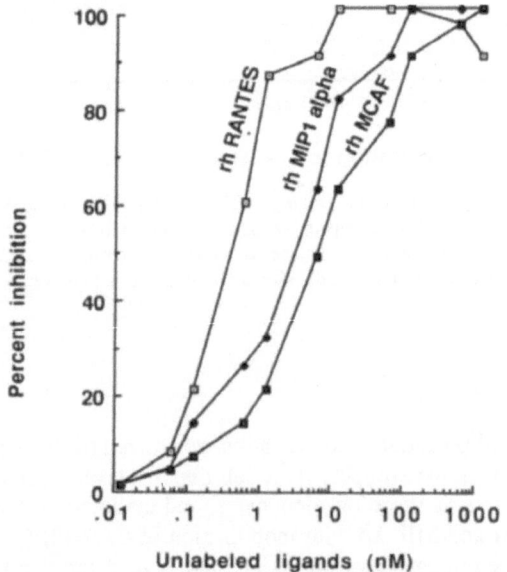

Figure 3. Displacement of ^{125}I-RANTES binding to THP-1 cells by unlabeled RANTES, MCAF and MIP-1α. Duplicate samples of 2 X 10[6] THP-1 cells in 200 μl binding medium were incubated with 0.1 nM ^{125}I-RANTES in the presence of increasing quantities of unlabeled cytokines. After incubation at 4°C for 4 hr, the cells were centrifuged through sucrose cushion and the radioactivity in cell pellets were measured.

150

Induction of Ca⁺² mobilization in THP1 cells by chemotatic cytokines

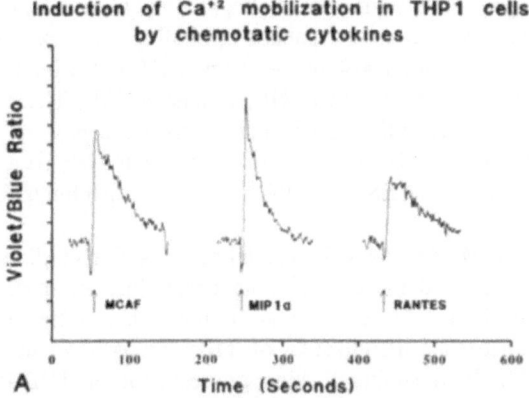

A

MCAF desensitizes THP1 to RANTES but RANTES does not desensitize THP1 to MCAF

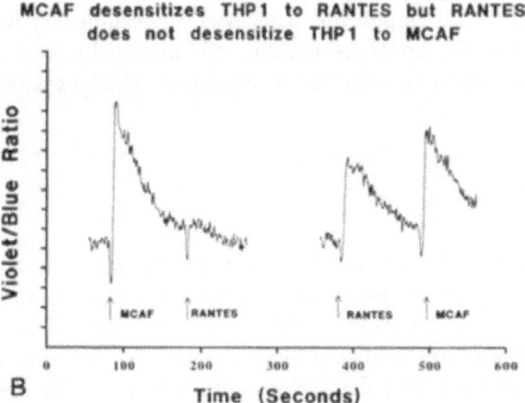

B

MIP1α desensitizes THP1 to RANTES but RANTES does not desensitize THP1 to MIP1α

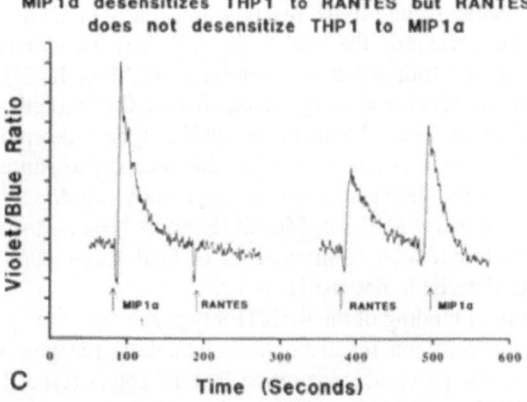

C

Figure 4. Desensitization of RANTES induced Ca⁺² mobilization by MCAF and MIP-1α. (A) The $[Ca^{+2}]_i$ of Indo-1 loaded THP-1 cells was monitored as they were stimulated with MCAF (500ng/ml), MIP-1α (500ng/ml), or RANTES (500ng/ml) where indicated. (B) THP-1 cells were stimulated with 500ng/ml MCAF followed 100 seconds later by 100ng/ml RANTES (left trace) or 500ng/ml RANTES followed 100 seconds later by 100 ng/ml MCAF (right trace). (C) THP-1 cells were stimulated with 500 ng/ml MIP-1α followed 100 seconds later by 100ng/ml RANTES (left trace) or 500 ng/ml RANTES followed by 100ng/ml MIP-1α (right trace).

151

1α binding was completely inhibited by both MIP-1α and MIP-1β (Fig. 2), whereas MCAF could partially compete for ^{125}I-MIP-1α binding. Competitive inhibition studies using ^{125}I labeled RANTES yielded yet another pattern of binding (Fig. 3); RANTES, MCAF, and MIP-1α could all compete for 100% of the specific RANTES binding, though RANTES competed with a greater affinity than MIP-1α and MCAF. These data taken together suggest the presence of at least three distinct receptors, a receptor for MCAF which cannot bind RANTES, MIP-1α or MIP-1β, a receptor which binds MIP-1α and MIP-1β but which does not bind MCAF or RANTES, and a third receptor which binds RANTES, MCAF and MIP-1α.

To investigate if MCAF and MIP-1α binding to the RANTES receptor has functional consequences, we examined the desensitizing capabilities of these various ligands for calcium mobilization in THP-1 cells. Figure 4 shows that pretreatment for 30 min with MCAF can desensitize Ca^{+2} mobilization of THP-1 cells to a subsequent challenge by MCAF or RANTES. In a similar fashion pretreatment of THP-1 cells with MIP-1α abolished Ca^{+2} mobilization to a subsequent challenge by RANTES or MIP-1α. RANTES on the other hand, only slightly reduced the Ca^{+2} mobilization by subsequent challenges to MIP-1α and MCAF. We interpret this data as an indication that RANTES, MCAF and MIP-1α can all bind to the RANTES receptor and functionally activate the receptor which results in the receptor being desensitized to subsequent challenges of RANTES or MCAF or MIP-1α.

DISCUSSION

MCAF, MIP-1α, MIP-1β and RANTES are all members of the chemokine beta subfamily and can all activate monocytes. These chemokines have specific receptors on monocytes and the human monocytic THP-1 cell line (8,9,10) (Table 1). In this report we examined the binding characteristics of these ligands and identified at least three distinct cell surface receptors on THP-1 cells. One of the receptors, which we will designate the RANTES type I receptor can bind RANTES, MCAF and MIP-1α, this receptor could not bind members of the chemokine alpha subfamily (e.g. IL-8, GRO, and IP10 data not shown), the second receptor, the MCAF receptor apparently binds MCAF but not MIP-1α and RANTES. The third receptor, the MIP-1α,β receptor will bind both MIP-1α and MIP-1β with apparently equal affinities but does not bind MCAF or RANTES. These receptors apparently are functional because each ligand can induce Ca^{+2} mobilization in THP-1 cells. The binding of MCAF and MIP-1α to the RANTES type I receptor is apparently also functional because these ligands can desensitize this receptor to subsequent challenges by RANTES. However, this desensitization is apparently unidirectional and RANTES presumably cannot desensitize MIP-1 or MCAF receptors because the number of RANTES receptors (500) is too low relative to the number of MIP-1α (~3000) and MCAF (18000) receptors to generate detectable desensitization.

The promiscuity of binding of the RANTES type I receptor for RANTES, MCAF and MIP-1α is probably a common feature of the chemokine receptor family because in a similar fashion one of the previously identified IL-8 receptors (type II) (11) can bind IL-8 and GRO while the IL-8 type I receptor (12) can bind IL-8 but not GRO (this volume). The identification of cDNAs for the chemokine beta receptors should prove useful in understanding the molecular basis of promiscuous and specific ligand binding.

ACKNOWLEDGEMENTS

The authors thank Drs. D. Longo, A. Lloyd, and D. Taub for helpful discussion and criticism on the manuscript, and Dr. M. Fivash for data analyses. JMW is recipient of the International Union Against Cancer-

American Cancer Society Roosevelt Cancer Research Award, and a fellowship from the Dainippon Pharmaceutical Company LTD. Osaka, Japan. We are grateful to Ms R. Unger and Ms L. Ridgell for the secretarial assistance in the preparation of the manuscript.

REFERENCES

1. Oppenheim, J.J., C.O.C. Zachariae, N. Mukaida, and K. Matsushima. 1991. Properties of the novel proinflammatory supergene "Intercrine" cytokine family, *Annu. Rev. Immunol.* 9: 617.
2. Schall, T.J. 1991. Biology of the RANTES/SIS cytokine family, *Cytokine* 3: 165.
3. Zachariae, C.O.C., A.O. Anderson, H.L. Thompson, E. Appella, A. Mantovani, J.J. Oppenheim, and K. Matsushima. 1990. Properties of monocyte chemotactic and activating factor (MCAF) purified from a human fibrosarcoma cell line, *J. Exp. Med.* 171: 2177.
4. Schall, T.J., K. Bacon, K.J. Toy, and D.V. Goeddel. 1990. Selective attraction of monocytes and T lymphocytes of the memory phenotype by the cytokine RANTES, *Nature* 347: 669.
5. Samanta, A.K., J.J. Oppenheim, and K. Matsushima. 1989. Identification and characterization of a specific receptor for monocyte-derived neutrophil chemotactic factor (MDNCF) on human neutrophils, *J. Exp. Med.* 169: 1185.
6. Munson, P.J., and D. Rodbard. 1984. Computerized analysis of ligand binding data: Basic principles and recent developments, Computers in Endocronology, pg 117-145.
7. Moser, B., C. Schumacher, V. von Tscharner, I. Clark-Lewis, and M. Baggiolini. 1991. Neutrophil-activating peptide 2 and gro/melanoma growth-stimulating activity interact with neutrophil-activating peptide 1/interleukin 8 receptors on human neutrophils, *J. Biol. Chem.* 266: 10666.
8. Wang, J-M., B. Sherry, D.J. Kelvin, and J.J. Oppenheim. Human recombinant macrophage inflammatory protein-1 alpha ans beta (MIP-1α and MIP-1β) and MCAF utilize common and unique receptors in the induction of monocyte chemotaxis, *J. Immunol.* in press.
9. Yoshimura, T., and E. Leonard. 1990. Identification of high affinity receptors for human monocyte chemoattractant protein-1 (MCP-1) on human monocytes, *J. Immunol.* 145: 292.
10. Valente, A.J., M.M. Rozek, C.J. Schwartz, and D.T. Graves. 1991. Characterization of monocyte chemotactic protein-1 binding to human monocytes, *Biochem. Biophys. Res. Comm.* 176: 309.
11. Murphy, P.M., and T.H. Lee. 1991. Cloning of complementary DNA encoding a functional human interleukin-8 receptor, *Science* 253: 1280.
12. Holmes, W.E., J. Lee, W.J. Kuang, G.C. Rice, and W.I. Wood. 1991. Structure and functional expression of human interleukin-8 receptor, *Science* 253: 1278.

STRUCTURAL AND FUNCTIONAL CHARACTERIZATION
OF THE INTERLEUKIN-8 RECEPTORS

M. Patricia Beckmann, Richard B. Gayle, Douglas Pat Cerretti,
Carl J. March, Subhashini Srinivasan, and Paul R. Sleath

Immunex Corporation
51 University Street
Seattle, WA 98101
USA

INTRODUCTION

Interleukin-8 (IL-8), growth regulatory gene/melanoma growth stimulatory activity (GRO/MGSA) and neutrophil activating peptide-2 (NAP-2) are members of a growing superfamily of cytokine polypeptides with sequence and structural homology (reviewed in Oppenheim et al., 1991). Four cysteines are conserved in these molecules which may serve to maintain their structural integrity. The three-dimensional structure of IL-8 has been solved by both NMR spectroscopy and x-ray crystallography. IL-8 exists as a homodimer which may undergo conformational changes upon receptor binding (6). Biological studies utilizing recombinant chemotactic peptides have illustrated several distinct functions for these cytokines. Neutrophils respond chemotactically by migrating toward IL-8, GRO/MGSA and NAP-2 (38, 25, 32). Additionally, neutrophil-endothelial cell adhesion has been observed to be enhanced by IL-8 (14), and both IL-8 and GRO/MGSA have been shown to induce a rise in cytosolic-free calcium ($[Ca^{++}]_i$) (25). Melanoma cells have also been reported to show autocrine growth regulation by GRO/MGSA (30).

Several investigators have determined that IL-8, GRO/MGSA and NAP-2 show cross-competition when binding to human cells (28, 32, 8, 25). Similarly, cells which have detectable surface IL-8 receptors show similar levels of GRO/MGSA receptors. These studies led to the hypothesis that the receptor for IL-8 might also bind GRO/MGSA.

In this paper a summary of the the molecular and biochemical characteristics of receptors for IL-8, GRO/MGSA and NAP-2 is presented. Additionally, through the characterization of chimeric receptors, localization of IL-8 and GRO/MGSA binding sites are deduced. Finally, preliminary evidence is presented showing that a portion of the IL-8 receptor molecule serves as an IL-8 antagonist. Through these studies a model for IL-8/IL-8R interactions is proposed.

The Chemokines, Edited by I.J.D. Lindley
et al., Plenum Press, New York 1993

CHARACTERIZATION OF IL-8 RECEPTORS

In order to characterize receptors for the chemokine molecules, we have radiolabeled recombinant IL-8 and GRO/MGSA with $Na^{125}I$ and have detected receptors on the surface of a variety of hematopoietic cells (2). Utilizing human neutrophils or the myelomonocytic cell line HL-60, we have observed both high affinity and low affinity receptors for IL-8 and GRO/MGSA. As shown in Figure 1, GRO/MGSA can bind to HL-60 cells with either a single class of high affinity binding sites (left panel), or in a biphasic manner (right panel). Similar results have been observed for IL-8 (not shown). The biphasic patterns of binding have been observed in ~30% of the studies undertaken in our laboratory. We have hypothesized that the activation state of the receptor molecule and its association with the heterotrimeric G protein complex may be responsible for the differences in the binding affinities observed. We have also observed that both IL-8 and GRO/MGSA undergo monomer-dimer transitions under varying conditions of pH and salt concentrations (data not shown). The different affinity classes observed may be due to transitions in the binding of monomer or dimer ligands to the receptor molecules. Moser et al. (1992) have also observed biphasic binding of GRO/MGSA to human neutrophils.

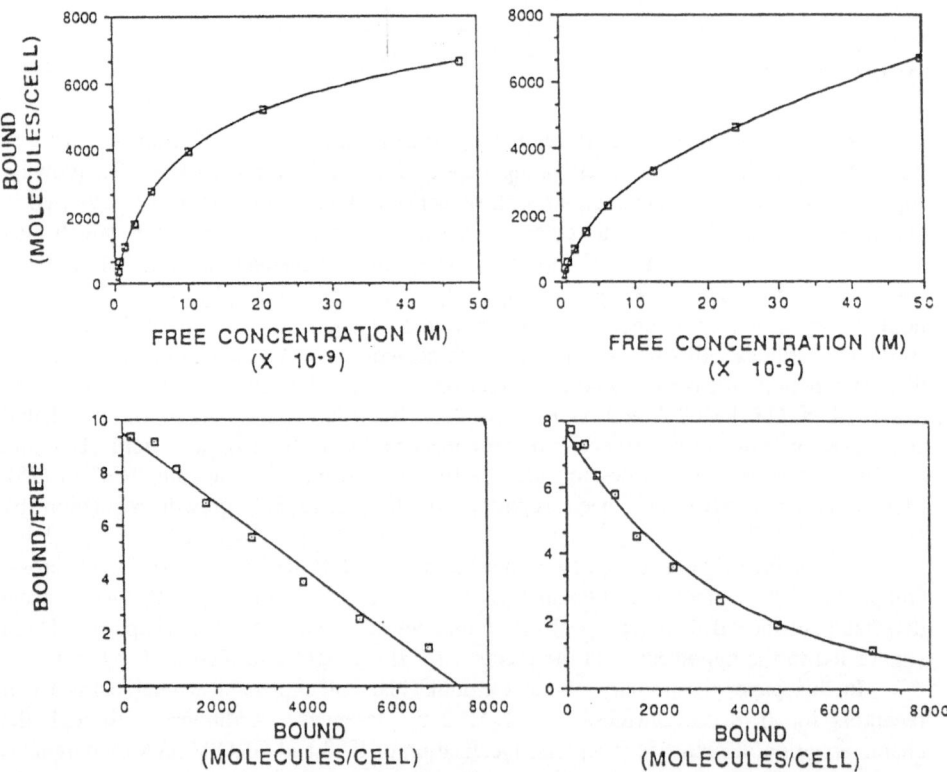

Figure 1. Equilibrium binding of ^{125}I-GRO/MGSA to HL-60 cells. Cells (2 x 10^6 cells/point) were incubated with various concentrations of radiolabeled GRO/MGSA. Free and cell-bound GRO/MGSA were separated over phthalate oil (2). Non-specific binding was determined in the presence of 100-fold molar excess of unlabeled GRO/MGSA.

Human neutrophils appear to be the cell source with the greatest number of receptors for IL-8 and GRO/MGSA. A wide range of receptor numbers is observed on neutrophils isolated from various human donors, and receptor numbers have varied from as great as 100,000 receptors to as low as 5,000 receptors per cell for IL-8. Generally 40-60,000 IL-8 binding sites can be detected on human neutrophils.

A variety of cytokines were tested for their ability to cross-compete for IL-8 and GRO/MGSA binding to human neutrophils. We have tested the macrophage inhibitory peptides (MIP-1a, MIP-1b) and the monocyte chemotactic activating factor (MCAF). Even at concentrations of 10^{-7} M, we observed no competition of IL-8 or GRO/MGSA binding. Other cytokines including IL-1, G-CSF, GM-CSF and IL-6 showed no competition at similar concentrations.

Conversely, utilizing ^{125}I-IL-8 we observed competition of binding with both unlabeled IL-8 and unlabeled GRO/MGSA as shown in Figure 2. These results have also been shown by Derynck et al., (1990); Moser et al. (1991); Oppenheim et al., (1991); and Sager et al., (1992). Experiments have also shown that unlabeled IL-8 and GRO/MGSA could compete with ^{125}I-GRO/MGSA binding to cells. This evidence suggests that IL-8 and GRO/MGSA may bind to the same receptor molecule on human neutrophils or other cell types that show binding to both molecules.

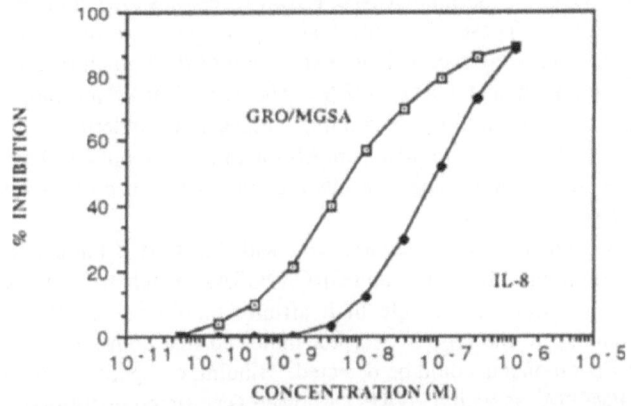

Figure 2. Inhibition of ^{125}I-IL-8 binding to human neutrophils. Purified human neutrophils (2 x 10^7 cells/ml) were incubated with ^{125}I-IL-8 (5 x 10^{-9}M) for one hour at 37°C with various concentrations of unlabeled GRO/MGSA or IL-8. Data were corrected for non-specific binding and analyzed as described (2).

MOLECULAR CHARACTERIZATION OF IL-8 RECEPTORS

Several years ago Thomas et al., (1990) isolated a cDNA, termed F3R, from a rabbit neutrophil cDNA library. The seven-membrane spanning receptor molecule was believed to be the receptor for a formylated peptide (N-formyl methionine-leucine-phenylalanine or fMLP). Because this receptor molecule appeared to be exclusively expressed on neutrophils, we isolated F3R and used it as a probe for the isolation of other related receptors (Beckmann et al., 1991). Upon expression of a cloned F3R in COS-7 cells, binding of ^{3}H-fMLP was not detected, but surprisingly specific ^{125}I-IL-8 binding was observed as shown in Table 1. While ^{125}I-IL-8 and GRO/MGSA binding has been detected on rabbit neutrophils, no ^{125}I-GRO/MGSA binding was observed on F3R expressed in COS-7 cells. The observations left open the possibility that a specific receptor existed for GRO/MGSA. Several groups have now documented that F3R is not the receptor for fMLP but the receptor for IL-8 (19, 36).

Table 1. Radiolabeled cytokine binding to rabbit IL-8 receptors.

Expression Vector	^{125}I-IL-8	^{125}I-GRO/MGSA	^{3}H-fMLP
Empty Vector	1901 ± 802	1103 ± 298	107 ± 31
Rabbit IL-8 Vector	31655 ± 2952	1294 ± 216	84 ± 23

COS-7 cells were transfected with pDC 302 containing the rabbit IL-8 receptor (2). Cells were incubated with radiolabeled ligands indicated (1.25 x 10^{-9}M) and binding determined.

Using the rabbit IL-8 receptor cDNA as a probe we isolated two related cDNAs from a human neutrophil cDNA library (5). These cDNAs encode proteins which show a high degree of amino acid similarity as described below. We have termed these receptors human IL-8 receptors type 1 and 2 while others have termed them A and B, respectively. They are identical to the receptors isolated by Holmes et al. (1991) and Murphy and Tiffany (1991). Overall, the human type 1 IL-8 receptor shows greater similarity to the rabbit IL-8 receptor. These similarities also correspond to functional similarities as described below. Following recombinant expression of the human IL-8 receptors, binding was assessed with IL-8 and GRO/MGSA. The type 1 IL-8 receptor showed specific binding of IL-8. When inhibition of binding studies were undertaken using radiolabeled IL-8 as a probe, IL-8 showed complete competition of binding but GRO/MGSA and NAP-2 did not compete for binding. These results are similar to those observed with the rabbit IL-8 receptor described above.

In contrast, when binding was assessed with the type 2 human IL-8 receptor, we observed direct binding of both IL-8 and GRO/MGSA. Scatchard analysis of IL-8 binding resulted in the detection of a single high affinity binding site, yet Scatchard plots of GRO/MGSA binding to the type 2 receptor indicated that both a high affinity and lower affinity binding components could be detected. Binding competition studies indicated that IL-8 and GRO/MGSA, as well as NAP-2, showed specific competition of IL-8 binding to the type 2 IL-8 receptors. Thus, in contrast to the type 1 IL-8 receptor, the type 2 receptor appears capable of showing promiscuity in binding to several ligands of the chemokine family.

ISOLATION OF IL-8 RECEPTOR GENOMIC CLONES

In order to identify the chromosomal location of the genes for the IL-8 receptors, human genomic clones were isolated (24). The type 1 IL-8 receptor genomic clones (*IL8R1*) varied in size from 13-17 kb and the type 2 (*IL8R2*) isolates from 4-16 kb in length. An additional class of IL-8 receptor genomic clones varying from 8-17 kb was also isolated. This third class of IL-8 receptor clones was identified as a pseudogene (*IL8RP*). The *IL8RP* has 77% and 88% identity with the type 1 and 2 IL-8 receptors, respectively. This gene cannot encode a functional protein, as six mutations in the sequence result in three stop codons and three frameshift mutations.

Utilizing cDNAs encoding the type 1 and 2 IL-8R and a 1.8 kb fragment of the IL-8 receptor pseudogene, Southern hybridizations of somatic rat-human cell hybrid DNAs indicated that all three genes were localized to human chromosome 2. Finer mapping utilizing fluorescence *in situ* hybridization (FISH) showed that all three genes reside on band q35 of chromosome 2.

Several human diseases have been linked to this region including van der Woude and Waardenburg syndromes (11) and rhabdomyosarcoma and leiomyomata (22). Waardenburg syndrome has characteristic pigmentary aberrations, and it is tempting to speculate that the IL-8R2 locus may be involved in this disease state through a defect in the interaction of this receptor with GRO/MGSA. Syntenic relationships of human chromosome 2q suggest that murine homologues are likely present on chromosome 1 (7). Proximal to this region is the Lsh-Ity-Bcg disease resistance locus (23, 20). This locus is thought to encode suspecptibility to infections including *Leishmania donovani* and *Salmonella typhimurium*. Genetic susceptibility to tuberculosis and leprosy in human populations may be analogous to this murine phenotype (33, 1). Further study in the IL-8R system is underway to examine these potential relationships.

CHARACTERIZATION OF CHIMERIC IL-8 RECEPTOR MOLECULES

The similarity in the IL-8 receptor sequences and the disparity in their binding characteristics led to construction of chimeric receptor molecules of type 1 and 2 (12). These chimeras allowed presentation of different combinations of amino acid residues to the ligands, thus defining regions responsible for binding to IL-8 or GRO/MGSA to be identified. Chimeras between the rabbit type 1 and the human type 2 IL-8 receptors were made as illustrated in Figure 3. Using conserved restriction enzyme sites in the molecules, different regions of receptor cDNAs were combined. Additionally, a recombinant rabbit IL-8 receptor was constructed which was devoid of the extracellular N-terminus (sRM) to assess the relative importance of the N-terminal portion of the receptor in IL-8 binding. All constructs contained an N-terminal peptide termed FLAG (containing the amino acid sequence DYKDDDDK) which provided an antibody epitope to be easily recognized on the proteins (17).

Following production of the human and rabbit IL-8 receptors in 293 cells, binding of IL-8 was assessed (Table 2). The equilibrium binding constants observed for the parental IL-8 receptors were similar to those observed on human neutrophils or for IL-8 receptors lacking the FLAG sequence following expression in COS-7 cells (5). Of the six chimeras produced, only #1, 2, 3, and 6 appear to bind IL-8. The binding affinity of IL-8 to these molecules was within 3-4 fold of the wild-type molecules indicating that the fusion of the two sequences likely had little effect on the overall structure of the protein. Chimeras #4, 5 and sRM did not display IL-8 binding, although IL-8 receptor protein was detected following immunoprecipitation of all of the IL-8 receptor proteins (data not shown). Some modulation of the affinities for IL-8 was demonstrated by the chimeric receptors. For instance, chimera #6, which has the N-terminal 212 amino acids of the type 2 IL-8 receptor fused to the last 146 amino acids of the rabbit type 1 IL-8 receptor, has slightly higher affinity for IL-8 than the reciprocal chimera #3 or the wild-type human type 2 receptor. Intracellular loops of receptors in this family are known to couple to G proteins and possibly dissociate from the receptor following ligand binding and signal transduction. The integrity of the G protein-receptor interactions is thought to be responsible for maintenance of high affinity binding. Thus, uncoupling of the G proteins from the receptor molecules may decrease the affinity of the ligand for the receptor (9), and be responsible for the variations in the binding affinities observed.

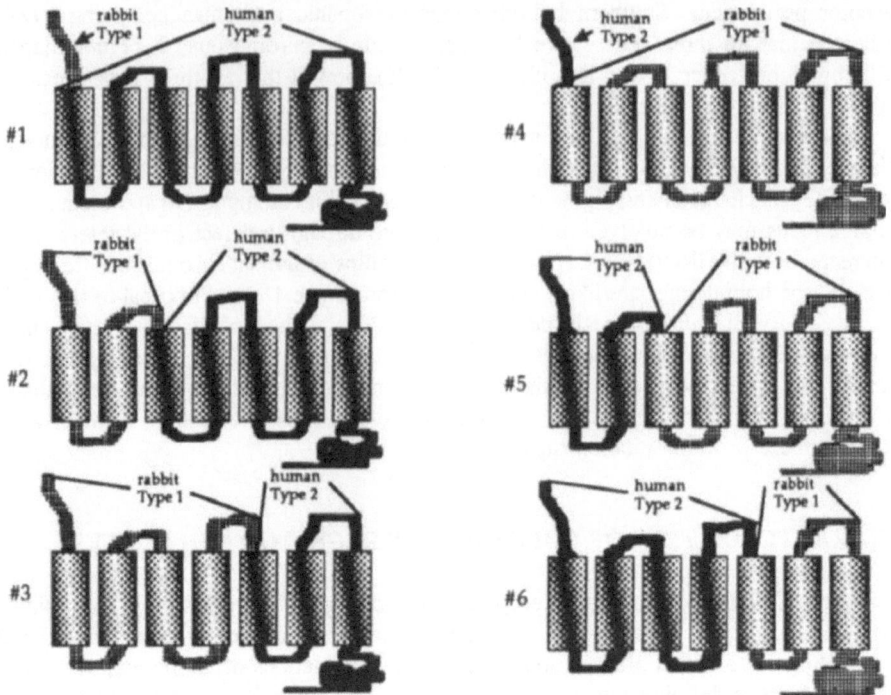

Figure 3. Schematic representation of chimeric rabbit-human IL-8 receptor molecules. Portions of human and rabbit IL-8 receptors present in chimeric receptors are shown. Transmembrane helices are depicted as rectangles. Cross-hatched lines represent rabbit type 1 sequences, while the black lines represent the human type 2 sequences.

In order to demonstrate which portions of the IL-8 receptor molecules are required for binding to GRO/MGSA, inhibition of ^{125}I-IL-8 binding with unlabeled IL-8 or GRO/MGSA was performed on the chimeric receptors. While the direct binding and inhibition by IL-8 resulted in similar K_I and K_D values (Table 1) the ability to bind GRO/MGSA differed markedly. Neither the rabbit nor human type 1 IL-8 receptor appears to bind GRO/MGSA (2, 5), while the human type 2 receptor binds GRO/MGSA effectively. The amino terminal section of the human type 2 IL-8 receptor is necessary for GRO/MGSA binding. Removal of this sequence, as evidenced by chimeras #1, 2, or 3, reduces or eliminates the binding to GRO/MGSA.

INHIBITION OF IL-8 BINDING BY PEPTIDES DERIVED FROM THE IL-8 RECEPTOR

Construction of chimeric IL-8 receptors using parts of the type 1 and type 2 IL-8 receptors has shown that the N-terminal portion of the receptor dictates its binding specificity. In addition, deletion of the N-terminal sequence from the receptor results in a loss of IL-8 binding. We therefore decided to synthesize peptides from the N-terminal regions of these receptors and to examine their ability to block IL-8 binding to HL60 cells (12). Shown in Table 3, section A, are the sequences of the N-terminal peptides for the human type 1 (il8r-5), human type 2 (il8r-13) and rabbit (il8r-1) IL-8 receptors, together with the inhibition constants (K_I) obtained for IL-8 binding. The rabbit and human type 1 N-terminal peptides show inhibition which appears to saturate (Figure 4) with K_I's of 1.7 x 10^{-5} M and 2.2 x 10^{-6} M respectively, whereas the human type 2 peptide shows a lower

Table 2. Summary of IL-8 and GRO/MGSA binding to chimeric IL-8 receptors.

IL-8 Receptor	KD IL-8 M	KI IL-8 M	GRO/MGSA
Rabbit IL-8R type 1	6.3×10^{-10}	1.4×10^{-9}	n.d.[a]
Human IL-8R type 2	2.9×10^{-9}	9.1×10^{-9}	1.5×10^{-8}
Chimera 1	7.0×10^{-10}	4.8×10^{-9}	4.3×10^{-7b}
Chimera 2	3.7×10^{-10}	2.2×10^{-10}	2.1×10^{-7b}
Chimera 3	8.3×10^{-9}	2.7×10^{-8}	n.d.
Chimera 4	n.d.	n.d.	n.d.
Chimera 5	n.d.	n.d.	n.d.
Chimera 6	6.7×10^{-10}	1.2×10^{-9}	3.4×10^{-9}
SRM	n.d.	n.d.	n.d.
Vector control	n.d.	n.d.	n.d.

L-8 receptor expression constructs were transfected into 293 cells and assessed for IL-8 or GRO/MGSA binding as described (12). For equilibrium binding studies, data was analyzed according to the method of Scatchard (1949). For inhibition of binding studies, cells were incubated with 8 x 10-10M 125I-IL-8 and various concentrations of unlabeled IL-8 or GRO/MGSA. Affinity constants (KD) and inhibition constants (K_I) were calculated on RS/1 run on a Microvax II under the VMS operating system.

[a] Not detected.
[b] Based on extrapolation of binding curve.

K_I (1.5×10^{-4} M by extrapolation) and the inhibition does not reach saturation. As a control sequence, the human type 1 IL-8 receptor peptide was scrambled, retaining the same amino acid composition, but arranged in a random sequence. Little or no inhibition was seen with this peptide (il8r-7) suggesting that the observed inhibition of IL-8 binding by the other N-terminal peptides is not due to a non-specific effect such as charge. It should be noted that all the N-terminal peptides bear an overall negative charge (il8r-1 (-5); il8r-5 (-8); and il8r-13 (-8)) at physiological pH whereas IL-8 bears an overall positive charge in the same pH range. The acidic nature of the N-terminal peptides limits their solubility in aqueous solution to less than 1×10^{-3} M unless the pH is greater than 8.0, above which oxidation and other uncharacterized reactions occur.

In addition to the full N-terminal peptides, a series of truncated peptides was synthesized from the human type 1 IL-8 receptor N-terminal region (Table 3, section B). Eleven C-terminal residues were removed from the sequence of il8r-5 to generate il8r-12, which had a K_I fifty-fold lower than il8r-5. Clearly, these residues contribute to IL-8 binding and suggest that the cysteine residue present in that sequence still contributes to binding even if it is involved in a disulfide bridge, possibly to the six-seven loop. This suggests a possible role for the disulfide bridge in constraining the conformation of the extracellular amino terminus of the IL-8 receptor. Further truncations from the C-terminal end of il8r-12 to generate il8r-14 and il8r-16 diminished the K_I still further. Removal of the first four N-terminal residues from il8r-5 appeared to have little effect on the K_I as shown from comparison of il8r-12 and il8r-11, but further truncations from the N-terminal end greatly diminished the inhibition of IL-8 binding.

In addition, for the rabbit IL-8 receptor, linear peptides were synthesized from the extracellular loops of the receptor, and we examined them for their ability to block IL-8 binding. These peptides demonstrated no ability to block IL-8 binding (Table 3, section C). This may be because the interaction of IL-8 with the loop peptides is relatively low in affinity in comparison to the N-terminal peptide or because these peptides are not adequately constrained to mimic the extracellular loops.

In order to understand the nature of the interaction between IL-8 and its N-terminal peptide, structural studies on both the peptide and the peptide-ligand complex are being pursued. Unfortunately, the low solubility of il8r-1 prevents its structure from being determined by NMR methods, but preliminary CD measurements suggest it exists as a partial helix. The apparent helical content of the three N-terminal peptides is greatest for il8r-1 followed by il8r-5 (where helix formation can be induced by the addition of trifluoroethanol to the solution) followed by il8r-13, which has no helical content. It is tempting to speculate that the helical segment of the N-terminal peptides is responsible for IL-8 binding as there is a direct correlation between the K_I observed for IL-8 binding and the amount of helical content in the peptide. Work is currently in progress to design peptides which exhibit higher K_I values for IL-8 binding, or which have increased solubilities and can be used for structural studies.

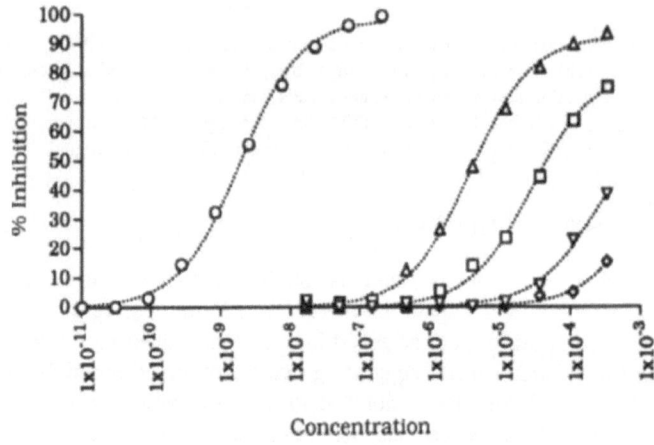

Figure 4. Inhibition of IL-8 binding by synthetic IL-8 receptor peptides. Inhibition of radiolabeled IL-8 binding to HL-60 cells was determined following incubation with IL-8 receptor peptides. The symbols represent peptides described in Table 3 including O, IL-8; ▲, il8r-1; □, il8r-5; ▼, il8r-13; ◊, il8r-7.

COMPARISONS OF IL-8 RECEPTOR SEQUENCES

The IL-8 receptor belongs to a superfamily of receptors known to signal by binding to G proteins. Members of this superfamily of receptors are characterized by the presence of seven hydrophobic segments separated by stretches of hydrophillic residues in extracellular and intracellular loops. These receptor molecules span the membrane seven times with the N-terminal portion extracellular, and the C-terminus cytoplasmic. The lengths of the transmembrane helices are conserved in the entire family as is a disulfide bond between the extracellular loops of helices two-three and four-five. The IL-8R has two additional cysteines in the extracellular segments, one in the N-terminal region and the other in the extracellular loop between the sixth and seventh transmembrane helices. These two cysteines are not conserved throughout the superfamily, but are found in the rat serotonin receptor. IL-8R has five cysteines present in the transmembrane region. An

Table 3. IL-8 receptor N-terminal peptide inhibition of IL-8 binding to HL60 cells.

Peptide sequence	K_I
A. amino terminal peptides	
rabbit Type 1	
il8r-1 H-MEVNVWNMTDLWTWFEDEFANATGMPPVEKDYSPCLVVTQTLNK-OH	2.2×10^{-6} M
human Type 1	
il8r-5 H-MSNITDPQMWDFDDLNFTGMPPADEDYSPCMLETETLNK-OH	1.7×10^{-5} M
human Type 2	
il8r-13 H-MESDSFEDFWKGEDLSNYSYSSTLPPFLIDAAPCEPE-OH	1.5×10^{-4} M[a]
scrambled human Type 1	
il8r-7 H-MPADYSSWMDENKNTTMGLNFDEPCLMTELDDPFPTIDNQ-OH	1.2×10^{-3} M[a]
B. truncated human Type 1 IL 8 receptor peptides	
il8r-11 H-MSNITDPQMWDFDDLNFTGMPPADEDYS-OH	1.1×10^{-4} M[a]
il8r-12 H-TDPQMWDFDDLNFTGMPPADEDYS-OH	1.5×10^{-4} M[a]
il8r-14 H-MSNITDPQMWDFDDLNFTGMPPA-OH	1.0×10^{-3} M[a]
il8r-16 H-MSNITDPQMWDFDD-OH	n. d.
il8r-17 H-TDPQMWDFDDLNFTGMPPA-OH	n. d.
il8r-10 H-DFDDLNFTGMPPADEDYS-OH	n. d.
il8r-9 H-DDLNFTGMPPADEDYS-OH	n. d.
C. extracellular peptides from rabbit IL-8 receptor	
il8r-2 Ac-KEKGWIFGTPLCKVVSLVKEVN-OH	n. d.
il8r-3 Ac-NNSSPVCYEDLGHNTAKWRMVLR-OH	n. d.
il8r-4 Ac-DTLMRTHVIQETCQRRNDIDR-OH	n. d.

Various concentrations of IL-8 receptor peptides were incubated with HL-60 cells (5×10^6 cells per point) in the presence of ^{125}I-IL-8 (4×10^{-10} M). Free and cell-bound IL-8 were separated and binding determined (12).

[a]   Based on extrapolation of binding curve
n. d. = no inhibition detected

```
HUDOPR: ----------------MDPLNLSWYDDDLERQNWSRPFNGSDGKAD------------RPHYNYYATLLTLLIAVIVFGNVLVCMAVSREKALQTT--
RTSERO: ---MARGSTIMTSPEHLNTSEASNWTIDAENRTNLSCEGYLPPTCLSILHLQEKN----------WSALLTVVIITIAGNILVIMAVSLEKKLQNAT--
RTOLFR: ----------------MTEENQTVISQFLLLFLPLPSEH----------QHVFYALFLSMYLITVLGNLIIIIILIHLDSHLHTPMY--
RTSBPR: ---------------MDNVLPMDSDLFPNISTNTSESNQFVQP----------TWQIVLWAAAYTVIVTSVVGNVVVIWIILAHKRMRTVTN--
HFPR**: ----------------METNSSLPTNISGGTPAVSAGY----------LFLDIITYLVFAVTFVLGVLGNGLVIWVAGFRMTHTVTT--
HUC5AR: ---------MNSFNYTTPDYGHYDDKDTLDLNTFVDKTSNTL----------RVPDILALVIFAVVFLVGVLGNALVVWVTAFEAKRT--
RBIL8R: ---1--MEVNVWNMTDLWTWFEDEFANATGMPFVEKDYSPCLVVT-3940----------QTLNRKYVVVIYALVFLLSLLGNSLVMLVILYSRSNRS---77--
HUBADR: ---1--MGQPGNGSAFLLAPNRSHAPDHDVTQ--2728----------RDEVWVVGMGIVMSLIVLAIVFGNVLVITAIAKFERLQT--66--
SQRHDO: ---------MGRDIPDNETWWYNPIYADIHPHWKQFDQ----------VPAAVYYSLGIFIAICGIIGCVGNGVVIYLFTKTSLQTP--
BORHDO: ---------MNGTEGPNFYVPFSNKTGVVRSPFEAPQYYL----------AEPWQFSMLAAYMFLLIMLGFPINFLTLYVTVQHKKLRTPL--
BRCORE: --------------------C----C----C----
PDB1BR: --------------MLELLPTAVEGVSQAQITGR----------PEWIWLALGTALMGLGTLYFLVKGMGVSDP--

HUDOPR: ----------NYLIVSLAVADLLVATIVMPWVVYLEVVGEWKFSRIH----------CDIFVTLDVMMCTASILNLCAISIDRYTAVAMPMLYNTR--
RTSERO: ----------NYFLMSLAIADMLLGFLVMPVSMLTILYGYRWPLPSKL----------CAIWIYLDVLFSTASIMHLCAISIDRYVAIQNPIHH--
RTOLFR: ----------LFLSNLSFSDLCFSSVTMPKLLQNMQSQVPSIPFAG----------CLTQLYFYLYFADLESFLLVAMAYDRYVAICFPLHYM--
RTSBPR: ----------YFLVNLAFAEACMAAFNTVVNFTYAVHNVWYYGLFY----------CKFHNFFPIAALFASIYSMTAVAFDRYMAIIHPLQP--
HFPR**: ----------ISYLNLAVADFCFTSTIPFFMVRKAMGGHWPFGWFL----------CKFLFTIVDINLFGSVFLIALIALDRCVCVLHPVWTQNH--
HUC5AR: ----------INAIWFLNLAVADFLSCLALPILFTSIVQHHHWPFGG----------AACSILPSLILLNMYASILLATISADRFLLVFKPIWCQNF--
RBIL8R: --78--VTDVYLLNNLAMADLFALTMPIWAVSKEKGWIFGT----112-------113------PLCKVVSLVKEVNFYSGILLLACISVDRYLAIVHATRTLTQ--153--
HUBADR: --67--VTNYFITSLACADLIVMGLAVVPFGAAHILMKMWTFGN----103-------104------FWCEFWTSIDVLCVTASIETLCVIAVDRYFAITSPFKYQ--142--
SQRHDO: ----------ANMFIINLAFSDFTFSLVNGFPLMTISCFMKYNVFGNAA----------CKVYGLICGIFGLMSIMTMTMSIDRYNVIGRPMSASKKMS--
BORHDO: ----------NYILLNLAVADLFMVFGGFTTTLYTSLHGYFVFGPTG----------CNIEGFFATLGGEIALWSLVVLAIERYVVVCKPMSNFRFGE--
BRCORE: ------C----C----CC----C--------------
PDB1BR: ----------DAKKFYAITTLVPAIAFTMYLSMLLG--YGLTMVPFGG----------EQNPIYWARYADWLFTTPLLLLDLALLV----DADQ--

HUDOPR: ----------YSSKKRRVTVMISIVWVLSFTISCPLLFGLNNADQNECIIANP----------AFVVSSIVSFYVPFIVTLLVVIKIYIVLRRRR--+142--
RTSERO: ----------SRFNSRTKAFLKIIAVWTISVGISMPIPVFGLQDDSKVFKEGSCLLA----------DDNFVLIGSFVAFFIPLTIMVITYFLTIKSLQKEATLCVSDL+50--
RTOLFR: ----------SIMSPRLCVSLVVLSWVLTTFHAMLTLLMARLSFCADNMIHFFCDISPLLKLSCSDTHVNEL--VIFVMGGLVIVIPFVLIIVSARVASILKVPSV--
RTSBPR: ----------RLSATATRKVFIFVIWVLALLLAFPQGYYSTETMPSRVVCMIEWPEHPNRTYEKAY----------HICVTVLIYFLPLLVIGYAYTVVGITLWASEIP---
HFPR**: ----------RTVS-LAKRVIIGPWVMALLLTLPVIIRVTTVPGKTGTVACTFNFSPWTNDPKERINVAVAMLTVRGIIRFIIGFSAPMSIVAVSYGLIATKIHKQGLIKSS--+16--
HUC5AR: ----------RGAGLAWIACAVAWGLALLTIPSFLYRVVREEYFPKVLCGVDYSHDKRR----------ERAVAIVRLVLGFLWPLLTLTICTFILLRTWSRRATRSTK--
RBIL8R: -154--KRHLVKFICLGIWALSLIILSLPFFLFRQVFSPNNSSPVCYEDIGHNTAKW-203-204-RMVLRILPHTFGFILPLLVMLFCYGFTLRTLFQAHMGQKH--243--
HUBADR: -154--SLLTKNKARVIILMVWIVSGLTSFLPIQMHWYRATHQEAINCYANETCCDFFT--195-196-NQAYAIASSIVSFYVPLVIMVFVYSRVFQEAKRQLQKIDKSEGRF-240+26
SQRHDO: ----------HRKAFIMIIFVWIWSTIWAIGPIFGWGAYTLEGVLCNCSFDYITRDTTTRSN----------ILCMYIFAFMCPIVVIFFCYFNIVMSVSNHEKEMAAMAKRLNAK-+16--
BORHDO: ----------NHAIMGVAFTWVMALACAAEPLVGWSRYIPEGMQCSCGIDYYTPHEETNN----------ESFVIYMFVVHFIIPLIVIFFCYGQLVFTVKEAAAQQQESATTQKAEKEV--
BRCORE: ------C----C----C----C----C----
PDB1BR: ----------GTIALVGADGIMIGTGLVGALT----------KVYSYRFV----------WWAISTAAMLYILYVLFFGFTS------KEASMRP--
```

```
HUDOPR:  -------KRATQMLAIVLGVFIICWLPFFITHILNIHCD------------------------CNIPPVLYSAFTWLGYVNSAVNPIIYTTFNIEFRK------+7-----------
RTSERO:  -------KVLGIVFFLFVVMWCPFFITNIMAVICKESCNE----------------------NVIGALLNVFVWIGYLSSAVNPLVTLFNKTYRS-------+83------
RTOLFR:  -------RGIHKIFSTCGSHLSVVSLFYGTIIGLYLCPSANNSTV-----------------KETVMAMYTVVTPMLNPFIYSLRNRDMKE----------+15------
RTSBPR:  -------VVKMMIVVVCTFAICWLPFHVFFLLPYINPDLYL--------------------KKFIQQVYLASMWNIAMSSTMVNPIIYCCLNDRFR-------+40------
HFPR**:  -------RPLRVLSFVAAAFFLCWSPYQVVALIATVRIRELL--------------------QGMYKEIGIAVDVTSALAFFNSCLNPMLYVFMGQDFRE------+40------
HUC5AR:  -------TLKVVAVVASFFIFWLPYQVTGIMMSFLEPSS----------------------PTFLLINKLDSLCVSFAYINCCINPIIVVAGQGFQG------+43-------
RBIL8R:  244----RAMRVIFAVVLIFLLCWLPYNIVLLADTLMRTHVIQETC--282-283-----QRRNDIDRALDATEILGFLHSCLNPIIYAFIGQNFR--318+35------
HUBADR:  267-KEHKALKTLGIIMGTFTLCWLPFFIVNIVHVIQDN-----301------302-----LIRKEVYILLNWIGYVNSGFNPLIYCRSPDFRIA--335+78-------
SQRHDO:  -------KISIVIVTQFLLSWSPYAVVALLAQFGPIE---------------------------WTPYAAQLPVMFAKASAIHNPMIYSVSHPKFRE-----+128-------
BORHDO:  -------TRMVIIMVIAFLICWLPYAGVAFYIFTHQGS------------------------DFGPIFMTIPAFFAKTSAVYNPVIYIMMNKQFRN-------+31------
BRCORE:  -------C---C--CC--C--CC--C------------------------------------C--C--CC--CC--C-------------------------
PDB1BR:  -------EVASTFKVLRNVTVVLWSAYPVVWLIG---------------------------SEGAGIVPL--NIETLLFMVLDVSAKVGFGLIILRS--------+23------
```

Figure 5. Sequence alignment for selected G protein-coupled seven-helix bundle receptor proteins. HUDOPR, human dopamine receptor (31); RTSERO, rat serotonin receptor (29); RTOLFR, rat olfactory receptor (4); RTSBPR, rat substance P receptor (37); HFPR**, human formyl peptide receptor (3); HUC5AR, human C5a receptor (13); RBIL8R, rabbit interleukin-8 receptor (2); HUBARD, human beta adrenergic receptor (18); SQRHDO, squid rhodopsin (15); BORHDO, bovine rhodosin (27); BRCORE, core residues in the three-dimensional structure of bacteriorhodopsin; PDB1BR, bacteriorhodopsin (10).(Note: C-terminal sequences are arbitrarily deleted after the seventh transmembrane helix and long loops are indicated by number of residues in the corresponding regions inserted in the sequences.) Helical regions in the three-dimensional structure of PDB1BR are shown with dotted lines.

alignment of the amino acid sequences of the members of this superfamily is shown in Figure 5 and reveals pairs of residues conserved in every transmembrane helix, suggesting similar topological arrangements of the seven helices inside the membrane for the superfamily.

From electron crystallographic techniques, the three-dimensional arrangement of the seven transmembrane helices are known for bacteriorhodopsin, a member of this superfamily. A sequence alignment between various species of rhodopsins reveals considerable divergence in the amino acid sequence across species. It is interesting to note that bovine rhodopsin shows more homology to IL-8R than to bacteriorhodopsin. This suggests that in spite of the lack of sequence homology between IL-8R and bacteriorhodopsin, they probably share a high degree of structural homology.

Based upon the three-dimensional structure of bacteriorhodopsin, we present a three-dimensional model of the IL-8R. Using the model building package FOLDER (34), a model for the IL-8R was constructed by constraining the extracellular loops by the two expected disulfide crosslinks as shown in Figure 6. The residue-by-residue alignment of the seven transmembrane helices between IL-8R and bacteriorhodopsin are not unambiguous at this stage of modelling. However, the lengths of the helices in IL-8R compared to those of bacteriorhodopsin do not allow for more than a two or three amino acid shift in the sequence alignment of the seven helices of these proteins. The model will be refined further once the disulfide topologies of the transmembrane cysteines are determined experimentally. Due to lack of experimental evidence for the disulfide topology in the transmembrane region, we have made no effort to crosslink nearby cysteines and have used only the lengths of the helices and the hydrophobic moments as constraints to build the model shown in Figure 6. Using this model, the potential disulfide links in the transmembrane region should be able to be determined.

By using synthetic peptides and chimeric receptors, we have shown that the N-terminal portion of the IL-8R is involved in the specific binding of IL-8. Numerous acidic amino acid residues are present on the N-terminal IL-8R and it could be speculated that a specific charge-charge interaction between this part of the receptor and the highly charged, basic IL-8 molecule is responsible for the ligand-receptor recognition. Based on the model it could also be hypothesized that the epitope for IL-8 on IL-8R could also consist of the C-terminal portion of the loop between transmembrane helices six and seven. There are a number of acidic amino acid residues in the six-seven loop which could form a continuous epitope with the N-terminal end of the IL-8R if the helices were linked into a ring structure as shown in the model. However, the N-terminal peptide data rules out a non-specific charge-charge interaction.

The relative sizes of IL-8 and the IL-8R shown in the model would prevent IL-8 from sinking into the transmembrane region of the receptor, (like the small organic molecules which bind to other receptors of this superfamily) without major disruption in the arrangement of the transmembrane helices of the receptor. However, by analogy to other members of this superfamily, it is hypothesized that a portion of IL-8 might interact with one or two charged amino acids in the transmembrane region of the receptor near the cell surface to trigger the biological response. In that case the role of the specific binding between the N-terminal segment of IL-8 and IL-8R might be required to retain the active orientations between the molecules. We have preliminary evidence that IL-8 and GRO/MGSA may exist as monomers under acidic or high salt conditions (Beckmann M.P., and Vanden Bos T., unpublished observations). Additionally, the IL-8 receptor binding properties of these ligands change based on the relative amounts of monomer or dimer in solution. Monomeric IL-8 and GRO/MGSA have higher affinity for the receptor. Recently Mantel et al. (1993) have observed that aggregation of another chemokine, MIP-1α modulates the myelopsuppresive effects of this factor. They conclude that the active form of MIP-1α is in the monomeric state.

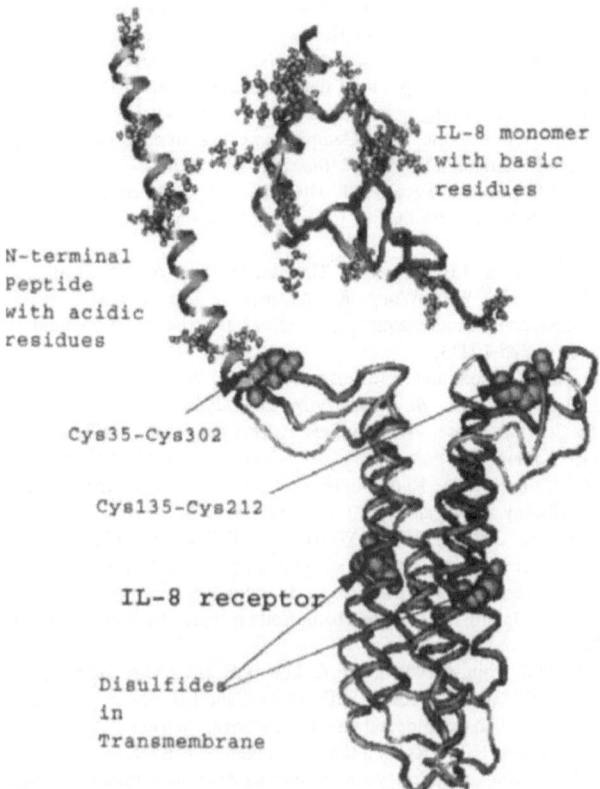

IL-8 monomer
with basic
residues

N-terminal
Peptide
with acidic
residues

Cys35-Cys302

Cys135-Cys212

IL-8 receptor

Disulfide
in
Transmembrane

Figure 6. Model of interaction of interleukin-8 with the IL-8 receptor. The model was built using FOLDER (34). The sequence alignment used in the model is shown in Figure 5. The seven hydrophobic segments in rabbit interleukin-8 receptor are aligned with that of bacteriorhodopsin such that four out of five cysteines in the transmembrane helices are in close proximity to form two disulfide bonds. Cys35-Cys302 and Cys135-Cys212 are two disulfides constraining the extracellular loops. Cys35-Cys302 is conserved in the entire family. The disulfide between Cys135-Cys212 is proposed in the model of interleukin-8 receptor and cysteines in similar loops are also found in rat serotonin receptor. Receptor model includes all the loops except the C-terminal residues.

Further research and experimentation is in progress in our laboratory to verify the proposed hypotheses. The studies should extend our understanding of the binding and structural correlations in an effort to identify therapeutically useful antagonists of IL-8 and other members of this chemotactic peptide family.

REFERENCES

1.   Abel L., and F. Demenais. 1988. Detection of major genes for susceptibility to leprosy and its subtypes in a Caribbean Island: Desirade Island. *Am. J. Hum. Genet.* 42: 256-266.
2.   Beckmann M.P., W.E. Munger, C. Kozlosky, T. Bos Vanden, V. Price, S. Lyman, N.P. Gerard, C. Gerard, and D.P. Cerretti. 1991. Molecular characterization of the rabbit IL-8 receptor. *Biochem. Biophys. Res. Comm.* 179: 784-789.
3.   Boulay F., M. Tardif, L. Brouchon, and P. Vignais. 1990. Synthesis and use of a novel N-formyl peptide derivative to isolate a human N-formyl peptide receptor cDNA. *Biochem. Biophys. Res. Commun.* 168: 1103-1109.

4.   Buck L. and R. Axel. 1991. A novel multigene family may encode odorant receptors: a molecular basis for odor recognition. *Cell* 65: 175-187.

5.   Cerretti D.P., C.J. Kozlosky, T. Bos Vanden, N. Nelson, D.P. Gearing, and M.P. Beckmann. 1993. Molecular characterization of receptors for human interleukin-8, GRO/melanoma growth stimulatory activity and neutrophil activating peptide-2. *Mol. Immunol.* 30: 359-367.

6.   Clore G.M., and A. Gronenborn. 1991. Comparison of the solution nuclear magnetic resonance and crystal structures of interleukin-8. *J. Mol. Bio.* 217: 611-620.

7.   Davisson M.T., P.A. Lalley, J. Peters, D.P. Doolittle, A.L. Hillyard, and A.J. Searle. 1991. Report of the comaprative committee for human, mouse and other rodents. *Cytogenet. Cell Genet.* 58: 1152-1189.

8.   Derynck R., E. Balentien, J.H. Han, H.G. Thomas, D. Wen, A.K. Samantha, C.O. Zacharaie, P.R. Griffing, R. Brachmann, W.L. Wong, K. Matsushima, and A. Richmond. 1990. Recombinant expression, biochemical characterization, and biological activities of the human MGSA/gro protein. *Biochemistry* 29: 10225-10233.

9.   Dohlman H.G., M.G. Caron, and R.J. Lefkowitz. 1987. A family of receptors coupled to guanine nucleotide regulatory proteins. *Biochemistry* 26: 2657-2664.

10.   Dunn R., J. McCoy, M. Simsek, A. Majumdar, S.H. Chang, U.L. RajBhandary, and H.G. Khorana. 1981. The bacteriorhodopsin gene. *Proc. Natl. Acad. Sci.* 78: 6744-6748.

11.   Frezal J. and A. Schinzel. 1991. Report of the committee of clinical disorders, chromosme aberrations and uniparental disomy. *Cytogent. Cell Genet.* 58: 986-1052.

12.   Gayle R., P.R. Sleath, C.W. Birks, K.S. Weerawarna, D.P. Cerretti, C.J. Kozlosky, N. Nelson, T. Bos Vanden, and M.P. Beckmann. 1993. The importance of the amono terminus of the interleukin-8 receptor in ligand interactions. *J. Biol. Chem.* In press.

13.   Gerard N.P., and C. Gerard. 1991. The chemotactic receptor for human C5a anaphylotoxin. *Nature* 349: 614-617.

14.   Gimbrone M.A., M.S. Obin, A.F. Brock, E.A. Luis, P.E. Hass, C.A. Herbert, Y.K. Yip, D.W. Leung, D.G. Lowe, W.J. Kohr, W.C. Darbonne, K.B. Bechtol, and J.B. Baker. 1989. Endothelial interleukin-8: a novel inhibitor of leukocyte endothelial enteractions. *Science* 246: 1601-1603.

15.   Hall M.D., M.A. Hoon, N.J.P. Ryba, J.D.D. Pottinger, J.N. Keen, H.R. Saibil, and J.B.C. Findlay. 1991. Molecular cloning and primary structure of squid (Loligo forbesi) rhodopsin, a phospholipase C-directed G-protein-linked receptor. *Biochem. J.* 274: 35-40.

16.   Holmes W.E., J. Lee, W.-J. Kuang, C.G. Rice, and W.I. Wood. 1991. Structure and functional expression of a human interleukin-8 receptor. *Science* 253: 1278-1280.

17.   Hopp T.P., K.S. Prickett, V.L. Price, R.T. Libby, C.J. March, D.P. Cerretti, D.L. Urdal, and P.J. Conlon. 1988. A short polypeptide marker sequence useful for recombinant protein identification and purification. *Bio/Technology* 6: 1204-1210.

18.   Kobilka B.K., T. Frielle, H.G. Dohlman, M.A. Bolanowski, R.A.F. Dixon, P. Keller, M.G. Caron, and R.J. Lefkowitz. 1987. Delineation of the intronless nature for genes for the human and hamster beta-2-adrenergic receptor and their putative promoter regions. *J. Biol. Chem.* 262: 7321-7327.

19.   Lee J., R. Horuck, G.C. Rice, and W.I. Wood. 1992. Characterization of complementary DNA clones encoding the rabbit IL-8 receptor. *J. Immunol.* 148: 1261-1264.

20.   Malo D., E. Schurr, D.J. Epstein, M. Vekemans, E. Skamene, and P. Gros. 1991. The host resistance locus Bcg is tightly linked to a group of cytoskeleton-associated protein genes that include villin and desmin. *Genomics* 10: 356-364.

21.   Mantel C., K.J. Young, S. Cooper, B. Kwon, and H.E. Broxmeyer. 1993. Polymerization of murine macrophage inflammatory peptide 1a inactivates its myelosuppressive effects in vitro: the active form is a monomer. *Proc. Natl. Acad. Sci.* 90: 2232-2236.

22.   Mitelman F., Y. Kankeo, and J. Trent. 1991. Report of the committee on chromosome changes in neoplasia. *Cytogenet. Cell Genet.* 58: 1053-1079.

23.   Mock B., M. Krall, J. Blackwell, A. O'Brien, E. Schurr, P. Gros, E. Skamene, and M. Potter. 1990. A genetic map of mouse chromosome 1 near the Lsh-Ity-Bcg disease resistance locus. *Genomics* 7: 57-64.

24.   Morris S.W., N. Nelson, M.B. Valentine, D.N. Shapiro, A.T. Look, C.J. Kozlosky, M.P. Beckmann, and D.P. Cerretti. 1992. Assignment of the genes encoding human interleukin 8 receptor types 1 and 2 and an interleukin 8 receprot pseudogene to chromosome 2q35. *Genomics* 14: 685-691.

25.   Moser B., C. Schumacher, V. von Tscharner, I. Clark-Lewis, and M. Baggiolini. 1991. Neutrophil-activitating peptide2 and gro/melanoma growth-stimulatory activity interact with neutrophil-activating peptide 1/ interleukin 8 receptors on human neutrophils. *J. Biol. Chem.* 266: 10666-10671.

26. Murphy P.M., and H.L. Tiffany. 1991. Cloning of complementary DNA encoding a functional interleukin-8 receptor. *Science* 253: 1280-1283.

27. Nathans J., and D.S. Hogness. 1983. Isolation, sequence analysis and intron-exon arrangement of the gene encoding bovine rhodopsin. *Cell* 34: 807-814.

28. Oppenheim J.J., O.C. Zacchariae, N. Mukaida, and K. Matsushima. 1991. Properties of the proinflammatory supergene "intercrine" cytokine family. *Annu Rev. Immunol.* 9: 617-648.

29. Pritchett D.B., A.W.J. Bach, M. Wozny, O. Taleb, R. Dal Toso, J.C. Shih, and P.H. Seeburg. 1988. Structure and functional expression of cloned rat serotonin 5HT-2 receptor. *EMBO J.* 7: 4135-4140.

30. Richmond A., E. Balentien, H.G. Thomas, G. Flaggs, D.E. Barton, J. Spiess, R. Bordoni, U. Francke, and R. Derynck. 1988. Molecular characterization and chromosmal mapping of melanoma growth stimulatory activity, a growth factor structurally related to b thromboglobulin. *EMBO J.* 7: 2025-2033.

31. Robakis N.K., M. Mohamadi, D.Y. Fu, K. Sambamurti, and L.M. Refolo. 1990. Human retina D2 receptor cDNAs have multiple polyadenylation sites and differ from a pituitary clone at the 5' non-coding region. *Nucl. Acids Res.* 18: 1299.

32. Sager R., A. Anisowicz, M.C. Pike, M.P. Beckmann, and T. Smith. 1993. Structural, regulatory and fuctional studies of the GRO gene and protein in *Cytokines:* Neutrophil-activating peptides and other chemotactic cytokines. Karger, Basel, in press.

33. Shields E.D., D.A. Russell, and M.A. Pericak-Vance. 1987. Genetic epidemiology of the susceptibility to leprosy. *J. Clin Invest.* 79: 1139-1143.

34. Srinivasan S., C.J. March, and S. Sudarsanam. 1993. An automated method for modeling proteins on known templates using distance geometry. *Protein Science* 2: 277-289.

35. Thomas K.M., H.Y. Pyun, and J. Navarro. 1990. Molecular cloning of the fMet-leu-phe receptor from neutrophils. *J. Biol. Chem.* 265: 20061-20064.

36. Thomas K.M., L. Taylor, and J. Navarro. 1992. The interleukin-8 receptor is encoded by a neutrophil-specific cDNA clone, F3R. *J. Biol. Chem* 266: 14839-14841.

37. Yokota Y., Y. Sasai, K. Tanaka, T. Fujiwara, K. Tsuchida, R. Shigemoto, A. Kakizuka, H. Ohkubo, and S. Nakanishi. 1989. Molecular characterization of a functional cDNA for rat substance P receptor. *J. Biol. Chem.* 264: 17649-17652.

38. Yoshimura T.K., K. Matsushima, J.J. Oppenheim, and E.J. Leonard. 1987. Neutrophil chemotactic factor produced by lipopolysaccharide (LPS)-stimulated human blood mononuclear leukocytes: partial characterization and separation from interleukin-1 (IL-1). *J. Immunol.* 139: 788-793.

ELUCIDATION OF STRUCTURE FUNCTION RELATIONSHIPS IN THE IL-8 FAMILY BY X-RAY CRYSTALLOGRAPHY

Manfred Auer, Susan R. Owens, Sabine Pfeffer[1],
Joerg Kallen[1], Erich Wasserbauer, Heinz Aschauer, Gerald Ehn,
Antal Rot, Jürgen Besemer, Charles Lam, and Ivan J. D. Lindley

Sandoz Research Institute
Brunnerstrasse 59
A-1235 Wien, Austria

[1]Preclinical Research
Sandoz Pharma Ltd.
CH 4002 Basel, Switzerland

INTRODUCTION

After its identification in 1987 as a novel neutrophil-activating cytokine, interleukin-8 (IL-8) was the center of interest of a large variety of interdisciplinary scientists for study of its action in vitro and in vivo (1,2,3,4). Soon it became clear that interleukin-8 plays a critical role in a series of diseases which are all characterized by strong neutrophil involvement. Rheumatoid artrithis, psoriasis, polytrauma, septicemia, adult respiratory distress syndrome (ARDS), asthma, emphysema, myocardial infarction, nephritis, inflammatory bowel disease, and gout are diseases which affect millions of people, and IL-8 is implicated in all these disease states. Therefore, inhibition of interleukin-8 itself, or blocking of the interaction with its seven transmembrane domain cellular receptor represents an attractive and logical approach for treatment of these inflammatory states. Possibilities for interference include neutralising antibodies, small molecules binding to the chemokine at critical positions, (e.g. domains involved in dimerization, heparin binding or receptor binding) or molecules which bind to the receptor. An additional approach would be the use of peptides based on the IL-8 structure or mutant IL-8 molecules to act as antagonists. As an alternative to screening for inhibitors in a series of biological assays, a rational way of finding an inhibitory principle involves the elucidation of structure/function relationships. This is usually done by production of mutant proteins of the target molecule and subsequent determination of biological activities. As only the primary structures of the mutants are known this might be called a "one-dimensional structure activity relationship". We are attempting to go one step further and actually determine the three-dimensional structure of mutant IL-8 derivatives with interesting and clear changes in biological activities and, based on structure/function relationships in three dimensions, get a more detailed insight into structural requirements for specific biological activities.

Besides molecular and cell biological work, IL-8 had also gained a lot of early attention in structural biology. The primary structure was determined in 1987 (5,6,7,8). Secondary structure information became available in 1989 from the first investigation by two-dimensional NMR spectroscopy (9), revealing a triple stranded antiparallel beta-sheet and a long C-terminal helix as the major structural elements of the monomer. One dimensional Nuclear Overhauser Enhancement (NOE) measurements showed that IL-8 is a dimer in solution. The full three dimensional structure of interleukin 8 in solution was reported in 1990 (10). With 2 symmetry-related antiparallel α-helices, (24 Å long and 14 Å apart) on top of a six stranded antiparallel β-sheet platform, the overall architecture is similar to the $\alpha 1/\alpha 2$ domains of the human class I histocompatibility antigen HLA-A2. At that time it was suggested that the two α-helices form the binding site for the cellular receptor and that their amphiphilic nature might be responsible for different specificities of IL-8 and related proteins. Crystallization of human recombinant IL-8 from saturated ammonium sulfate at pH 8.5 (11) and from three different lengths of polyethylene glycol at pH 6.5 (12) by vapour diffusion techniques was reported in the same year. Structural work on IL-8 also includes a new technical variation in crystallography. For the first time the crystal structure was determined by molecular replacement techniques using as a model the solution structure derived from NMR experiments (13). A comparison between solution and crystal structure and sequence alignments with murine macrophage inflammatory protein 2 (mu MIP-2) and human melanoma growth-stimulating activity (MGSA, GRO-α) revealed that the regions at the amino terminus (Glu-4 through Cys-9) and the β-bend at His-33 (Gly-31 through Glu-38) projecting into solution may play a role in receptor interaction. A more detailed investigation of NMR and crystal structure suggested that differences in the distance between the two helices (open and closed form, corresponding to 14 and 12 Å distance respectively) might be of functional significance (14). It was suggested that blocking of this potential conformational change might provide a possibility for rational design of inhibitors of the binding of IL-8 to its receptor(s).

The conclusions drawn from comparison of conserved amino acids within the CXC chemokine family and the results of the X-ray crystallographic analysis were confirmed recently by biochemical experiments, using IL-8 mutants (15, 16) and chemically synthesized analogs (17). In addition to C-terminal truncations (reduced, but did not totally abolish activity in neutrophil elastase release, chemotaxis and receptor binding assays), and N-terminal mutants, (Glu-4, Leu-5 and Arg-6 are absolutely necessary for activity) we are also interested in three-dimensional structure/activity relationship of mutants in the so called "adhesion region" (S^{44} DGR) and of chimeric proteins containing C- or N-terminal parts of other members of the chemokine family.

We report here the production, purification, biological characterization, crystallization and preliminary crystallographic analysis of 4 mutants out of 3 classes of IL-8 derivatives described above.

1. IL-8 lacking the last 6 amino acids on the C-terminus, [IL-8/1-66], (Fig. 1a)
2. IL-8 with amino acid 54-72 (KENWVQRVVEKFLKRAENS) replaced by the corresponding amino acids of MGSA (Gro α), (ASPIVKKIIEKMLNSDKSN), [IL-8/Chi1], (Fig. 1b).
3. IL-8 with amino acid 45-47 (DGR), ("adhesion sequence") replaced by (AGH), [IL-8/Adh1], (Fig. 1c).
4. IL-8 with amino acid 44-47 (SDGR) replaced by the rotated sequence (RGDS), IL-8/Adh.2], (Fig. 1d)

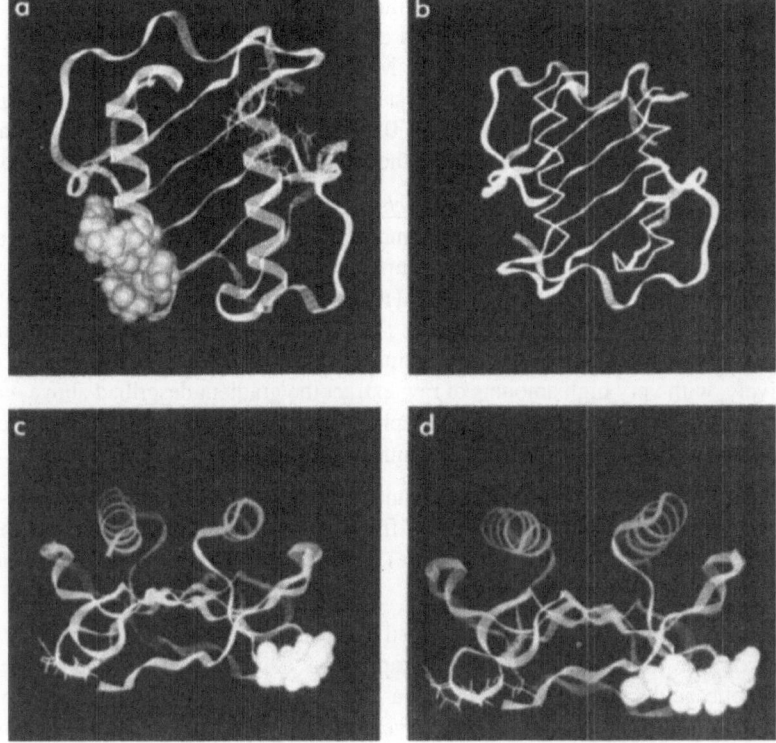

Figure 1. Three dimensional model of mutation sites using a space filling representation for changed aminoacids within the IL-8 wild type structure shown in ribbon plot. a: IL-8/1-66, b: IL-8/Chi1, c: IL-8/Adh1, d: IL-8/Adh2.

DESIGN AND PRODUCTION OF MUTANTS

All mutants were produced by direct manipulation of the synthetic IL-8 gene in the expression plasmid construct p(NAF)6T3, which carries ampicillin resistance and has the IL-8 gene under the control of the E. coli Tryptophan promoter and an artificial transcription terminator downstream of the gene (8). Portions of the gene were removed by cutting with restriction nucleases and separation on an agarose gel. Subsequently, new synthetic double stranded DNA with the desired sequence and with "sticky ends" to complement the restriction sites was ligated into the plasmid, creating the altered gene ready for expression.

The new constructs were sequenced using the Sequenase enzyme (United States Biochemical, Cleveland, Ohio) to confirm the correct sequence before expression and purification of the mutant protein (8, 15).

PURIFICATION

A general purification scheme was followed for all mutants.

Cell extraction: Frozen E. Coli cells were stirred in 2ml/g of 50 mM NaCl, 20 mM Tris-HCl, pH 8.0 at room temperature for 20 minutes and the pellet resuspended in 4 ml/g of the same buffer. The sample was then ultrasonicated on ice for 5x20 seconds on/off cycles at 100Watt and the E. coli envelopes collected by centrifugation at 15000g for 15 minutes. After resuspension in 10 ml/g of 6 M guanidine hydrochloride, 50 mM morpholinoethanesulphonic acid MES, pH 6.5, the sample was stirred for 1 hour at 4^0C and then dialysed (MWCO 1000) against 5 litres of 0.5% acetic acid overnight. The extract was centrifuged at 20000 g for 20 minutes and the supernatant collected for purification.

Primary purification by ion-exchange chromatography: Depending on the amount of raw extract three different purification schemes were followed. FPLC-system: The crude supernatant diluted 1:5 with equilibration buffer 50 mM MES pH 6.5 and then 1:1 with distilled water, was loaded onto a Mono-S HR 16/10 Pharmacia FPLC column. Proteins were eluted with a linear gradient from 0 to 1M NaCl at a flow of 5 ml/minute. Fractions of 5 ml were collected and their absorption at 280 nm monitored. Gravity column: Alternatively, with very high amounts of raw extract the gradient described above was used on either a gravity column filled with S-Sepharose FF material (Pharmacia) or on a S-Sepharose 35/100 Fast Flow Pharmacia column mounted on a Bio-Pilot system.

Refolding: To ensure that IL-8 mutant proteins were present in its correct conformation, the pool of protein containing fractions from the ion-exchange columns was adjusted to 2 mM GSH (reduced glutathione), 0.02 mM GSSG (oxidised glutathione) pH 8.0 and stirred overnight at 4^0C.

Reversed Phase HPLC: The refolded pool of IL-8 mutant containing fractions obtained from the ion-exchange chromatography was adjusted to 10% acetonitrile, 0.1% trifluoro acetic acid (TFA) and applied to a Brownlee Aquapore BU 300 column. The column was equilibrated with a starting solvent of 10% acetonitrile, 0.1% TFA in water (HPLC grade) and a linear gradient programmed with a limit buffer of 90% acetonitrile, 0.1% TFA. Flow rate was 10 ml/minute and 13 ml fractions monitored at A280 were collected.

Final purification by ion-exchange chromatography: IL-8 mutant fractions from the reversed-phase separations were pooled and diluted 1:4 with 50 mM MES buffer, pH 6.5. The sample was applied to a Pharmacia Mono-S HR 5/5 column, equlibrated with the same buffer as above and eluted with a linear gradient from 0 - 1 M NaCl. Flow rate was 1 ml/minute and 1 ml fractions were collected.

Analytical methods: SDS-PAGE analysis: Fractions from each chromatographic run were checked for the presence and purity of IL-8 mutant protein by SDS-PAGE as described by Laemmli (18).5-20% polyacrylamide gels were used with a discontinuous Tris/glycine buffer using the Bio-Rad Protean II Slab Cell system. Samples (20-40 ul) were mixed with equal volumes of final sample buffer (62.5 mM Tris-HCl pH 6.8, 10% glycerol, 2% SDS, 5% -mercaptoethanol), boiled at 95^0C for 5 minutes and applied to the 4% poly acrylamide stacking gel. Silver staining of the gels was carried out according to the method of Merril et al. (19) or Ansorge et al. (20) followed by immediate counter-staining with Coomassie Blue (25% methanol, 7.5% acetic acid, 0.01% CB R-250) and destain with 25% methanol, 7.5% acetic acid overnight.

Analytical HPLC: The final protein pool was analyzed for purity and concentration by comparison with a 1 mg/ml standard of native IL-8 on a Beckman System Gold HPLC system. Native and mutant proteins had a retention time of approximately 7.5 minutes under a 20 minute linear gradient of 10% to 90% acetonitrile, 0.1% TFA in water, on a Vydac protein C4 column. All proteins used for crystallization experiments had > 99% purity.

Concentration of the final purified pool was achieved using Filtron Microsep centricons 3000 at a speed of $\leq$ 5000 rpm. Volumes larger than 10 ml were initially reduced using an Amicon concentration cell 8010, YM2 (diameter 25 mm).

BIOLOGICAL ACTIVITIES

Purified mutant proteins were tested for biological activity in three different assay systems: Receptor binding, stimulation of elastase release from fresh human neutrophils (PMN), and chemotactic effect of IL-8 mutants on freshly isolated human neutrophils. Cleavage at position 66 (IL-8/1-66) did not lead to a reduction in affinity to the receptor. However, only 50% enhancement of elastase release compared to control were measured, and chemotactic activity was reduced to approximately 10% relative to wild type protein. The chimeric protein IL-8/Chi1 had lost 46% in receptor binding activity, whereas IL-8/Adh1 showed 83% relative affinity in the receptor binding assay, and 87% activity in the elastase release experiment. The chemotactic effect of this adhesion region mutant was approximately 30% of the wild type activity.

CRYSTALLIZATION

IL-8 mutants were crystallized from concentrated solutions (5-25 mg/ml) using the hanging drop vapour diffusion method in multi well tissue culture plates (Linbro model FB-16-24-TC). Drops containing 2 to 25 µl of protein/buffer mixtures were equilibrated against a 1 ml chamber of precipitating solution. Dialysis plates constructed and provided for trials by Dr. Gebhard Schertler (MRC Cambridge) were also used successfully for crystallization of IL-8/1-66. 60 ul of protein solution was dialysed against 4.5 ml of precipitating solution.

IL-8/1-66

88 g of E. coli yielded a total of 15.58 mg of crystallization grade protein. It was concentrated to 3 batches containing 7.93 mg/ml (200 µl), 24.7 mg/ml (500 µl) and 16.1 mg/ml (240 µl).

Dialysis: 30 ul of a 24.7 mg/ml protein solution was mixed with an equal volume of 50 mM MES pH 6.5, 0.32 M NaCl, 5 mM NaN_3, 60 % $[NH4]_2SO_4$ (AS) and equlibrated against 44%, 46% and 48% AS in the dialysis chamber. The first two sets of conditions yielded single orthorhombic crystals within 8 days, the largest of which formed against 46% AS (1.4 mm x 0.32 mm) (fig.2a).

Vapour diffusion: 8 µl of the above solution was mixed with an equal volume of 50 mM MES pH 6.5, 0.32 M NaCl, 5 mM NaN3, 60% AS and equilibrated against the same buffer containing 31% to 39% AS. A shower of crystals (< 0.1 mm all dimensions) formed against 39% AS within 9 days which were used to seed the other drops. Growth of these crystals occured in the 34% - 38% AS drops, the largest to a size of 0.32 mm^2 against 34% AS within the next 45 days. (fig. 2b)

IL-8/Chi1

The chimeric protein between IL-8 and GRO-α was crystallized by vapour diffusion from a 10.1 mg/ml solution. 3 µl of this solution was mixed with an equal volume of 50 mM MES, pH 6.5, 5 mM NaN_3 , 0.32 M NaCl, 20% AS and equilibrated against the same buffer containing 23 - 27% AS. 6 µl of the protein solution was mixed with an equal volume of 50 mM MES, pH 6.5, 5 mM NaN_3 , 10% Polyethylene glycol (PEG) 8000 and equilibrated against 50 mM MES, pH 6.5, 5 mM NaN_3 , 0.32 M NaCl, 20% AS. All drops yielded crystals suitable for X-ray diffraction. The largest crystals (0.32 x 0.48 x 0.4 mm), from drops containing 27% ammonium sulphate, were obtained in 4 days (fig. 2c). The PEG 8000 drop yielded 3 single crystals (0.5 x 0.4 x 0.4 mm) and a cluster of 7 crystals, the largest of which measured (1.4 x 0.7 x 0.6 mm) after 8 days (fig. 2d).

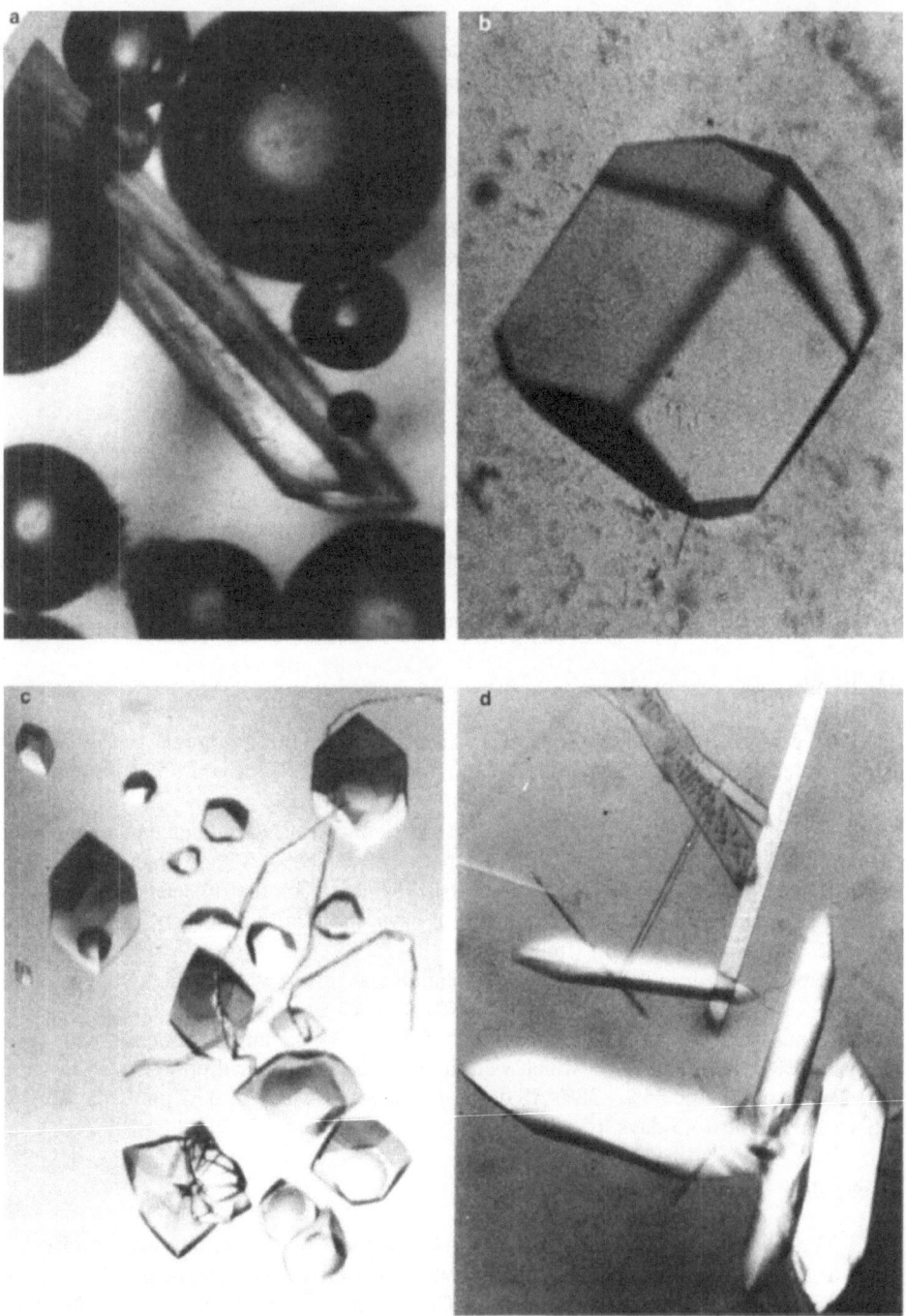

Figure 2. IL-8 mutant crystals. a: IL-8/1-66 crystals obtained through dialysis against 46% AS, pH 6.5. b: IL-8/1-66 crystal grown by vapour diffusion from 34% AS, pH 6.5. c: IL-8/Chi1 crystals obtained by vapour diffusion from 27% AS, pH 6.5. d: IL-8/Chi1 crystals obtained from 25% PEG 8000, pH 6.5.

Figure 2. e: IL-8/Adh1 needles grown from 30% AS, pH 6.5 by vapour diffusion. f: Long needles (up to 3 mm in length) of IL-8/Adh2 grown from 42% AS, pH 7.5 using the hanging drop vapour diffusion technique.

IL-8/Adh1

The first protein bearing a mutation in the SDGR region was crystallized by vapour diffusion from 7 μl drops containing 5 μl of 22.74 mg/ml protein solution in 50 mM MES, pH 6.5, 0.52 M NaCl, mixed with 2 μl of precipitating solution. Long thin needles (0.2 x 0.01 mm) were observed in drops containing 28% and 30% ammonium sulphate after a period of 14 days, which grew to a length of 0.3 mm after a further 27 days (fig. 2e).

IL-8/Adh2

The SDGR → RGDS mutant protein was crystallized by vapour diffusion from a 7.3 mg/ml solution of 50 mM HEPES, 0.53 M NaCl, pH 7.5 when equilibrated against the same buffer containing 5 mM NaN_3 and 40% or 42% ammonium sulphate. 2 μl of precipitating solution was mixed with 5 μl protein solution and needles (0.8 x 0.01 mm) formed at room temperature under the above conditions within 20 days (fig 2f).

PRELIMINARY X-RAY CRYSTALLOGRAPHIC STUDY

Crystals obtained from IL-8/1-66 (dialysis and vapour diffusion) and IL-8/Chi1 were suitable for structure determination. Native data sets were collected on a FAST area detector using CuK_α-radiation (40 kV, 80 mA) produced by a rotating anode X-ray generator (FR571). The program MADNES (21) was used for evaluation of measured X-ray intensities.

Data collection of the cubic IL-8/1-66 crystals obtained from dialysis was performed with synchrotron radiation and the image plate detector at the European Molecular Biology Laboratory (EMBL) Hamburg. The MOSFLM program package was used to process the data. The following table is a listing of the parameters, statistics and completeness for all data collections and shows the comparison to parameters obtained in our former study on wild type IL-8 (12), which is in agreement with data published by Baldwin et al. (11).

	IL-8(wt)	IL8/1-66 A	IL-8/1-66 B	IL-8/Chi1
Crystallization method	Vapour diffusion	Vapour diffusion	Dialysis	Vapour diffusion
Crystal form	Trigonal	Cubic	Orthorhombic	Tetragonal
Point group	321	432	222	422
Space group	P3(1)21 P3(2)21	P4(1)32	P2(1)2(1)2(1)	P4(1)2(1)2 P4(3)2(1)2
Cell dimensions (Å)	a = 40.86 b = 40.86 c = 90.26	a = 123.01 b = 123.01 c = 123.01	a = 36.65 b = 55.56 c = 62.29	a = 55.7 b = 55.7 c = 119.9
	$\alpha = \beta = 90^0$ $\gamma = 120^0$	$\alpha = \beta = \gamma = 90^0$	$\alpha = \beta = \gamma = 90^0$	$\alpha = \beta = \gamma = 90^0$
Number of unique refl.	3099	8128	3951	5579
Resolution (Å)	2.4	2.6	2.55	2.7
Completeness		92%	90.4	99.3
R [sym]	3.4%	8.5%	8.6%	7.3%

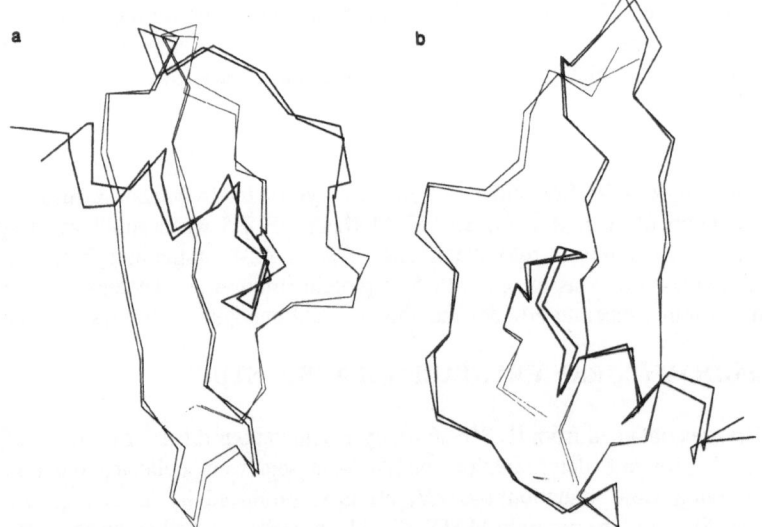

Figure 3. a: Overlay of a C-α plot of molecule A of the IL-8/1-66 structure with C-α plot of molecule "A" in IL-8 wild type protein. b: Same overlay for B and "B" molecule of IL-8/1-66 and IL-8 (wt), respectively.

STRUCTURE OF IL-8/1-66

The orthorhombic crystal form of IL-8/1-66 with complete data to 2.55 Å resolution was used for solving the three-dimensional structure. There are two molecules in the asymmetric unit and the packing density is rather tight with Vm=2.1 Å*3/Da. For the analysis of local symmetry elements a self rotation function was calculated in the resolution range of 8.0-3.0 Å. The received peak indicates a 2-fold axis relating the two molecules in the asymmetric unit. The crystal structure of IL-8/1-66 was solved by Patterson search techniques on the basis of the monomeric molecular model, on which the mutated residues were truncated. The correct orientation of the molecules was determined using the rotation function within the XPLOR program package and the SEARCH program of CCP4 program package for the translation function. Crystallographic refinement was carried out with XPLOR. Use was made of the rigid body refinement, molecular dynamics and conventional energy-restrained least squares refinement procedures on positional parameters and B-factors. Model building was performed on an Evans & Sutherland interactive display system with the program "O". With the correctly orientated and positioned molecule, a Fourier map with 2Fo-Fc model phases was calculated and the model checked and built into that. The current R factor is 17.5% for all atoms 1-66 and 89 water molecules, and data between 2.65-10 Å resolution. The chain fold of IL-8/1-66 molecule consists of three antiparallel beta-strands connected with loops and one long helix made of the carboxyl-terminal residues 56-66. (See figure 3a,b for a comparison between IL-8/1-66 molecule A and B.) Two molecules in the asymmetric unit form a dimer via a pseudo two fold axis. It is stabilized by hydrogen bonds between the first beta-strand (23-29) in each molecule and by additional side chain interactions. In that way the two molecules form a conformation with a six stranded beta-sheet and two antiparallel helices. The crystal structure of IL-8/1-66 and the wild type structure has been compared. When the dimers were superimposed, the rms deviation for 4-66 of C-alpha atoms was 0.975 Å. The main difference between the IL-8/1-66 and the wild type structure is the center-to-center distance between the helices. The distance between C-alpha residues 63-A and 63-B is 4.47 Å in IL-8/1-66 and 8.32 Å in wild type IL-8. The helices are much closer to each other (fig. 4). The two disulphide bridges, between Cys-7 and Cys-34 and between Cys-9 and Cys-50 are located in clear defined electron density and have not changed. His 33 has also not changed, and the NE2 atom forms a hydrogen bond to carbonyl oxygen of Glu-29.

DISCUSSION

The role of chemokines as diffusable mediators between cells implies that a whole series of different biological activities must be contained within a small protein of usually less than hundred amino acids. Although it is generally accepted, and supported by a long series of experiments, that IL-8 is a major inflammatory mediator *in vivo*, it is not fully proven that IL-8 is proinflammatory, and there are even experiments suggesting that IL-8 may act as an anti-inflammatory agent. At the moment it is not known which cells respond to IL-8 in haptotactic and which ones in a chemotactic fashion, and a series of high and low affinity receptors for IL-8 have been described on different cell types over the last years. Cross reactivities of chemokines on different receptors are not fully understood. It might also be the case that different N-terminally processed IL-8 (and generally chemokine) forms, (77, 72, 70, 69 aminoacids) are used for triggering of different molecular signals (22).

To answer this vast number of questions concerning biological activities of IL-8, more understanding on the molecular level of structures involved is urgently needed. Although in case of IL-8 a crystal and a NMR structure was solved, information on the structural changes within the N-terminus on receptor binding or changes within the α-helical region on binding to heparin or other extracellular matrix proteins is not available. Because of the molecular nature of the known receptors (seven helix membrane spanners) and other ligands like heparin (difficult to produce homogenious in length), structural work on complexes will probably be a long term project. It is, however, possible to obtain limited, but useful, understanding about structural features responsible for a certain biological effect by solving the three dimensional structure of a mutant protein with well characterized biological activities. This approach was taken in this work. As the chemokine family offers the possibility to characterize derivatives with mutations or truncations in

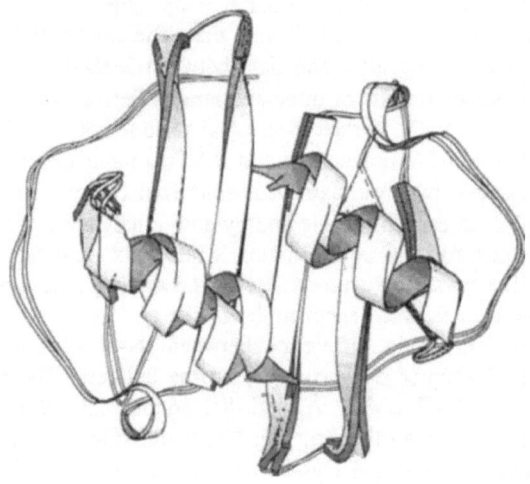

Figure 4. Schematic Richardson diagram of IL-8(wt) structure truncated by 6 aminoacids overlayed with IL-8/1-66 structure. IL-8/1-66 is the plot with about 50% reduced distance between the two α-helices.

regions essential for different activities not in one, but in a series of different assays, a fairly clear picture of the structure activity relationsship on the primary sequence level is obtained. With more and more structures (also of mutant proteins) solved, it becomes clear that even most single point mutations or simple truncations cause severe changes in the original wild type 3D-structure of a protein. These changes have to be taken into consideration when changes at certain position of a peptide sequence are related to biological activity. The first results on such a study on IL-8 are now available. Already at the stage of crystallization different crystal forms were obtained for all mutants characterized so far. Whereas the IL-8 wild type protein crystallized in a trigonal space group, IL-8/1-66 gave cubic and orthorhombic crystals depending on the conditions. The chimeric protein, IL-8/Chi1, revealed a tetragonal space group. The two mutants with a substitution in the adhesion region produced very long needles not seen in other IL-8 crystallization trials so far.

On a molecular level, the first three dimensional structure of an IL-8 mutant protein solved, IL-8/1-66, showed a surprisingly strong change in distance, (4 Å), between the remaining portion of the two α-helices (lacking 6 amino acids). This finding, in combination with the biological result of practically unchanged receptor binding but only 10% of chemotactic activity remaining, allows a better experimental understanding of interaction processes. The orientation and distance between the two carboxy-terminal helices is very critical for the binding interactions necessary for chemotaxis. The strong conformational change at the helical region however, would suggest that the N-terminus alone mediates receptor binding, with not even indirect effects coming from the C-terminus. Although it must be kept in mind that conformational changes on binding to the various effector molecules are also likely to occur in mutant proteins, those changes cannot exceed a certain limit. A careful interpretation of the structural results on IL-8 mutants together with exact biological characterization should therefore provide us with a better molecular understanding of thermodynamic and kinetic stability of the chemokine proteins and of their interaction with target molecules.

REFERENCES

1.   Baggiollini M., A. Walz, and S.L. Kunkel. 1989. Neutrophil-activating peptide-1/Interleukin 8, a novel cytokine that activates neutrophils. *J. Clin. Invest.* 84: 1045-1049.
2.   Oppenheim J.J., C.O.C. Zachariae, N. Mukaida, and K. Matsushima. 1991. Properties of the novel proinflammatory supergene "intercrine" cytokine family. *Annu. Rev. Immunol.* 9: 617-648.
3.   Matsushima, K. and J.J. Oppenheim. 1989. Interleukin 8 and MCAF: Novel inflammatory cytokines inducible by IL1 and TNF. *Cytokine,* Vol.1, 1:2-13.
4.   Van Damme, J. Interleukin-8 and related molecules. 1991. *The Cytokine Handbook,* 201-214.
5.   Yoshimura, T., K. Matsushima, S. Tanaka, E.A. Robinson, E. Appella, J.J. Oppenheim, and E. Leonard. 1987. Purification of a human monocyte-derived neutrophil chemotactic factor that has peptide sequence similarity to other host defence cytokines. *Proc. Natl. Acad. Sci. U.S.A.* 84:9233-9237.
6.   Walz, A., P. Peveri, H. Aschauer, and M. Baggiolini. 1987. Purification and amino acid sequencing of NAF, a novel neutrophil-activating factor produced by monocytes. *Biochem. Biophys. Res. Commun.* 149/2: 755-761.
7.   Gregory, H., J. Young J.M. Schroeder, U. Mrowietz, and E. Christophers. 1988. Structure determination of a human lymphocyte derived neutrophil activating peptide. *Biochem. Biophys. Res. Commun.* 151(2): 883-890.
8.   Lindley I., H. Aschauer, J.M. Seifert, C. Lam, W. Brunowsky, E. Kownatzki, M. Thelen, P. Peveri, B. Dewald, V. von Tscharner, A. Walz, and M. Baggiolino. 1988. Synthesis and expression in Escherichia coli of the gene encoding monocyte-derived neutrophil-activating factor: biological equivalence between natural and recombinant neutrophil-activating factor. *Proc. Natl. Acad. Sci. U.S.A.* 85(23): 9199-9203.
9.   Clore, G.M., E. Apella, M. Yamada, K. Matsushima, and A.M. Gronenborn. 1989. Determination of the secondary structure of interleukin-8 by nuclear magnetic resonance spectroscopy. *J. Biol. Chem.* 264: 18097-18911.
10.   Clore, G.M., E. Appella, M. Yamada, K. Matsushima, and A.M. Gronenborn. 1990. Three dimensional structure of interleukin 8 in solution. *Biochemistry* 29:1689-1696.
11.   Baldwin, E.T., K.A. Franklin, E. Appella, M. Yamada, K. Matsushima, A. Wlodawer, and I.T. Weber. 1990. Crystallization of Human Interleukin-8. *J. Biol. Chem.* 265: 6851-6853.
12.   Auer, M., J. Kallen, S. Schleischitz, M.D. Walkinshaw, E. Wasserbauer, G. Ehn, and I.J.D. Lindley. 1990. Crystallization and preliminary X-ray crystallographic study of interleukin-8. *FEBS Lett.* 265(1,2): 30-32.
13.   Baldwin, E.T., I.T. Weber, R. St. Charles, J.-C. Xuan, E. Appella, M. Yamada, K. Matsushima, B.F.P. Edwards, G.M. Clore, A.M. Gronenborn, and A. Wlodawer. 1991. Crystal strucure of interleukin 8: Symbiosis of NMR and crystallography. *Proc. Natl. Acad. Sci. U.S.A.* 88: 502-506.
14.   Clore, G. M., and A.M. Gronenborn. 1991. Comparison of the solution nuclear magnetic resonance and crystal structures of interleukin-8. Possible implications for the mechanism of receptor binding. *J. Mol. Biol.* 217: 611-620.

15. Lindley, I.J.D., H. Aschauer, C. Lam, J. Besemer, and A. Rot. 1990. In vitro and in vivo activity of mutagenised versions of recombinant human NAP-1/IL-8 and identification of functionally important domains. *Molecular and Cellular Biology of Cytokines*, 345-350. Eds. Oppenheim, J.J., M.C. Powanda, M.J. Kluger, C.A. Dinarello.

16. Hebert, C.A., R.V. Vitangcol, and J.B. Baker. 1991. Scanning mutagenesis of interleukin-8 identifies a cluster of residues required for receptor binding. *J. Biol. Chem.* 266: 18989-18994.

17. Clark-Lewis, I., C. Schumacher, M. Baggioloni, and B. Moser. 1991. Structure-activity relationships of interleukin-8 determined using chemically synthsized analogs. *J. Biol. Chem.* 266: 23128-23134.

18. Laemmli, U.K. 1970. Cleavage of structural proteins during the assembly of the head of bacteriophage T_4. *Nature.* 227: 680-685.

19. Merril, C.R., D. Goldman, S.A. Sedman, and M.H. Ebert. 1981. Ultrasensitive stain for proteins in polyacrylamide gels shows regional variation in cerebrospinal fluid proteins. *Science* 211: 1437-1438.

20. Ansorge, W. 1985. *J. Biochem. Biophys. Methods.* 11: 13-20.

21. Messerschmidt, A. and J.W. Pflugrath. 1987. *J. Appl. Cryst.* 20: 309-315.

22. Miller, M.D., and M.S. Krangel. 1992. Biology and biochemistry of chemokines: A family of chemotactic and inflammatory cytokines. *Crit. Rev. Immun.* 12: 17-46.

OVERVIEW OF CHEMOKINES

Joost J. Oppenheim

Laboratory of Molecular Immunoregulation
National Cancer Institute
Building 560, Room 21-89A
Frederick, U.S.A

It has been very difficult for investigators involved in studies of the chemotactic factors to agree on terminology; hence the plethora of strange names for these cytokines. It was therefore especially gratifying that this intensely involved and partisan group of researchers could meet and after heated discussion reach an almost unanimous agreement to endorse the chemokine α and β nomenclature for the subfamilies of chemoattractant cytokines whose genes are located on human chromosome 4q and 17q regions respectively.

My overview of this meeting was facilitated immensely by the pivotal questions posed in Dr. Marco Baggiolini's elegant introduction of the chemokine α subfamily. Although only some of his questions were answered, and additional questions became apparent in the course of the workshop, by dealing with the questions and answers the meeting can be put in the proper perspective.

Dr. Baggiolini challenged the theme of the meeting by asking whether all the chemokines are actually chemoattractants and whether this difficult bioassay really provides the best means of characterizing the biological activity of these chemokines. IP 10 was the only remaining chemokine, not known to have chemotactic activities, until Dr. Dennis Taub reported this cytokine to be a moderately potent *in vitro* chemoattractant for monocytes and T cells. Thus, although chemotaxis assays have many shortcomings, they still provide the most sensitive and generally applicable means of evaluating this cytokine family. The rapid mobilization of calcium provides an attractive alternative. However, although this may provide a necessary signal, induction of calcium mobilization may not be sufficient to initiate subsequent cell responses and may therefore not be as indicative of subsequent cell activation events, as chemotaxis. Perhaps the polarization assay reflecting shape change described by Dr. Peter Wilkinson, assays of actin polymerization, or effects on adhesion proteins as suggested by Dr. Timothy Springer may eventually provide more quantitive and reproducible indicators of chemokine activities. Until then the chemotaxis assay remains the best *in vitro* correlate of the capacity by chemokines to mobilize cells *in vivo*.

Dr. Marco Baggiolini amply documented the capacity of chemokine α members, namely IL-8, GRO, NAP-2 and ENA-78, to act on neutrophils, presumably based on the utilization of the type II "rhodopsin-like" receptor for IL-8. He questioned the relevance of the lymphocyte effect of IL-8. In response, Dr. Kouji Matsushima documented that at least a half dozen reports in the literature confirm the observation that IL-8 has modest in vitro chemoattractant effects on T lymphocytes. However, this T cell chemoattractant effect may be indirect and mediated by agents produced by IL-8 stimulated neutrophils. Alternatively, Dr. Peter Wilkinson pointed out that the ability of IL-8 to rapidly influence assays of light polarization and chemotaxis is enhanced in activated T cells, but to detect this effect of IL-8 requires the presence of large numbers of monocytes in mononuclear cell preparations. In addition to RANTES as previously reported by Dr. Tom Schall (1), three other chemokines, namely IP-10, MIP 1α and MIP 1β, are considerably more potent chemoattractants according to Dr. Taub of various T cell subsets than IL-8. However, it is true that the contribution of these chemokines in attracting T cells to *in vivo* inflammatory sites remains to be established.

The major effects of the chemokines have been shown to be differentiative. Do chemokines influence cell growth? MIP 1α was reported by Dr. Gerard Graham to inhibit the replication of early hematopoietic stem cells and thus can inhibit cell growth. Both MIP 1α and MIP 1β are reported by Dr. Hal Broxmeyer and coworkers to costimulate the growth of later hematopoietic progenitor cells (2). GRO was codiscovered by Ann Richmond as a melanoma growth stimulating activity (MGSA) (3) and IL-8 has been reported to be a modest comitogen for keratinocytes (4). Overall, until now chemokines have been shown to have only limited effects on cell growth.

Do the chemokines act largely on leukocytes or do they also have significant effects on somatic cells? The major effects ascribed to platelet factor 4 and β thromboglobulin are fibroblast chemotaxis and activation (5,6). As previously mentioned, IL-8 is a comitogen for keratinocytes (4) and GRO/MGSA stimulates melanoma cell lines (3). Consequently to date chemokines act largely on leukocytes and appear to act on nonleukocytic cells only if engaged in host defence and reparative processes.

Why the redundancy in activities of the chemokines? In addition to four chemokines acting on neutrophils, four on T cells, and two on fibroblasts (CTAP-III and PF4), six chemokines are now known to act on monocytes, namely MCP-1, IP-10, MIP 1α, MIP 1β, RANTES and I-309. Dr. Clemens Dahinden described two chemokines, MCP-1/MCAF and RANTES, as inducers of basophil histamine release. According to Dr. Jens Schröder, RANTES as well as MIP 1α chemoattract eosinophils. However, the *in vitro* assays fail to indicate the time of *in vivo* appearance and compartmentalization of these activities. We have insufficient information at present to assign accurate pathophysiological roles to the chemokines. I predict that these chemokines probably will be sufficiently specialized not to be redundant *in vivo*. On the other hand structurally homologous variants of some of the chemokines may represent redundant gene products. There are three distinct molecular species of GRO that exhibit greater than 90% homology in amino acid sequence. These variants of GRO presumably represent mutated gene products. However, it has not been clearly established whether the variants use the same receptors and are duplicative in their activities. Although MIP 1α and MIP 1β exhibit only about 70% homology in their amino acid sequence, yet they were reported by Dr. David Kelvin to bind to the same receptor on monocytes with equal affinity. According to Dr. Taub, MIP 1α and MIP 1β preferentially attract CD8 and CD4 T cell subsets respectively and each of them is reported to have different effects on hematopoietic stem cells and eosinophils. Consequently, these two chemokines have acquired divergent capabilities and in retrospect perhaps should have been given more distinct names. At this meeting Dr. Jo Van Damme presented his elegant detective studies that led to the identification and purification of two additional members of the chemokine β subfamily with chemotactic activity for monocytes. Since both these molecules show about 72% homology with MCP-1, he has named them MCP-2 and MCP-3. This may turn out to be appropriate, but as with MIP 1α and MIP 1β, these variants may also have some distinctive activities and may be more than just redundant duplicates.

One question that is of particular interest to me is the role of chemokines in the cytokine cascade. Chemokines are readily induced by proinflammatory cytokines such as IL-1, TNF, PDGF, IL-2 and IFNγ. Therefore like IL-6, chemokines appear to be secondary "effector" cytokines that are more specialized in their activities and are not themselves potent promulgators of the cytokine cascade. In that regard, recent reports that chemokines such as MIP-1α can induce IL-1 and TNF (7) need to be confirmed.

Dr. Antal Rot challenged the paradigm that chemokines function largely in host defense and repair processes. Based on assays of the chemokine content in his colleagues' sweat, Dr. Rot deduced that sweat glands are a very rich source of IL-8. Provided this does not represent a response by sweat gland cells to bacterial organisms present on the skin, this may represent the first evidence that these chemokines can be produced constitutively or in response to stress and may therefore participate in maintaining homeostasis. In addition, Dr. Ed Leonard discussed his data showing that most normal humans have circulating free IgG antibody to both MCP-1 and IL-8, as well as immune complexes of these cytokines. This leads him to suggest that the antibody functions as a molecular trap that prevents interaction of chemoattractant with target cells in the circulation. This data also suggests that these chemokines may be continually produced either spontaneously or in response to frequent environmental microbial invaders.

A number of the participants asked with some trepidation how many more chemokines will be discovered? At the risk of being rapidly superannuated, based on the logarithmic shape of the curve for the rate of chemokine discovery, I would predict that most of the chemokines have been identified, and excluding variants, that we will end up with about a dozen.

How many chemokine receptors are there and will they all be members of the rhodopsin superfamily? Many of the chemokines can be shown to mobilize calcium within seconds in the appropriate target cells, as can all the neutrophil chemoattractants that use the two rhodopsin-like receptors that have been identified for IL-8. This suggests that the other chemokines will also use the seven transmembrane domain G-protein coupled receptors. In addition, the fact that the responses to some of the chemokine β members are inhibited by pertussis toxin also suggests they utilize rhodopsin-like receptors (unpublished observation). The cross-binding and desensitization data presented by Drs. Silvano Sozanni and Kelvin suggest the existence of distinct monocyte receptors for MIP-1α and β and MCP-1 and that only the RANTES receptor can bind all three of these ligands. There appear to be distinct receptors for MIP-1α, MIP-1β, RANTES and IP-10 on various T cell subsets, and presumably there are unique receptors for PF-4, βTG and I-309. These observations predict considerable chemokine receptor heterogeneity as well as cross-utilization.

Many questions remain completely unanswered, such as the identification of antagonists of chemokines. Hopefully Dr. Patricia Beckman and others attempting to design antagonists using mutational approaches or Dr. Auer using crystallographic methods will succeed in generating chemokine inhibitors. The therapeutic capabilities of chemokines as anticancer agents has begun to be evaluated by transfecting tumor cells by Dr. Alberto Mantovani using MCAF and Tom Schall using RANTES. The latter seems more promising in that, unlike MCAF, it may induce tumor immunity.

The mechanism of signal transduction presumably utilizes G proteins and phosphokinases since chemokines have been shown to induce the phosphorylation of intracellular substrates (8). However the identification of the specific kinases, and the means by which chemokines regulate genes remain unknown. The mechanisms of chemokine regulation of adhesion proteins is also crucial for understanding how they can cause cells to infiltrate and migrate through tissues. Hopefully we will have more data to bear on these issues at the next meeting. By then we may also hear more on the subject of transgenic mice that overexpress chemokines, or that may be made deficient in chemokines through, homologous recombination. In the meanwhile I must express my gratitude at having been granted the privilege of working in such a vital and exciting field with a group of dynamic colleagues that, as evidenced by this meeting, have been able to make excellent progress in increasing our understanding of the contribution of chemokines to the integrity of the host.

REFERENCES

1. Schall, T.J., K. Bacon, K.I. Toy, and D.V. Goedell. 1990. Selective attraction of monocytes and T lymphocytes of the memory phenotype by the cytokine RANTES. *Nature* 34: 669.

2. Broxmeyer, H.E., B. Sherry, L. Lu, S. Cooper, K.O. Oh, P. Tekamp-Olson, B.S. Kwon, and A. Cerami. 1990. Enhancing and suppressing effects of recombinant murine macrophage inflammatory proteins on colony formation *in vitro* by bone marrow myeloid progenitor cells. *Blood* 76: 110.

3. Richmond, A., E. Balentien, H.G. Thomas, G. Flaggs, D.E. Barton, J. Spiess, R. Bordoni, U. Francke, and R. Derynck. 1988. Molecular characterization and chromosomal mapping of melanoma growth stimulatory activity, a growth factor structurally related to β-thromboglobulin. *EMBO J.* 7: 2025.

4. Krueger, G., C. Jorgensen, C. Miller, J. Schroeder, M. Sticherling, and E. Christophers. 1990. Effects of IL-8 on epidermal proliferation. *J. Invest.Dermatol.* 94: 545 (Abstract).

5. Senior, R.M., G.L. Griffin, J.S. Juang, D.A. Walz, and T.F. Deuel. 1983. Chemotactic activity of a platelet alpha protein for fibroblasts. *J.Cell.Biol.* 96: 382.

6. Castor, C.W., J.W. Miller, and D.A. Walz. 1983. Structural and biological characteristics of connective tissue activating peptide (CTAP-3), a major human platelet-derived growth factor. *Proc.Natl.Acad.Sci. USA* 80: 765.

7. Fahey, T.J. III., K.J. Tracey, P. Tekamp-Olsen, L.S. Cousens, W.G. Jones, T. Shires, A. Cerami, and B. Sherry. 1992. Macrophage inflammatory protein 1 modulates macrophage function. *J.Immunol.* 148: 2764.

8. Oppenheim, J.J., C.O.C. Zachariae, N. Mukaida, and K. Matsushima. 1991. Properties of the novel proinflammatory supergene "intercrine" cytokine family. *Ann.Rev.Immunol.* 9: 617.

REGULATION OF CYTOKINE SECRETION BY POXVIRUS ENCODED PROTEINS

G.J. Kotwal and S. Jayaraman

Division of Molecular Virology and Division of Clinical Virology, James N. Gamble Institute of Medical Research, Cincinnati, U.S.A.

Poxviruses encode proteins in the telomeric regions of their genome which are non-essential for virus replication in tissue culture but play an important role *in vivo* in virus-host interactions. One of the best characterized of such proteins is the major secretory protein which is significantly similar to the human C4b-binding protein (Kotwal and Moss. 1988. *Nature.* 335: 176-178). Poxviruses also encode genes which are structurally similar to the serine protease inhibitor (Serpin) superfamily (Kotwal and Moss. 1989. *J. Virol.* 63: 600-606). A 38 kDa Serpin encoded by cowpox virus (CPV) has been shown to inhibit cellular infiltration and recently shown to block intracellular processing of IL-1 Beta mediated by a cysteine protease. We have overexpressed the 38 kDa Serpin of CPV and its homolog encoded by vaccinia virus using the T7-EMC vaccinia expression system. Previous studies indicate that the poxviral serpins from within the right end of the genome have structure, localization and biosynthetic pathway similar to placental plasminogen activator inhibitor (Kotwal and Moss, unpublished). Employing an IL-1 Beta-sensitive T cell clone, we determined levels of IL-1 Beta secreted in the culture media from a human monocyte cell line U937 infected with wildtype poxvirus or recombinant viruses; either lacking the ability to express the 38 kDa protein or overexpress this and other serpins. Our results suggest that there is a significant reduction in the level of IL-1 Beta secretion in the CPV infected cells in comparison to the recombinant lacking the 38 kDa protein. Thus, inhibition of the inflammatory response reported during infection in the chorio allantoic membrane may be due, at least in part, to the down regulation of IL-1 Beta secretion by the CPV 38 kDa protein.

INTERLEUKIN-8 EXPRESSION IN GASTRIC CANCER

J.E. Crabtree[1], J.I. Wyatt[1], P. Peichl[2], L.K. Trejdosiewicz[1], N. Ramsay[1], P.H. Nichols[1], J.N. Primrose[1], and I.J.D. Lindley[2]

[1]Departments of Clinical Medicine and Pathology, St. James's University Hospital, Leeds, U.K. and [2]Sandoz Research Institute, Vienna, Austria

We have previously demonstrated that interleukin-8 (IL-8) is expressed by epithelial cells in normal and gastritic antral mucosa. The aims of this study were to investigate IL-8 expression in gastric carcinoma and in gastric cancer epithelial cell lines. Tumour tissue from 16 patients undergoing surgery for gastric carcinoma was rapidly frozen. Cryosections were immunolabelled with a mouse monoclonal antibody to IL-8 followed by FITC-labelled goat anti-mouse IgG antibody. Double immunofluorescence with rabbit anti-cytokeratin antibody was used to identify infiltrating carcinoma cells. A rabbit anti-fibronectin antibody was used as a negative control. Specificity of IL-8 immunolabelling was confirmed by competitive inhibition with exogenous recombinant IL-8. Strong IL-8 expression was observed in the neoplastic cells of all 16 patients, expression in diffuse (n = 6) and intestinal (n = 8) type gastric cancer (Lauren classification) being equally positive. IL-8 expression was also observed in one subject with a mixed tumour type and one unclassified gastric carcinoma. IL-8 positive cells infiltrating the stroma were cytokeratin positive, fibronectin negative. Constitutive expression of IL-8 mRNA was observed in 3 gastric cancer cell lines, Kato 3, MKN 45 and ST 42 using PCR. Expression of antigenic IL-8 in the cell lines was confirmed by immunofluorescence. These results demonstrate that IL-8 is expressed constitutively in gastric cancer cell lines and that gastric cancer cells *in vivo* show positive IL-8 expression. The relevance of IL-8 expression to the peri-tumoural inflammatory reaction and the possible contribution to diagnosis require further study.

SYNERGISTIC IL-8 SYNTHESIS BY HUMAN PERITONEAL MESOTHELIAL CELLS (HPMC) FOLLOWING IL-1β AND TNFα TREATMENT

N. Topley, Z. Brown, A. Jörres, J. Westwick, G.A. Coles, and J.D. Williams

[1]Institute of Nephrology, Cardiff, U.K., [2]Pharmacology, University of Bath, U.K., and [3]Nephrology, Charlottenburg, Berlin, Germany

The initiation of peritonitis in patients on continuous ambulatory peritoneal dialysis is accompanied by the influx of neutrophils into the peritoneal cavity. The current study was initiated to examine the HPMC as a possible source of neutrophil chemotactic activity and to characterise the released product and examine its regulation by other cytokines.

Unstimulated HPMC under growth arrested conditions released IL-8 in a constitutive and time dependent manner. Stimulation of HPMC with IL-1β or TNFα resulted in a time and dose dependent IL-8 generation; after 24 h the levels induced by IL-1β and TNFα (both at 1000 pg/ml) were (mean $\pm$ SEM, n=5) 101 $\pm$ 26.6 (p<0.01) and 35 $\pm$ 8.09 (p<0.01) respectively. This release was inhibited following co-incubation with the relevant anti-cytokine antibody or preincubation with either cycloheximide or actinomycin D. Treatment of HPMC with IL-1β or TNFα resulted in increased expression of IL-8 specific mRNA.

Stimulation of HPMC with combinations of IL-1β and TNFα resulted in a synergistic increase in IL-8 release. This effect was significant at combined doses of IL-1β (50 pg/ml) and TNFα (500 pg/ml) and above, when the release of IL-8 was 88 ± 27% above the additive IL-8 release values (z=2.201; p<0.01).

Western blot analysis using specific anti-IL-8 antibody demonstrated the presence of two major immunoreactive bands between 9 and 10 kD, in HPMC culture supernatants.

These data demonstrate that HPMC synthesise IL-8 and that its release can be regulated as a result of induction of mRNA expression and de-novo protein synthesis by other cytokines.

DETECTION OF A POTENT CHEMOTACTIC ACTIVITY IN SUPERNATANTS FROM HUMAN MONOCYTES STIMULATED WITH ENDOTHELIN-1 (ET-1)

E. Helset, T. Sildnes, and Z. Konopski

Institute of Clinical Medicine and Institute of Medical Biology, University of Tromse, Tromse, Norway

A role for the endothelial derived, potent vasoactive peptide Endothelin-1 (ET-1) in inflammatory reactions in the microvasculature has been suggested (1). We have earlier shown that ET-1 causes lung injury in an interaction with leukocytes in isolated rat lungs (2). In this work we have studied the chemotactic activity and release of cytokines in supernatants from human monocytes (MC) stimulated in vitro with different concentrations of ET-1 for 1, 4, 8, and 24 hours. For the study of chemotactic activity, a Boyden chamber was employed, and the cytokines TNF, IL-1, and IL-6 were analysed by bioassays. We found that supernatants from MC stimulated with ET-1 contained a potent chemotactic activity against MC and polymorphonuclear cells (PMN). Chemotactic activity against PMN is produced already after 1 hour of stimulation with ET-1. The effect of ET-1 is dose-dependent. The chemotactic activity against PMN does not correspond to the levels of TNF, IL-1 or IL-6 detected in the supernatants. The chemotactic activity against MC appears in the supernatants first after 24 hours of stimulation with ET-1. This chemotactic activity is also dependent on dose of ET-1 used for stimulation. The chemotactic activity against MC corresponds to the increased levels of IL-1 detected in the supernatants. To conclude, our data suggests that ET-1 may take part in inflammatory reactions, both by inducing production of cytokines and chemotactic factors.

1.   Murch, S.H. et al. 1992. *The Lancet* 339: 381-385.
2.   Helset, E. et al. 1992. (In press). *Circulatory Shock.*

GLIOBLASTOMA SECRETE MCP-1, A MONOCYTE CHEMOTACTIC FACTOR

I. Desbaillets, A-C. Diserens, N. de Tribolet, and E. Van Meir

Laboratoire de Neurochirurgie, CHUV, CH-1011 Lausanne, Switzerland

Immunohistological studies on glioblastoma tumor sections have demonstrated the presence of variable amounts of infiltrating macrophages. To understand how these macrophages are recruited from the circulation through the blood brain barrier, we have tested for the presence of macrophage chemotactic factors such as MCP-1. MCP-1 is a human monocyte chemoattractant protein produced by several cell types such as fibroblasts, PMA stimulated PBMC and human glioma cell-line U-105MG. We found various constitutive levels of different specific MCP-1 mRNAs in 10/13 cell lines. The inflammatory cytokines IL-1β and TNFα could induce or increase the MCP-1 mRNA in all cell lines analyzed. To demonstrate *in vivo* MCP-1 production we showed that ex vivo extracted tumors contained MCP-1 mRNA, both in low and high grade astrocytomas. In situ hybridization is actually performed on glioblastoma tissue sections to determine the *in vivo* cellular origin of MCP-1.

We are also currently analyzing the cell line supernatants, CSF and cyst fluids of patients to check the presence of soluble MCP-1 protein. These data will further be correlated with immunostaining on frozen sections to identify the MCP-1 producing cells.

NEUTROPHIL CHEMOTACTIC ACTIVITY IN THE BRONCHOALVEOLAR LAVAGE FLUID OF ASTHMATICS

L.M. Teran, T. Begishvili, M.G. Campos, R. Djukanovic, M.K. Church, D.E. Davies, and S.T. Holgate

University of Southampton, Southampton, U.K.

Neutrophil chemotactic activity (NCA) has been widely described in the serum of asthmatics after antigen challenge. We now report NCA in the bronchoalveolar lavage (BAL) fluid of asthmatic and non asthmatic subjects. BAL (about 50 ml) was obtained from 12 subjects (7 asthmatics and 5 normals) aged 26 to 40 years at bronchoscopy. Separation of NCA was achieved by FPLC cation exchange chromatography (Mono S column). Twenty five fractions were collected from each subject and neutrophil chemotaxis was assayed in a microBoyden chamber using 5 μm polycarbonate filter and stained for counting with Hema Gurr stain.

Neutrophil chemotactic activity was detected in three different fractions in 4 out of 7 asthmatics. In 1 out of 5 normal subjects had detectable NCA but showed a different profile from that of the asthmatics. Thus, asthmatic subjects secrete neutrophil chemotactic factors into the bronchial lumen in this chronic disease.

THE CHEMOATTRACTANT PROPERTIES OF INTERLEUKIN-4

M.T.A. Villar, J.A. Douglass, C. Heusser, P. Bradding, S.T. Holgate, and
M.K. Church

University of Southampton, Southampton, U.K.

Interleukin-4 (IL-4) is an important cytokine in B-cell differentiation and has been recently found to have a role in the selective recruitment of eosinophils via up-regulation of V-CAM. We have studied the effects of IL-4 *in vitro* as a chemoattractant for both eosinophils and neutrophils. Neutrophils were isolated on Ficoll-Hypaque to a purity of 98%, while eosinophils from atopic donors were isolated on discontinuous Percoll gradients to a purity of $68 \pm 8\%$. Chemotaxis was assayed in multiwell modified micro Boyden chambers using polycarbonate filters with 5 μm and 8 μm pore size respectively. Chambers were incubated for one hour at 37° C, and filters removed, fixed and stained with Haema-Gurr. 10 high power fields were counted per duplicate experiment and results expressed as cells per high power field (cells/HPF). The mean $\pm$ SEM of 5 separate experiments are presented. Concentrations of IL-4 ranging from 0.1-1000 ng/ml were tested. Neutrophils demonstrated a dose-response curve with an optimum response of 63.4 ± 14.5 cells/HPF at 10 ng/ml which was significantly different to the response to buffer of 30.0 ± 9.2 cells/HPF (p=0.032). FMLP (10^{-8}M) as positive control gave a response of 121 ± 23.6 cells/HPF. Eosinophils also showed a dose-response curve with an optimum response at 100 ng/ml IL-4 of 9.9 ± 1.86 cells/HPF compared to buffer response of 4.7 ± 1.1 cells/HPF (p=0.055). The peak response to PAF (10^{-7}M) was 12.6 ± 2.5 cells/HPF. We conclude that IL-4 appears to be a chemoattractant for both neutrophils and eosinophils.

IL-8 INDUCED CHEMOTAXIS OF PERIPHERAL NEUTROPHILS IN CHILDREN WITH CYSTIC FIBROSIS (CF)

Y. Dai, T.P. Assadullahi, M.K. Church, J.K. Shute, and J.O. Warner

University of Southampton, Southampton, U.K.

Inflammation associated with neutrophil infiltration is a commonly observed feature of children with CF. Production of the major neutrophil chemotactic cytokine interleukin-8 (IL-8) by alveolar macrophages or supporting cells is potentially of great importance in the pathology of CF. We have previously shown that both sputum and BAL IL-8 concentrations are higher in children with CF compared with asthmatics, normal children or children with lung infection. We have therefore investigated the chemotactic effects of IL-8 on peripheral neutrophils obtained from a CF group (n=18, aged 4-20 yr) and a control group of normal children (n=13, aged 5-12 yr). Cells were isolated by dextran sedimentation followed by separation on Lymphoprep. Chemotaxis was assayed using multiwell microchemotaxis chambers and 5 μm polycarbonate filters. Filters were fixed and stained with Haema-Gurr for counting. Results are expressed as number of neutrophils per high power field (HPF). Optimal chemotaxis of control and CF neutrophils were observed at an IL-8 concentration of 1×10^{-8}M. At the optimum concentration the number of cells migrating were 150 ± 11/HPF for controls and 140 ± 14/HPF for the CF group (p=n.s.). At lower IL-8 concentrations, the number of neutrophils migrating were significantly lower for the CF group (p=0.042 at $1x10^{-9}M$ and p=0.026 at $5x10^{-9}M$). These results show that neutrophils from both control and CF groups have receptors for IL-8, but that the CF group have a different degree of responsiveness to this cytokine. The mechanism underlying this is currently under investigation.

UNUSUAL PROPERTIES IN VITRO AND IN VIVO OF 198, AN ANTIBODY THAT RECOGNIZES RABBIT CD11b

A. Shock[1], J. Rutter[1], D. Howat[1], T. Minion[1], S. Lane[1], E. Ball[1], P. De Baetsellier[2], and M. Robinson[1]

[1]Celltech Ltd., 216 Bath Road, Slough, Berkshire, U.K. and [2]Institut voor Moleculaire Biologie, Vrije Universiteit, Brussels, Belgium

The leukocyte integrin CR3 (CD18/CD11b) is thought to be important in the migration of neutrophils (N) into tissues during inflammation. The aim of the present study was to compare the properties of 6.5E, a mab with cross-reactivity against human and rabbit leukocyte integrin β chain CD18, with those of 198 and KIM 118, antibodies against the α chain of CR3 (CD11b) in rabbit and human cells, respectively. This was accomplished by assessing the ability of antibodies to block the adhesion of PMA-stimulated N (human N or glycogen elicited rabbit peritoneal N) to protein-coated plastic or to endothelial cells (from human umbilical vein (HUVEC) or rabbit ear (REEC) sources). In the human system, KIM 118 and 6.5E were found to be potent at blocking adhesion of N to both plastic and HUVEC. IC_{50} values for 6.5E and KIM 118 on plastic were 0.30 and 0.23 µg/ml, respectively; comparable values on HUVEC were 0.24 and 0.37 µg/ml. In the rabbit system, 6.5E also blocked adhesion of rabbit N to plastic and to REEC (IC_{50} = 0.3 and 0.1 µg/ml, respectively); however whilst 198 blocked adhesion of N to plastic (IC_{50} 1.0 µg/ml) it did not block adhesion of N to REEC. These results suggest that 198 recognizes an unusual epitope on CR3 or that LFA-1 and/or p150,95 are important regulators of adhesion in the rabbit. Data will also be presented to show the unusual effects of 6.5E and 198 *in vivo*.

EFFECT OF MONOCYTE CHEMOTACTIC CYTOKINE GENE TRANSFER ON MACROPHAGE INFILTRATION, GROWTH AND METASTATIC BEHAVIOUR OF A MURINE MELANOMA

B. Bottazzi, S. Walter, D. Govoni, F. Colotta, and A. Mantovani

"Mario Negri" Institute for Pharmacological Research, 20157 Milan, Italy

Tumor-derived chemotactic factors have been identified and suggested to play a role in the regulation of macrophage infiltration in neoplastic tissues. Aim of this study was to investigate the *in vivo* relevance of a tumor-derived chemotactic factor molecularly identified as monocyte chemotactic protein (MCP) by gene transfer in a murine melanoma. MCP-producing melanoma clones showed a twofold increase in the percentage of tumor-associated macrophages (TAM) compared to control clones and to the parent line B78/HI. MCP-producing cells exhibited a slower growth rate *in vivo* with a prolongation of survival time while the *in vitro* growth rate of melanoma clones was unaffected by MCP gene transfer. Similar differences between MCP-producing and control cells, in terms of macrophage infiltration and growth rate, were detected after implantation in athymic mice. After inoculation of small cell numbers (100 cells/mouse), MCP-producing cells were slightly, but significantly, more tumorigenic. Finally the metastatic behaviour of MCP-producing and control clones was studied: after i.v. injection MCP-expressing cells were more metastatic than control cells in terms of lung involvement and of the occurrence of extrapulmonary lesions. These results demonstrate that a tumor-derived chemotactic cytokine can indeed play a role in the regulation of mononuclear phagocyte recruitment in neoplastic tissues.

EXPRESSION OF MONOCYTE CHEMOTACTIC PROTEIN-1 (MCP-1) BY MONOCYTES AND ENDOTHELIAL CELLS EXPOSED TO THROMBIN

F.L. Sciacca, F. Colotta, M. Sironi, W. Luini, and A. Mantovani

"Mario Negri" Institute for Pharmacological Research, 20157 Milan, Italy

Thrombin, in addition to being a key enzyme in haemostasis, affects a series of endothelial and leukocyte functions, and thus may be involved in the regulation of inflammatory reaction. Since leukocyte recruitment and activation are important events in inflammatory and thrombotic processes, in this study we have examined the possibility that thrombin can induce the production of a cytokine chemotactic for mononuclear phagocytes. Human peripheral blood mononuclear cells (PBMC) exposed *in vitro* to thrombin expressed transcript of monocyte chemotactic protein-1 (MCP-1, alternative acronyms: JE, monocyte chemotactic and activating protein, tumor derived chemotactic factor). Among circulating mononuclear cells, monocytes were identified as the cells expressing MCP-1 in response to thrombin. Thrombin was two to three fold less effective than endotoxin in inducing MCP-1 transcript in PBMC. In addition to monocyte, also endothelial cells (EC) expressed MCP-1 in response to thrombin, although at lower levels compared to monocyte. Actinomycin D experiments indicated that induction of MCP-1 by thrombin in PBMC and in EC was gene transcription dependent. The protein synthesis inhibitor cycloheximide blocked thrombin-induced MCP-1 expression in PBMC, whereas it superinduced both constitutive and thrombin-induced expression of MCP-1 in EC, indicating different mechanism of regulation of this gene in mononuclear phagocytes versus other cell types. Thrombin stimulated mononuclear cells and endothelial cells to release chemotactic activity for monocytes that could be at least partially inhibited by absorption with anti-MCP antibodies. It is well known that thrombin is involved in inflammation: it induces a series of functional modifications in endothelial cells (secretion of PGI2 and IL-1), it is chemotactic for monocytes and it degrades some components of basal membrane; moreover endothelial cells are strategically located at the interface between blood and tissue and upon exposure to inflammatory signals (LPS, IL-1) monocytes and endothelial cells express cell associated procoagulant activity: hence, the induction of MCP-1 by thrombin may be an important mechanism in the recruitment of monocytes at sites of vasculary injury. These data further indicate that blood coagulation, inflammatory reaction and immunity interact closely.

IL-8 PRODUCTION DURING URINARY TRACT INFECTION

W. Agace[1], S. Hedges[1], M. Ceska[2], and C. Svanborg[1]

[1]Dept. of Medical Microbiology, Clinical Immunology, Lund University, Lund, Sweden and [2]Sandoz Research Institute, Vienna, Austria

Infection of the urinary tract leads to a local inflammatory response mediated at the mucosa. In patients with natural urinary tract infection (UTI) and in patients deliberately colonized with *Escherichia coli* in the urinary tract there is a rapid recruitment of neutrophils into the urine. We have shown previously that urinary tract epithelial cell lines produce a range of cytokines including IL-8 following stimulation with *E. coli* (1). In this study we analyzed the urine of patients following deliberate colonization of the urinary tract with three *E. coli* strains, for the influx of PMNs and presence of IL-8. IL-8 was found in the urine of all patients following UTI. There was a strong correlation of 0.7 between levels of IL-8 and numbers of PMNs in the urine. IL-8 was not found in the serum of patients during the period of colonization. The J82 bladder and A498 kidney epithelial cell lines and neutrophils collected from fresh human blood produced large amounts of IL-8 in response to the same *E. coli* strains. These results suggest that IL-8 may play an important role in neutrophil influx during UTI and that local epithelial cells as well as activated neutrophils may be involved in the production of this protein at the site of infection.

1. Agace, W., S. Hedges, U. Andersson, J. Andersson, M. Ceska and C. Svanborg. Selective cytokine production by epithelial cells following exposure to *E. coli*. Submitted.

NAP-1/IL-8 RECEPTOR EXPRESSION ON HUMAN NEUTROPHILS

W. Schnitzel, U. Monschein, and J. Besemer

Sandoz Research Institute, Vienna, Austria

NAP-1/IL-8 receptor expression by human neutrophils was studied by binding and chemical cross-linking experiments using iodinated NAP-1/IL-8 preparations. NAP-1/IL-8 associated with and dissociated from its receptor(s) with more than one rate, indicating expression of more than one receptor and/or different affinities. Chemical cross-linking of radioactive NAP-1/IL-8 to neutrophils labeled four protein bands with apparent molecular masses between 55 and 81 kDa. In addition, a protein band with molecular mass corresponding to the NAP-1/IL-8 dimer was seen. Also mRNAs for two NAP-1/IL-8 receptors (1,2) could be detected in human neutrophils. Thus, our data are consistent with expression of at least two NAP-1/IL-8 receptors on neutrophils.

1. Holmes, W.E. *et al*. 1991. *Science* 253: 1278.
2. Murphy, P.M., and H.L. Tiffany. 1991. *Science* 253: 1280.

MACROPHAGE-DERIVED NEUTROPHIL CHEMOTACTIC FACTOR (MNCF) INDUCES NEUTROPHIL MIGRATION THROUGH LECTIN-LIKE ACTIVITY

M. Dias-Baruffi[1], M.C. Roque-Barreira[1], F.Q. Cunha[2], and S.H. Ferreira[2]

Departments of [1]Immunology and [2]Pharmacology, Faculty of Medicine, 10049, Ribeirao Preto, Sao Paulo, Brazil

We have previously reported that rat peritoneal macrophages stimulated with LPS release a factor (MNCF) which induces neutrophil migration which is not blocked by glucocorticoids. The initial characterization of MNCF showed that it did not correspond to any of the current known chemotactic cytokines or to known chemotactic agents. Recent studies have shown that neutrophil migration to the inflammatory focus may involve protein molecules with carbohydrate-binding sites, the LEC-CAMs. This fact prompted us to investigate the possible lectin-like nature of MNCF. Thus, the supernatants of macrophage monolayers stimulated with LPS were submitted to affinity chromatography on immobilized sugar columns. We observed that the D-galactose (D-gal) binding fraction retained the MNCF activity. This fraction, consisting of four protein components, was submitted to chromatography on Sephadex G-75, yielding a homogeneous preparation of the active component. MNCF has a MW of 54 KDa (gel filtration and SDS-PAGE) and a pI of 2-3 (isoelectrofocusing and chromatofocusing). A direct mechanism of action of MNCF on neutrophil migration initially suggested by the absence of blockade of the phenomenon *in vivo* by glucocorticoids, was confirmed by the observation that a protein-carbohydrate interaction is involved in this activity. Homogeneous preparation of MNCF induced neutrophil migration *in vivo* and *in vitro*. The *in vitro* effect was blocked by D-gal which did not interfere with the chemotactic effects of known interleukins (IL-1β, TNF-α, IL-6 and IL-8). The present results strengthen our previous suggestion that MNCF is yet-undescribed monokine which induces neutrophil migration through a direct mechanism involving the D-gal binding site of the molecule.

INTERLEUKIN 8 RELEASE FROM HUMAN PMN IS MODIFIED BY AIRBORNE PARTICULATED MATTER

B. Hitzfeld, K.H. Friedrichs, and H. Behrendt

Med. Institute of Environment, Hygiene at the Heinrich-Heine-University, Auf'm Hennekamp 50, 4000 Düsseldorf, Germany

Purified polymorphonuclear leukocytes (PMN), obtained from blood of healthy, non-allergic donors were exposed to organic dust extracts of airborne particulated matter. The dust was collected on glass fibre filter sheets using high volume samplers HVS150 during the winter season 1989/90 in a highly industrialized region of northwestern Germany. The

PMN were incubated either with the organic dust extracts alone or subsequently stimulated with either Ca-Ionophore A23187 or with opsonized zymosan. IL-8 release was measured in the supernatants using an ELISA technique with a monoclonal antibody against IL-8.

The results showed that dust extract alone was able to induce IL-8 release from the PMN of at least half of the donors. The other donors showed a strong inhibition. Ca-Ionophore led in almost all donors to an increased release of this cytokine, being on average twofold higher than in controls. Again half of the donors showed a stimulation of Ca-Ionophore release after preincubation with organic dust extracts. Incubation with opsonized zymosan stimulated IL-8 release; prestimulation with dust extract led in the same donors to a further increase in cytokine release.

Organic extracts of airborne particulated matter can therefore directly release Interleukin 8 and can also prime human PMN for release of stimulus-induced release. Since these pollutants also activate and prime human basophils for anti-IgE-stimulated histamine release and since we have observed interactions between basophils and neutrophils in such preparations, IL-8 release from PMN may act as a HRF in these basophils. Organic dust extracts can thus affect the cells involved in allergic inflammation as well as their interaction.

Supported by grant 07ALL04 of the BMFT.

INTERLEUKIN-8 INDUCES THE EXPRESSION OF HLA-DR ANTIGEN ON HUMAN KERATINOCYTES

A. Sz. Kenderessy[1], L. Kemény[1], G. Michel[2], A. Beetz[2], T. Ruzicka[2], and A. Dobozy[1]

[1]Department of Dermatology, Albert Szent-Györgyi Medical University, Szeged, Hungary and [2]Department of Dermatology, University of Munich, Germany

Recent results suggest that interleukin-8 (IL-8) may directly influence several functions of epidermal cells, such as chemotaxis and Candida albicans killing activity. Since human keratinocytes possess the ability to express cell surface molecules characteristic of immune cells, we were interested in whether IL-8 was capable of modulating immune-associated surface markers of human keratinocytes. A transformed human epidermal cell line and cultured normal human keratinocytes were used for the studies. Cells were incubated with 10 nM IL-8 for 24 or 48 h, and the expression of HLA-DR, CD11a, CD18, OKM5 (CD36) and ICAM-1 was investigated by flow cytometry. We have found that IL-8 induced the MHC-II antigen expression on human keratinocytes. On the other hand, there was no effect on the expression of the other cell surface markers studied, suggesting that the IL-8-induced HLA-DR expression is a rather specific effect of this cytokine. The modulation of immune-associated cell surface markers of human keratinocytes by IL-8 may be of significance in the pathogenesis of inflammatory skin diseases.

CLONING, SEQUENCING AND EXPRESSION OF OVINE NEUTROPHIL-ATTRACTANT PROTEIN-1/IL-8 AND MONOCYTE CHEMOATTRACTANT PROTEIN-1

H-F. Seow[1], T. Yoshimura[2], P. Wood[1], and I.G. Colditz[3]

[1]CSIRO Division of Animal Health, Parkville and [3]Armidale, Australia and [2]National Cancer Institute, Frederick, U.S.A.

A cDNA library was constructed from mRNA purified from con A-stimulated ovine spleen cells and probed with cDNAs of human NAP-1 and human MCP-1. Ovine NAP-1 cDNA comprised 1434 base pairs with an open reading frame that encodes a 101 amino acid protein with 78.2% similarity to human NAP-1. Ovine MCP-1 cDNA comprised 719 base pairs with an open reading frame that encodes a 99 amino acid protein with 66.6% similarity to human MCP-1. The cDNAs were expressed as fusion proteins in E. coli using the pGEX 2T vector. Biological activity was assessed in an *in vitro* chemotaxis assay and *in vivo* following intradermal injection. Following thrombin cleavage from GST, ovine NAP-1 had greater chemotactic activity for ovine peripheral blood neutrophils than human NAP-1. Additional results will be presented.

INHIBITION OF LPS-MEDIATED RESPONSES IN HUMAN WHOLE BLOOD BY RECOMBINANT BACTERICIDAL/PERMEABILITY INCREASING PROTEIN

P. Conlon, K. Mészáros, J. Watkins, S. Aberle, L. Grinna, A. Horwitz, T. Parsons, G. Theofan, and J. Parent

XOMA Corporation, Berkeley, CA, U.S.A.

BPI is a LPS-binding protein with bactericidal and LPS-neutralizing activity that is found in the azurophilic granules of neutrophils. Since an N-terminal proteolytic fragment of BPI displays the potent biological properties of the 55 kD intact protein (BPI_{55}), we have produced, purified, and characterized, a recombinant protein, $rBPI_{23}$, corresponding to the N-terminal half of human BPI. $rBPI_{23}$ binds specifically and with high affinity to highly purified smooth LPS isolated from a variety of clinical isolates via $rBPI_{23}$ interaction with the lipid A region. We investigated the ability of $rBPI_{23}$ to inhibit LPS-mediated, cellular responses in whole human blood, a complex environment which more closely resembles the *in vivo* situation. We observed that $rBPI_{23}$ inhibited the LPS-dependent production of a variety of inflammatory cytokines, including TNF, IL-1β, IL-6, and IL-8. This response was specific in that $rBPI_{23}$ had no effect on the ability of irrelevant stimuli (i.e., zymosan) to induce these cytokines. The LPS dependent production of oxygen-derived free radicals by human blood cells, resulting in chemiluminescence, was also inhibited by $rBPI_{23}$. Again, this inhibition was specific in that $rBPI_{23}$ had no effect on the ability of irrelevant stimuli (PMA or zymosan) to induce chemiluminescence. In the complex setting of whole human blood, LPS-dependent chemiluminescence was found to be independent of the presence of TNF, as antibodies to TNF had no effect in this system. The profound effects of rBPI$_{23}$ on the LPS-mediated, cellular responses in whole blood suggest that BPI has promise as a therapeutic agent in Gram-negative infections.

THE ACTIVITY OF IL-8 IN COMBINATION WITH GMCSF AND TNFα

P.J. Roberts[1], A.R. Pizzey[1], A. Khwaja[1], J. Carver[1], A. Mire-Sluis[2], and D.C. Linch[1]

[1]Haematology Department, University College London, U.K. and [2]National Institute for Biological Standards, Potters Bar, Herts., U.K.

The effect of cytokines on phagocytes can be either direct e.g. increasing cell adhesion molecule (CD11b) expression, or indirect e.g. enhancing or 'priming' an agonist-stimulated activity. rh-IL-8 (Sandoz) (500 ng/ml) upregulated CD11b expression to 255 ± 47% of control (n = 5). When neutrophils were incubated with IL-8 + GMCSF (1 - 100 ng/ml) the effect on CD11b expression was additive. Incubation of either purified or whole blood neutrophils with IL-8 rapidly primed fMLP-dependent superoxide production by 297 ± 40%, (n = 17). The amount of IL-8 required to give maximal priming was 100 ng/ml in whole blood, cf 10-50 ng/ml for purified cells, concentrations that did not stimulate a respiratory burst in the absence of an agonist. IL-8 was a weaker primer than GMCSF (10 ng/ml) or TNFα (100U/ml), its effect was transient, with loss of activity by 2 hours. When IL-8 and GMCSF were co-incubated in whole blood assays of the respiratory burst, the priming effect at low GMCSF concentrations (10 pg - 1 ng/ml) was additive, whereas at high GMCSF concentrations (10 ng/ml) the effect was less than with GMCSF alone. Ligand binding studies showed that high concentrations of IL-8 (500 ng/ml) did not alter the binding of ^{125}I-GMCSF to neutrophils. In parallel experiments IL-8 increased FMLP receptor expression from 18,000 to 28,000 receptors/cell with little change in receptor affinity, and co-incubation of IL-8 with GMCSF (10 ng/ml) further increased the number of fMLP receptors to 39,000/cell. Respiratory burst priming by combinations of IL-8 with TNFα was at least additive throughout the whole concentration range. The data suggest that in inflammatory lesions IL-8 has the potential to increase neutrophil activity, especially in the presence of other cytokines such as GMCSF and TNFα.

INDUCTION OF IL-8 IN INFLAMMATORY BOWEL DISEASE

S.J.H. van Deventer[1,2], S.A. Radema[1], K. Lammers[1], D. Hommes[1], M. Ceska[3], J. Jansen[1], and GNJ Tytgat[2]

[1]Center for Thrombosis, Hemostasis, Atherosclerosis and Inflammation Research, and [2]Department of Gastroenterology, Academic Medical Center, Meibergdreef 1105, AZ Amsterdam, The Netherlands and [3]Sandoz Research Institute, Vienna, Austria

Ulcerative colitis (UC) is an inflammatory disease of unknown etiology that is characterized by an influx of inflammatory cells, particularly neutrophils, into the mucosa of the large bowel. Using immunohistochemistry and in situ hybridization, we have previously shown that IL-1 is predominantly expressed by enterocytes in experimental colitis and UC. In a prospective clinical trial, biopsy specimens from 20 consecutive patients with UC and 5 controls, were obtained by endoscopy. IL-8 and IL-1 levels were assessed by ELISA, either in immediately homogenized biopsy specimens or after 23 hour organ culture. Although the diseased bowel mucosa contained more IL-8 than the normal mucosa (200 pg/mg), considerable amounts of IL-8 were also detected in the normal mucosa (98 pg/mg; p=0.004). Moreover, specimens obtained from macroscopically normal and from diseased mucosa from UC patients contained similar amounts of IL-8 (200 vs 180 pg/mg p=0.43). We have shown that, in UC, IL-8 is transcribed and translated by inflammatory cells as well as enterocytes. In HT-29/19a enterocytic cells, IL-8 production can be dose-dependently induced by IL-1 and TNF, and these stimuli are able to induce a stimulus-directed polarized IL-8 secretion that is dependent on rearrangement of the cytoskeleton. In conclusion, large amounts of IL-8 are present in the normal bowel mucosa, suggestive of a role in normal neutrophil and T-cell trafficking. In the diseased bowel, increased levels of IL-8 seem to be important for neutrophil activation in UC, presumably in concert with other stimuli.

ELEVATED CIRCULATING AND TISSUE IL-8 IN ALCOHOLIC HEPATITIS

N. Sheron[1], G. Bird[1], M. Sticherling[2], M. Ceska[3], I.J.D. Lindley[3], and R. Williams[1]

[1]Institute of Liver Studies, King's College Hospital, London SE5 9RS, U.K., [2]Dept. of Dermatology, University of Kiel, Kiel, Germany and [3]Sandoz Research Institute, Vienna, Austria

Acute alcoholic hepatitis is characterised by a unique degree of liver neutrophil infiltration, often accompanied by marked peripheral neutrophilia. We examined plasma (n = 164) and tissue (n = 50) levels of the neutrophil activator and chemotaxin, interleukin-8, in patients with a spectrum of alcoholic liver disease, and in normal and disease control subjects. Levels of circulating IL-8 were undetectable in normals but highly elevated in patients with alcoholic hepatitis particularly in those who died (gM 600 ng/l, Cl 323-1120 v 184 ng/l, Cl 114-309 in survivors). Levels correlated with biochemical indicators of severe disease (bilirubin R = 0.38, INR R = 0.28, WCC R = 0.35, creatinine R = 0.34), and with TNFα (R = 0.43) and soluble TNF receptors (p55 R = 0.59). In contrast moderate elevations in the level of circulating IL-8 were seen in alcoholic cirrhosis (gM 93 ng/l, Cl 40-213) and in alcoholics undergoing alcohol withdrawal (gM 137 ng/l, Cl 72-259). Levels in non-alcohol related inflammatory liver disease were comparatively low (gM 17 ng/l, Cl 10-29). In liver tissue from patients with alcoholic liver disease, local levels of IL-8 correlated with the degree of neutrophil infiltration (R = 0.71, n = 15) and levels were much higher in alcoholic hepatitis (987 pg/mg, Cl 351-1623) compared with other alcoholic liver disease (103 pg/mg, Cl 0-220), normal liver (20 pg/mg, Cl 0-61) and non-alcohol related liver disease (219 pg/mg, Cl 142-295). Synthesis of IL-8 within the liver may play a role in the pathogenesis of alcoholic liver disease.

EXPRESSION OF MONOCYTE CHEMOATTRACTANT PROTEIN-1 IN BASAL KERATINOCYTES OF PSORIATIC LESIONS

R. Gillitzer[1], K. Wolff[2], G. Stingl[2], and R. Berger[2]

[1]Department of Dermatology, University of Würzburg Medical School, Würzburg, Germany and [2]Department of Dermatology, University of Vienna Medical School, Vienna, Austria

To further our understanding of psoriasis (PS) pathogenesis, it is necessary to analyze the factors responsible for the abnormal cutaneous topobiology of infiltrating leukocytes, in particular the juxtaposition of proliferating basal keratinocytes (KC) to immunocompetent cells. Surprisingly, our immunohistological analysis of psoriatic lesions (n = 11) revealed that, as opposed to neutrophils and T cells, basically mobile dermal macrophages/dendrocytes (DM/DD) are almost exclusively restricted to the dermal compartment with only few cells penetrating the basal membrane. They are encountered in the papillary dermis and are frequently arranged along the basement membrane of the cubbing rete ridges, thus having the closest spatial proximity to proliferating KC. Recently, the potent monocyte chemoattractant protein-1 (MCP-1) was isolated and sequenced. Since T cells, fibroblasts, endothelial cells, smooth muscle cells and KC have the potential to secrete MCP-1 *in vitro* and all cell types are integral parts in lesions of PS, we performed in situ hybridization studies with ^{35}S-labelled MCP-1 RNA probes to detect and localize MCP-1 mRNA expression. Using MCP-1 antisense but not sense probes, we consistently detected highly abundant silver grain precipitates along the basal epidermal layer of the tips of the rete ridges and to a lesser extent in cells residing in the papillary dermis. In contrast, all suprabasal epidermal cells and (except few single cells) the resident and passenger cells in the dermal compartment were quiescent. Thus, the highest concentration of MCP-1 is most likely achieved at the dermal-epidermal junction and may explain the particular distribution of DM/DD in PS. This suggests that a dialogue between proliferating KC and DM/DD, mediated by MCP-1, is one of the dominating regulatory events in PS pathogenesis.

IL-8 CONCENTRATION IN β-THALASSAEMIA AND AFTER BONE MARROW TRANSPLANTATION

M. Uguccioni[1], R. Meliconi[2], S. Nesci[3], G. Lucarelli[3], G. Gasbarrini[2], M. Ceska[4], and A. Facchini[1]

[1]Lab. Immunologia e Genetica, Istituto di Ricerca Codivilla-Putti, IOR, Bologna, Italy, [2]Patologia Medica I University of Bologna, Italy, [3]Divisione Ematologica e Centro Trapianto, Midollo Osseo, Pesaro, Italy and [4]Sandoz Research Institute, Vienna, Austria

Several immunological defects can be found in patients with β-thalassaemia; among these the impairment of neutrophil phagocytic and killing functions are of utmost relevance. In alternative to blood transfusions and chelating therapy bone marrow transplantation

(BMT) had become an accepted treatment for β-thalassaemia even if graft versus host disease (GvHD) remain an important adverse condition that may occur after BMT. During GvHD neutrophil chemotaxis and phagocytic defects are well recognized. In order to evaluate the role of IL-8 in the neutrophil defects in β-thalassaemia and GvHD, we determined IL-8 serum concentrations in 30 patients with β-thalassaemia before and after BMT. IL-8 was measured by ELISA. Patients with β-thalassaemia had higher serum IL-8 concentrations than age and sex matched normal controls. Patients with severe liver siderosis and fibrosis had the highest IL-8 serum concentration. After BMT, IL-8 serum concentration fell significantly in patients with successful engraftment. Conversely in patients with acute GvHD IL-8 serum concentrations were not statistically different from the concentrations found before BMT and were higher than in patients with no complications and patients with graft rejection. IL-8 may play a role in the immune dysregulation which occurs in β-thalassaemia and it may be involved in the immune mechanisms leading to GvHD.

(Grant from IOR Ricerca corrente - Area 2).

CYTOKINE PRODUCTION IN FELTY'S SYNDROME

R. Meliconi[1,2], M. Uguccioni[3], F. Chieco-Bianchi[1,] G. Kingsley[1], C. Pitzalis[1], G. Gasbarrini[2], M. Ceska[4], A. Facchini[3], and G.S. Panayi[1]

[1]Rheumatology Unit, Dept. of Medicine, UMDS Guy's Hospital, London SE1, U.K., [2]Patologia Medica I University of Bologna and [3]Lab. Immunologia e Genetica, IOR, Bologna, Italy and [4]Sandoz Research Institute, Vienna, Austria

The pathogenesis of Felty's Syndrome (FS) consisting of rheumatoid arthritis (RA), splenomegaly and neutropenia is unknown. 93% of FS are HLA DR4Dw4+; this association implies a role for T cells. FS is mainly characterized by a deficiency in neutrophil number and activity. Various cytokines are directly or indirectly involved in neutrophil maturation and activation. Therefore we studied the potential for cytokine production by peripheral blood mononuclear cells (MNC) from FS (11 cases), RA without neutropenia (10 patients) and from 10 normal controls (NC). MNC were cultured in unstimulated and OKT3 stimulated conditions. The following cytokines were evaluated by ELISA in all supernatants: GM-CSF, TNFα, IL-1β, IL-8. Unstimulated MNC from the 3 groups produced similar amounts of cytokines. All groups showed a similar rate of stimulated GM-CSF and TNFα production. Stimulated RA MNC produced significantly higher amounts of IL-1β than NC. FS MNC production of IL-1β was in between that of RA and NC. IL-8 was produced in greater amounts by FS and RA MNC compared with NC (FS vs. NC = $p<0.01$, Ra vs. NC = $p<0.05$). In addition, we evaluated IL-8 serum concentration in FS, RA and NC. Significantly increased levels of IL-8 were found in FS and RA compared to NC sera ($p<0.001$). Our results show no intrinsic abnormality in the ability to produce cytokines in FS. Since no significant difference in cytokine production was found between RA and FS, cytokines cannot account for the decreased neutrophil count in FS.

(Grant from IOR Ricerca corrente - Area 2).

THE LYMPHOCYTE ATTRACTANT EFFECTS OF INTERLEUKINS 1α AND 8 ARE NOT DUE TO THE RELEASE OF A SECONDARY CHEMOATTRACTANT FACTOR

R. Pleass and R. Camp

St. John's Institute of Dermatology, St. Thomas' Hospital, London, U.K.

The *in vitro* peripheral blood lymphocyte (PBL) attractant effects of cytokines, including interleukin (IL)-1α and IL-8 which have been proposed to play a role in skin disease, may require further characterisation. We have therefore determined whether these effects of IL-α and IL-8 are direct or due to *in vitro* production by PBL of secondary chemoattractant substance(s) during the migration assay. In preliminary experiments the concentration of specific neutralising antibody (NAb) needed to neutralise maximal responses to IL-1α and IL-8 in the PBL migration assay were determined. Subsequently, PBL (2×10^5 in 0.1ml) were incubated with a range of concentrations of IL-1α or IL-8 for 30 min at 37°C. Neutralising concentrations of corresponding NAb were then added after 30 min, supernatants harvested and tested in the PBL migration assay after serial dilution[1]. Responses were compared with those due to the corresponding IL alone[2], and NAb-free supernatant from PBL incubated in the absence of IL (background, Bg[3]), in the same assay. Results were as follows:

IL-8[2]	IL-8 Nab[1]	Bg[3]	IL-1α[2]	IL-1α NAb[1]	Bg[3]
2.07+0.09	1.42+0.12	1.37+0.03	1.97+0.03	1.31+0.12	1.36+0.09

(Results expressed as a mean maximal migration index + S.E.M., n=3).

The lack of attractant activity above Bg in supernatants to which NAb had been added following incubation of PBL with IL-1α and IL-8 for 30 min, indicates that the *in vitro* PBL attractant effect of these cytokines is direct and not due to the release of a secondary chemoattractant factor.

SERIAL CHANGES IN CIRCULATING INTERLEUKIN-8 AND NEUTROPHIL ELASTASE AFTER MAJOR SURGERY

A. Murata[1], H. Toda[1], N. Matsuura[2], Y. Oka[1], and T. Mori[1]

[1]Dept. Surgery II, Osaka Univ. Med. Sch., Osaka, Japan and [2]Dept. Pathol. II Wakayama Med. Sch. Wakayama, Japan

Interleukin 8 (IL-8) is a cytokine known as a potent neutrophil activator. Neutrophil elastase (NE) is released by activated neutrophil. In this study, we tried to examine serial changes in serum immunoreactive IL-8 and NE after major surgeries and elucidate the interaction between cytokine response and surgical trauma. We studied 10 patients undergoing radical surgery against thoracic esophageal cancer, 5 patients undergoing elective total gastrectomy and 3 patients suffering from severe acute pancreatitis. Serum IL-8 levels were determined by EIA and plasma NE levels were also determined by EIA. Furthermore we also examined serum IL-6 levels in these patients simultaneously as an index of severity of surgical trauma (1).

Serum IL-8 levels were 118-616 pg/ml (esophageal cancer), 47.9-530 pg/ml (total gastrectomy) and 430-2405 pg/ml (acute pacreatitis). Peak of IL-8 levels in each group were shown at the 1st postoperative day. Meanwhile, peak plasma NE levels were shown at 3rd or 4th postoperative day in each group and their values were in wide range. Serial changes in serum IL-6 levels were paralleled by IL-8 levels after surgery.

These data suggested that serum IL-8 along with IL-6 were released very early after major surgery and preceded the increase of plasma NE.

1)   Murata, A., et al. 1990. *Immunol. Invest.* 19: 271-278.

ROLE OF INTERLEUKIN-8 IN THE PATHOGENESIS OF NEUTROPHIL-MEDIATED ACUTE LUNG INJURY

N. Matsuura[1], A. Murata[2], H. Imanaka[3], H. Toda[2], H. Nakagawa[3], N. Tomiyama[1], N. Takeuchi[1], N. Taenaka[3], and K Kakudo[1]

[1]Dept. Pathol. II, Wakayama Med. Sch., Wakayama, Japan, [2]Dept. Surg. II, Osaka Univ. Med. Sch., [3]Intensive Care Unit, Osaka Univ. Hospital, Osaka, Japan

Neutrophils play an important role in the pathogenesis of acute lung injury, suggesting that neutrophil accumulation should precede acute lung injury. Interleukin 8 (IL-8) is known to be a major stimulant of neutrophil chemotaxis. We investigated the effect of intravenous administration of IL-8 into rabbits to clarify the possible role of IL-8 in acute lung injury. Recombinant IL-8 was administered systemically by a bolus injection in rabbits which were placed on a high frequency oscillation. Arterial blood gas and counts of circulating WBC were analysed sequentially. Rabbits were autopsied to examine the changes in the lung histologically. After IL-8 injection, PaO_2 and the count of circulating WBC were decreased rapidly, until the value of PaO_2 and circulating WBC nadired 70% and 20% of control, respectively. Histological examination revealed massive accumulation of neutrophil into the lung. PaO_2 was recovered within 30 min. and also circulating WBC were increased above the level of control after 30 min. These data demonstrated that IL-8 administration resulted in mild respiratory insufficiency accompanied with prominent circulating WBC decrease due to the accumulation of neutrophil into lung. To generate severe acute respiratory failure observed clinically, the second stimulus may be required to activate neutrophil to damage lung tissues.

INTERLEUKIN-8 IN ACUTE MENINGOCOCCAL INFECTIONS; CORRELATION WITH SEVERITY OF DISEASE

M. van Deuren, J. vd Ven-Jongekrijg, J.W.M. van der Meer

Department of Internal Medicine, University Hospital Nijmegen, The Netherlands

Endotoxin (ET) and ET stimulated pro-inflammatory cytokines such as TNF, IL-1β and IL-6 play a pivotal role in the genesis of severe Gram-negative infections, e.g. meningococcal meningitis or sepsis. High plasma concentrations of these cytokines correlate with the severity of disease. Since ET, TNF, as well as IL-1β induce IL-8 production, we measured the plasma concentration of IL-8 in 12 patients with acute meningococcal infections at admission at the Intensive Care Unit, by using an ELISA (Quantikine™, R and D Systems, Minneapolis, USA).

Patients were divided in 3 subgroups. Group A (n=6) had meningitis (leucocytes in CSF > 100/mm^3) without hemodynamic involvement; Group B (n=3) had meningitis with hypotension or shock and group C (n=3) had shock without meningitis (leucocytes in CSF < 100/mm^3). One patient - in Group C - died shortly after admission.

Median values of IL-8 in group A were 63 pg/ml (range 32-250), in Group B 2900 pg/ml (325-3050) and in Group C > 6000 pg/ml (2200 - 6000). In 3 patients, each belonging to one of the subgroups, IL-8 was followed during 8 days. In these patients IL-8 decreased to < 100 pg/ml at 50 hours after start of antibiotics and remained low during reconvalescence. The IL-8 plasma concentration probably reflects the ET mediated activation of the cytokine network and may - next to TNF, IL-1β and IL-6 - be used as a prognostic indicator in acute meningococcal disease.

IL-6 AND IL-8 LEVELS IN PATIENTS WITH RHEUMATOID ARTHRITIS

T. Takakuwa[1], S. Endo[1], K. Inada[2], H. Yamashita, M. Hoshi[3], M. Komagamine[3], S. Hoshi[1], and M. Ceska[4]

[1]Critical Care and Emergency Center and [2]Department of Bacteriology of Iwate Medical University, [3]Morioka Yuai Hospital, 19-1 Uchimaru, Morioka 020, Japan and [4]Sandoz Research Institute, Vienna, Austria

IL-6 and IL-8 levels in the synovial fluid and plasma in patients with rheumatoid arthritis (RA) and osteoarthritis (OA) were measured. Fourteen patients with RA and 12 patients with OA were subjected in this study. IL-6 and IL-8 levels were measured by ELISA. The measurable limit of these 2 levels was 0.156 ng/ml and 0.01 ng/ml. IL-6 level in the synovial fluid in all patients with RA was 5.05 ± 0.93 ng/ml (mean ± SD) exceeding detectable limit. IL-6 level in 8 of 12 patients with OA was 1.12 ± 0.73 ng/ml exceeding detectable limit. IL-8 level in synovial fluid in all patients with RA was 9.30 ± 2.85 ng/ml exceeding measurable limit. IL-8 level in 4 of 12 patients with OA was 0.33 ± 0.38 ng/ml exceeding measurable limit. Both IL-6 and IL-8 levels in the synovial fluid were significantly higher in patients with RA than those in patients with OA. The significant correlation was observed between IL-6 and IL-8 levels in the synovial fluid in patients with RA (r=0.5864, p<0.05). Both IL-6 and IL-8 levels in the synovial fluid in volunteers were less than detected limit. IL-6 in the plasma was detected in 11 of 14 patients with RA and 2 of 12 patients with OA, and its level was low. IL-8 in the plasma was detected in 4 of 14 patients with RA and 1 of 12 patients with OA, and its level was low in all cases. From the above mentioned results, it was suggested that production of high levels of IL-6 and IL-8 in the synovial fluid in patients with RA may be related to the morphological formation of the arthritis.

PLASMA INTERLEUKIN 8 LEVELS IN PATIENTS WITH SEPTIC SHOCK

K. Inada[1], S. Endo[2], T. Takakuwa[2], A. Yamada[2], Y. Kuwata[2], S. Hoshi[2], H. Yamashita[1], M. Yoshida[1], and M. Ceska[3]

[1]Department of Bacteriology and [2]Critical Care and Emergency Center, Iwate Medical University, 19-1 Uchimaru, Morioka 020, Japan and [3]Sandoz Research Institute, Vienna, Austria

IL-8, polymorphonuclear leukocyte elastase (PMNE) and endotoxin at the development of septic shock were measured. IL-8 and PMNE were measured by ELISA. Endotoxin level was measured by endotoxin-specific Endospecy test (Seikagaku Corporation, Ltd., Tokyo). The pre-procedure of the plasma was carried out by using new PCA method developed by us. Its normal level was less than 9.8 pg/ml. IL-8 level at the development of the septic shock was 6.28 ± 9.00 ng/ml (mean $\pm$ SD, n=29), and significantly higher (p<0.001) than 0.35 ± 0.35 ng/ml (n=40) in the sepsis without shock. IL-8 level in volunteers was less than measurable limit. The significant correlation was observed between IL-8 and PMNE levels at the development of septic shock (r=0.6916, p<0.001). No correlation was observed between endotoxin and IL-8 levels at the development of septic shock (r=0.1839, p=0.3395). It was suggested that IL-8 may be related to the production of PMNE at the septic shock, and substances except for endotoxin may also be related to the stimulation of the production of IL-8.

INHIBITORY EFFECT OF PROTEASE-INHIBITOR FOR PRODUCTION OF INTERLEUKIN 8 (IL-8) AND POLYMORPHONUCLEAR LEUKOCYTE ELASTASE (PMNE)

S. Endo[1], K. Inada[2], T. Takakuwa[1], S. Hoshi[1], H. Yamashita[2], and M. Yoshida[2]

[1]Critical Care and Emergency Center and [2]Department of Bacteriology, Iwate Medical University, 19-1 Uchimaru, Morioka 020, Japan

We have reported inhibitory effect (inhibitory effect of BRM) of Ulinastatin (Mochida Pharmaceutical Co., Ltd., Tokyo) and Nafamostat Mesilate (FUT-175, Babyu Pharmaceutical Co., Ltd., Tokyo) being protease-inhibitors for production of TNFα, IL-1β and IL-6 in human peripheral mononucleocytes due to endotoxin-stimulation. This time, we studied inhibitory effect due to Ulinastatin and FUT-175 for production of IL-8 in human peripheral granulocytes and human vasoendothelial cells by endotoxin-stimulation. When endotoxin-stimulation was carried out after adding Ulinastatin and FUT-175 in human peripheral granulocytes and human vasoendothelial cells, production of IL-8 was almost dose-dependently inhibited. When human peripheral granulocytes were stimulated by endotoxin after addition of Ulinastatin and FUT-175, production of elastase was dose-dependently inhibited. When human peripheral granulocytes were stimulated by endotoxin after addition of anti-IL-8 antibody, production of PMNE tended to be inhibited. It was suggested that IL-8 is related to the production of PMNE. The production of IL-8 in human peripheral granulocytes and human vasoendothelial cells due to endotoxin-stimulation was inhibited by Ulinastatin and FUT-175 is effective for prevention of damage of organs developing by activation of neutrocytes due to IL-8.

EFFECT OF DILUENTS AND CAPTURE ANTIBODIES ON DETECTION OF NAP-1 BY SANDWICH ELISA

I. Sylvester, A.J. Suffredini, R.L. Danner, and E.J. Leonard

National Institutes of Health, Frederick and Bethesda, MD, U.S.A.

This study was undertaken because of a problem that arose during an investigation of the effects of i.v. LPS in normal human subjects. It was reported that LPS caused a rapid and striking increase in serum NAP-1. However, in subsequent assays of these sera, peak serum NAP-1 concentrations were considerably lower with an ELISA developed in our laboratory than with a commercially available sandwich ELISA kit (R and D Systems, Minneapolis, MN). We now describe the cause of this discrepancy, which related to the RD6 diluent used in the R & D ELISA for the NAP-1 standard curve. Standard curves were compared for NAP-1 in different diluents, including buffered BSA, pooled human serum, and RD6 (a presumed animal serum, specified by R & D only for dilution of NAP-1 for the standard curve). With the R & D ELISA plate monoclonal capture antibody, the standard curve for NAP-1 diluted in RD6 was depressed 4 to 6-fold relative to the curve for NAP-1 diluted in buffered BSA or in pooled human serum with undetectable NAP-1. It follows that serum NAP-1 concentrations calculated from the RD6 standard curve are incorrectly high, since the RD6 curve is depressed by something in RD6 that is not present in pooled human serum. The effect of diluents also depends on the specific capture antibody. With a different capture antibody, dose-response curves for NAP-1 diluted in buffered BSA or RD6 were identical, whereas the curve for NAP-1 diluted in pooled human serum was depressed. Since diluents or serum do not affect different monoclonal antibodies in the same way, careful evaluation of these interactions is required in the development or application of an ELISA system.

DO THE TWO TYPES OF IL-8 RECEPTORS ON HUMAN NEUTROPHILS MEDIATE DIFFERENT CELLULAR RESPONSES?

C. Lam, A. Tuschil, and I.J.D. Lindley

Sandoz Research Institute, Vienna, Austria

Although two distinct calcium mobilizing IL-8 receptors have recently been cloned (1,2), it is unclear whether coupling of both receptors is necessary for IL-8 to stimulate neutrophil responses other than calcium transients. In an attempt to address this issue, we used NAP-2 at concentrations shown recently to mobilize intracellular calcium (3) presumably by binding to common receptors on human neutrophils with two quite different affinities (3,4). Addition of IL-8 or NAP-2 to human neutrophils resulted in a transient mobilization of cytosolic calcium with minimum effective dose observed at 0.1 and 6 nM respectively. Preactivation of the neutrophils with an optimal amount of NAP-2 (>12 nM) significantly suppressed the calcium transients that could be stimulated by IL-8 at < 1 nM but not at higher concentrations. Similarly, pre-exposure of the cells to IL-8 at 0.1 nM blocked the stimulatory effect of NAP-2 at < 100 nM but not at higher concentrations. These effects may be ascribed to a phenomenon of tachyphylaxis. However, when neutrophil activation was monitored by the release of elastase, NAP-2 was only active at > 100 nM whereas IL-8 was effective at 1 nM. Exposure of the neutrophils to NAP-2 for 2 min. markedly attenuated exocytosis in response to IL-8 dose-dependently. Similar effects were observed when NAP-2 was added simultaneously with the IL-8, suggesting that NAP-2 interfered with its proper binding. Together, these findings suggest that the coupling of any one type of IL-8 receptor results in calcium mobilization whereas coupling of both receptors may be a prerequisite for exocytosis. The existence of NAP-2 in a form that differs from IL-8 would partly explain the apparent antagonistic effect of NAP-2 seen here.

1. Holmes, W.E., J. Lee, W-J. Kuang, G.C. Rice, and Wood. 1991. *Science*. 253: 1278-1280.
2. Murphy, P.M., and H.L. Tiffany. 1991. *Science*. 253: 1280-1283.
3. Moser, B., C. Schumacher, V. von Tscharner, I. Clark-Lewis, and M. Baggiolini. 1990. *J. Biol. Chem.* 266: 10666-71.
4. Schnitzel, W., B. Garbreis, U. Monschein, and J. Besemer. 1991. *BBRC*. 180: 301-307.

INTERLEUKIN 8 PLAYS A FUNDAMENTAL ROLE IN INFLAMMATORY PROCESSES IN VIVO

M. Bolanowski, M. Baganoff, C. Deppeler, D. Meyer, D. Widomski[1], D. Fretland[1], Y. Zhang[2], B. Jakschik[2]

[1]Monsanto Company, St. Louis, MO, U.S.A., G.D. Searle, Skokie, IL, U.S.A., and [2]Washington University Medical School, St. Louis, MO, U.S.A.

Several lines of evidence suggest that human interleukin-8 (IL-8) is a potent mediator of neutrophil (PMN) infiltration, inflammation, and tissue damage associated with several diseases (e.g. psoriasis, rheumatoid arthritis, and inflammatory bowel disease). To more fully define the role of IL-8 in human disease, we developed a mouse monoclonal antibody (DM/C7) which neutralized the biological activity of human IL-8. *In vitro* DM/C7 specifically inhibited IL-8 binding to receptors on human PMNs (IC_{50} = 0.4 µg/ml vs. 1 nM IL-8) and chemotaxis of human PMNs toward IL-8 (IC_{50} = 0.9 µg/ml vs. 2 nM IL-8). As expected, DM/C7 had no effect on the *in vitro* activities of IL-1β, fMetLeuPhe, C5a, or LTB$_4$. In a mouse dermal chemotaxis assay, DM/C7 (10 µg, i.d. or 100 µg, i.v.) specifically blocked 100% of PMN influx toward sites of IL-8 injection (100 ng, i.d.). In this assay, DM/C7 also inhibited all PMN influx toward sites of IL-1β injection but had no effect on influx toward sites of LTB$_4$ or C5a injection. In normal mice, PMN infiltration of the skin induced by deposition of IgG immune complexes was reduced 60-75% by DM/C7 (100 µg, i.v.). Surprisingly, DM/C7 was totally ineffective vs. IgG complex induced PMN influx in W/Wγ mast cell deficient mice. Following reconstitution of mast cells from a syngeneic donor, W/Wγ mice acquired a DM/C7 sensitive PMN influx which was indistinguishable from that observed in normal animals. Together, these data strongly support three conclusions: 1) local or systemic administration of DM/C7 neutralizes IL-8; 2) DM/C7 neutralizes the biological activity of an endogenous mouse antigen; and 3) IL-8 plays a fundamental role in at least one PMN-dependent inflammatory process *in vivo*. Furthermore, the inability of DM/C7 to inhibit PMN infiltration of W/Wγ mouse skin suggests that mast cells may mediate the response to or production of IL-8 in this model. The protective effect of DM/C7 in mice implies that small molecule IL-8 receptor antagonists will be therapeutically useful in humans.

ENDOTHELIAL CELL BINDING OF NAP-1/IL-8: ROLE IN NEUTROPHIL EMIGRATION

A. Rot

Sandoz Research Institute, Vienna, Austria

Experimental evidence indicates that neutrophil emigration induced *in vivo* by intradermal injections of various attractants, including NAP-1/IL-8, is regulated by mechanisms involving hypothetical attractant receptors present in skin sites. Using an in situ binding assay on human and animal skin samples, we identified NAP-1/IL-8 binding sites on the surface of venular endothelial cells. We argue that these binding sites are instrumental for NAP-1/IL-8-induced neutrophil emigration. Under the conditions of venous blood flow only the endothelial cell surface-bound NAP-1/IL-8 can effectively stimulate the second, integrin-mediated step of neutrophil adhesion to the endothelium. Conversely, soluble NAP-1/IL-8 should inhibit neutrophil-endothelial cell adhesion. Thus the endothelial binding sites for NAP-1/IL-8 could be involved in regulation of neutrophil-endothelial cell adhesion.

In addition, we show that NAP-1/IL-8, but not formyl peptide, induces the *in vitro* neutrophil migration in Boyden-type chamber when bound to the surface of the polycarbonate filter. By analogy with *in vitro* haptotaxis we also suggest that neutrophil transmigration across the endothelial cell barrier can be induced by NAP-1/IL-8 bound to the endothelial cell surface.

HEPARIN BINDING OF NAP-1/IL-8: THE EFFECT OF HEPARIN AND RELATED GLYCOSAMINOGLYCANS ON IN VITRO NEUTROPHIL RESPONSES TO NAP-1/IL-8

L.M.C. Webb[1], B. Moser[2], M. Baggiolini[2], and A. Rot[1]

[1]Sandoz Research Institute, Vienna, Austria and [2]Theodor Kocher Institut, University of Bern, Bern, Switzerland

NAP-1/IL-8, by its structural homology to PF4, was postulated to bind to heparin and related glycosaminoglycans. We confirmed the affinity of NAP-1/IL-8 for heparin and using chemically synthesised analogs of NAP-1/IL-8 found that this interaction occurred via the C-terminal alpha helix of NAP-1/IL-8 i.e. a site of the molecule distinct from the putative neutrophil receptor-binding domain. We also found that heparan sulfate enhanced neutrophil migration in response to NAP-1/IL-8. This enhancement was not seen with heparin, chondroitin sulfate A or chondroitin sulfate B. Neutrophil migration induced by formyl peptide was unaffected by heparan sulfate and related molecules. Conversely, the *in vitro* activity of elastase released by neutrophils in response to either NAP-1/IL-8 or formyl peptide was significantly inhibited by both heparin and heparan sulfate.

These observations imply that heparan sulfate, present *in vivo* on the endothelial cell surface and in the basement membrane, can enhance the chemotactic activity of NAP-1/IL-8 for neutrophils yet inhibit the activity of released elastase. This may explain why NAP-1/IL-8-induced neutrophil transmigration across the endothelial cell barrier results in minimal damage to endothelial cells and surrounding tissue.

THE EFFECT OF 5-LIPOXYGENASE INHIBITORS ON THE ACTIVITY OF IL-8

P.J. Roberts, A.R. Pizzey, and D.C. Linch

Department of Haematology, UCHMS, University College, London, U.K.

The cytokine interleukin-8 (IL-8) has direct effects on neutrophils such as increasing the expression of CD11b adhesion molecules or chemotactic peptide receptors and also enhances agonist-mediated responses such as the generation of superoxide. The metabolism of endogenous ^{3}H-arachidonate can be estimated by the release of radioactive metabolites from the cell. Incubation of purified neutrophils with IL-8 (10-500 ng/ml) for up to 30 min. did not directly stimulate this activity, however IL-8 at 10 and 100 ng/ml enhanced the activity of cells stimulated with calcium ionophore (1 μM) by 126 $\pm$ 16% (n = 5) and 144 $\pm$ 11% (n = 12), respectively. To examine the role of leukotriene metabolites such as 5-HETE or LTB_4 in the signal transduction pathways of IL-8, we employed two lipoxygenase inhibitors, piriprost (1) at 87 μM and MK886 (2) at 100 nM, concentrations that inhibited the release of radioactivity from ionophore-stimulated cells by 94 $\pm$ 1% (n = 5) and 89 $\pm$ 2% (n = 11), respectively. The inhibitory activity of these compounds on leukotriene synthesis under identical conditions was confirmed by reverse phase HPLC. MK886 inhibited the release of leukotrienes from IL-8-primed cells stimulated with ionophore by 96 $\pm$ 1% (n = 4), but did not inhibit the upregulation of CD11b stimulated by IL-8, nor did it inhibit the priming of fMLP-stimulated superoxide production by IL-8. Similar data were obtained with piriprost. We conclude therefore that leukotriene synthesis does not mediate the direct activation of phagocytes by IL-8, nor the priming of the fMLP-stimulated respiratory burst.

1.   Bach, M.K., J.R. Brashler, and H.W. Smith *et al.* 1982. *Prostaglandins* 23: 759-771.
2.   Gillard, J. *et al.* 1989. *Can. J. Physiol., Pharmacol.* 67: 456-464.

CHEMOTACTIC CYTOKINE GENERATION BY HUMAN LUNG MICROVASCULAR ENDOTHELIAL CELLS: DIFFERENTIAL REGULATION BY GAMMA INTERFERON

Z. Brown[1], M.E. Gerritsen[2], R.M. Strieter[3], S.L. Kunkel[3], and J. Westwick[1]

[1]Department of Pharmacology, University of Bath, Avon, U.K., [2]Miles Laboratories, New Haven, Connecticut, U.S.A. and [3]Department of Pathology, University of Michigan, Ann Arbor, U.S.A.

The migration of selective populations of leucocytes from the intravascular lumen to extravascular sites in response to an inflammatory stimulus involves a series of co-ordinated events which include leucocyte adhesion, diapedesis and chemotaxis. The primary vascular cell which initiates these events is probably the microvascular endothelial cell. In this study we have examined human lung microvascular endothelial cells (HLME) for their ability to express and secrete IL-8 and MCP-1 in response to pro-inflammatory cytokines. Resting HLME did not express detectable amounts of IL-8 or MCP-1, as determined by Northern analysis for mRNA or as antigenic levels by ELISA. Treatment of these cells with either IL-1 (0.01-10 units/ml) or TNFα (0.1-10 ng/ml) produced a dose related secretion of antigenic IL-8; 2.1 ± 0.13 to 12.3 ± 0.5 ng and 5.7 ± 0.33 to 18.9 ± 1.5 ng with IL-8 and TNF respectively. Similarly, IL-1 and TNF-induced a dose related generation of MCP-1 2 ± 0.08 to 31 ± 0.27 ng/ml or 2.1 ± 0.2 to 11.9 ± 0.71 ng/ml with IL-1 and TNF respectively. Northern analysis for IL-8 and MCP-1 mRNA confirmed the above results. The addition of gamma interferon to resting HLME produced a dose related expression of MCP-1 mRNA, without increase in IL-8 mRNA. Interferon gamma (0.5 to 500 units/ml) induced a dose related of MCP-1, secretion of 0.85 ± 0.06 to 9.46 ± 0.22 ng/ml, while antigenic IL-8 remained below the level of detection (25 pg/ml).

Thus gamma interferon - a product of activated T cells - has the potential of dictating leucocyte traffic by selective induction of MCP-1 by microvascular endothelial cells.

THE EFFECT OF HAEMOFILTRATION ON CIRCULATING CYTOKINES IN CHILDREN UNDERGOING CARDIOPULMONARY BYPASS: A PRELIMINARY STUDY

L. Armstrong[1], N. Moat[3], J. van der Linden[3], Z. Brown[2], R.L. Robson[2], J.C.R. Lincoln[3], J. Westwick[2], and A.B. Millar[1]

[1]Schools of Postgraduate Medicine and [2]Pharmacy and Pharmacology, University of Bath, Bath, U.K. and [3]Royal Brompton National Heart and Lung Hospital, U.K.

Pulmonary dysfunction still occurs following cardiopulmonary bypass (CPB) and remains a major cause of morbidity and mortality especially in the paediatric age group. This is consequent upon the so called systemic inflammatory response to CPB with an increase in inflammatory mediators. Haemofiltration (HF) may be able to attenuate the effects of this response by elimination of some or all of these mediators. We undertook a prospective randomised study to investigate the effect of HF on plasma levels of TNF alpha, IL-6, and IL-8, in 18 infants and children undergoing deep hypothermic CPB. Serial plasma samples were taken before, during and up to 24 hrs post CPB. TNF alpha was assayed in the plasma samples using a specific bioassay (WEHI cell line). IL-6 and IL-8 were detected by using highly specific ELISAs. In all cases there was an increase in TNF alpha reaching a maximum between 2 and 3 post CPB of 0.31 ± 0.57 iu/ml in those haemofiltered and 6.98 ± 20.4 iu/ml in those not haemofiltered. There were wide variations in these levels but at all time points, TNF alpha was lower in those patients who were haemofiltered. IL-6 levels increased in 15 of the 18 subjects after CPB ($0.03- > 2$ ng/ml) but there were no consistent differences with HF. IL-8 was detected in 12 patients, 7 of whom had HF (1.12 - 3.24 ng/ml).

We conclude that TNF alpha and IL-6 are induced by CPB and may contribute to the systemic inflammatory response in this condition. By contrast, IL-8 production was variable in relation both to CPB and HF. This preliminary study suggests a potential role for HF to alter cytokine production in response to CPB.

INTERLEUKIN 8 (IL-8) IS PRODUCED BY THE BRONCHOALVEOLAR LAVAGE CELL POPULATION FROM PATIENTS WITH ACTIVE BUT NOT INACTIVE SARCOIDOSIS

A. Millar[1], Z. Brown[2], M. Spiteri[3], and J. Westwick[2]

[1]Schools of Postgraduate Medicine and [2]Pharmacy and Pharmacology, University of Bath, Bath, U.K. and [3]Royal Free Hospital, London, U.K.

Pulmonary sarcoidosis results in fibrotic changes within the lungs in some patients but the mechanisms by which this occurs are poorly understood. Interleukin-8 (IL-8) is a potent chemoattractant cytokine which has been proposed as playing a role in the development of lung injury and subsequent fibrosis in other forms of pulmonary disease. We considered that IL-8 might have a similar role in pulmonary sarcoidosis. In order to investigate this, bronchoalveolar lavage was performed on three groups of subjects: 6 symptomatic patients with biopsy proven sarcoidosis prior to any treatment, 6 asymptomatic patients with biopsy proven sarcoidosis and no treatment for at least 6 months, and 6 normal volunteers. Whole cell populations from this lavage were cultured for 24 hours in serum free media. The resultant supernatants were assayed for the presence of IL-8 using a highly specific double ELISA. No IL-8 was detected in the cell culture supernatants of the normal volunteers nor the asymptomatic patients with sarcoidosis, but was present $(11.93 \pm 0.71$ ng/ml, SD $\pm$ SEM) in the supernatants from those with active sarcoidosis. These findings suggest that IL-8 production may have a role in the acute inflammatory phase of sarcoidosis, although the mechanisms by which subsequent fibrosis develops remain unclear.

GASTRIC INTERLEUKIN-8 AND INTERLEUKIN-8 IgA AUTOANTIBODIES IN *HELICOBACTER PYLORI* INFECTION

J.E. Crabtree[1], P. Peichl[2], J.I. Wyatt[1], U. Stachl[2], I.J.D. Lindley[2]

[1]Departments of Clinical Medicine and Pathology, St. James's University, Hospital, Leeds, U.K. and [2]Sandoz Research Institute, Vienna, Austria

Gastric infection with *Helicobacter pylori* (HP), which causes chronic gastritis, is characterised by neutrophil infiltration. Whilst HP derived neutrophil chemotactic factors have been identified, little is known of the potential of the gastric mucosa to produce chemotactic cytokines. Using *in vitro* organ culture techniques, this study has investigated 1) IL-8 production by human antral endoscopic biopsies and 2) gastric secretion of autoantibodies to IL-8. 24 hour IL-8 secretion in HP+ patients with active gastritis (i.e. intraepithelial neutrophil infiltration) (mean $\pm$ SEM, 89.1 $\pm$ 12.7 ng/mg biopsy protein, n = 17) was greater than patients with inactive gastritis (28.8 $\pm$ 8.01, n = 10, p < 0.002) (8/10 HP+) and HP negative subjects with normal mucosa (21.7 $\pm$ 4.1, n = 19, p < 0.0001) or reactive gastritis (19.1 $\pm$ 6.7, n = 4, p < 0.01). IgA autoantibodies to IL-8 were present in 19 patients (13 active gastritis, 4 inactive gastritis) and concentrations were correlated with IL-8 production (p < 0.001, r = 0.51, n = 50). Immunolocalisation demonstrated that IL-8 was present in the epithelium of histologically normal antral mucosa (n = 7) and that HP infection was associated with increased epithelial and lamina propria IL-8 expression. The local synthesis of IL-8 is likely to be an important factor in regulating mucosal neutrophil infiltration and activation in patients with HP infection. The pathological significance of mucosal IL-8 IgA autoantibodies and the expression of this cytokine in gastric epithelium requires further investigation. The IgA anti-IL-8 antibodies may however represent a down-regulatory response of the host to limit damage associated with a chronic bacterial infection.

EOSINOPHILS FROM ALLERGIC INDIVIDUALS, BUT NOT NORMAL INDIVIDUALS, SHOW CHEMOTACTIC RESPONSIVENESS TO NAF/IL-8

P.L.B. Bruijnzeel[1], R.A.J. Warringa[2], and L. Koenderman[2]

[1]Department of Pulmonary Diseases, University Hospital Utrecht, Utrecht, The Netherlands and [2]MBL/TNO Department of Pharmacology, Rijswijk, The Netherlands

Eosinophils from peripheral blood of normal individuals do not show chemotactic responsiveness to NAF/IL-8. In contrast eosinophils from allergic asthmatics or atopic dermatitis patients do show a significant chemotactic response. *In vitro* experiments have indicated that eosinophils from normal individuals become responsive toward NAF/IL-8 after pretreatment with pM concentrations of the cytokines IL-3, IL-5, and GM-CSF. The magnitude of this response is almost similar to the one observed in allergic individuals. This suggests that eosinophils from allergic individuals have been primed *in vivo* and that the above mentioned cytokines could be responsible. NAF/IL-8 becomes chemotactic at a concentration of 0.1 nM and is optimally chemotactic at 10 nM. The number of cells mobilised amounts to about half of that mobilised by optimal PAF concentrations. Moreover, when mild allergic asthmatic individuals were challenged with allergen a significant increase of the NAF/IL-8-induced chemotactic response of eosinophils occurred 3 hours after challenge. This suggests that NAF/IL-8 may contribute to eosinophil mobilisation in allergic inflammatory reactions and more particular in the late phase allergic reaction. This idea is further substantiated by the fact that NAF/IL-8 also is capable of mobilising eosinophils across endothelial cell monolayers.

MONOCYTE CHEMOTACTIC PEPTIDE-1 (MCP-1) AND NAP-1/IL-8 FORMATION IN BABOON SEPTICEMIA ARE PARTIALLY TNF DEPENDENT

H. Redl[1], G. Schlag[1], E. Paul[1], B. Strieter[2], S. Kunkel[2], M. Ceska[3], J. Davies[4], M. Bodmer[5], and R. Foulkes[5]

[1]Ludwig Boltzmann Institute Exp. Clin. Traumatol, Vienna, Austria and [2]Dept. Pathol., University of Michigan, Ann Arbor, U.S.A. [3]Sandoz Research Institute, Vienna, Austria. [4]Roodeplaat Res. Lab., Pretoria, S.A. and [5]Celltech Ltd., Berkshire, U.K.

Cytokines like IL-1 and IL-6 are released in sepsis in a TNF dependent manner. Therefore we asked the question, whether TNF might be responsible for MCP-1 and NAP-1 production in baboon septicemia (2 hours infusion of live E. coli 2 x 10[9] CFU/kg). We performed sepsis experiments with and without pretreatment with humanized TNF-AB (CDP571, Celltech) 1 mg/kg, which completely neutralizes circulating TNF. MCP-1 was increased from 0.6 ng/ml at baseline up to 6.8 ng/ml at 4 hours.

(median Q_1-Q_3) ng/ml	MCP-1 (4)	NAP-1 (4h)
Control	6.8 (5.5 - 7.7)	18.2 (16.7 - 23.4)
TNF AB 1 mg/kg	* 3.9 (2.2 - 5.9)	* 7.3 (5.7 - 12.9)

* Signif. different.

Despite high circulating levels of endotoxin ($\approx$ 200 ng/ml peak), neutralizing TNF leads to a diminished production of MCP-1 and NAP-1. Therefore we conclude that MCP-1 plasma levels in baboon septicemia are increased and partially dependent on TNF, despite the presence of a high LPS plasma concentration.

INCREASED EXPRESSION OF MONOCYTE CHEMOATTRACTANT PROTEIN-1 (MCP-1) IN GLOMERULI FROM RATS WITH ANTI THY 1.1 GLOMERULONEPHRITIS

R.A.K. Stahl[1], M. Disser[1], K. Hora[2], and D. Schlöndorff[2]

[1]Departments of Medicine, Divisions of Nephrology, University of Frankfurt am Main, Germany and [2]Albert Einstein College of Medicine, New York, U.S.A.

The infiltration of monocyte-macrophages in the glomerulus is one of the hallmarks of glomerulonephritis and may play an important pathogenetic role. Monocyte chemoattractant protein-1 (MCP-1) and colony stimulating factor-1 (CSF-1) are monocyte-specific cytokines with chemoattractant and activating activities for monocytes. MCP-1 and CSF-1 can be generated by several cell types including glomerular mesangial cells and can be stimulated by cytokines and immune complexes. To study the expression of CSF-1 and MCP-1 in a model of proliferative glomerulonephritis we used Northern blot analysis and immunohistochemistry. The glomerular lesion was induced in rats by the i.v. injection of a heterologous antibody, directed against the Thy 1.1 antigen which is localized on glomerular mesangial cells. Northern blot analysis revealed comparable amounts of CSF-1 in glomeruli isolated from control, untreated rats, and from rats after 30 min to 3 wks of injection of Thy 1.1. In contrast, control glomeruli contained low mRNA levels for MCP-1, which markedly increased 30 min after the induction of the nephritis, were then reduced at 24 hours and increased again at 5 and 21 days after induction of the disease. These time points following antibody injection are associated with mesangial immune complex formation (30 min), mesangiolysis (24 hours) and proliferative glomerulonephritis (5 and 21 days). By immunohistology the presence of MCP-1 was demonstrated only in glomeruli with a predominant mesangial distribution. The mesangial immunofluorescence for MCP-1 followed a pattern similar to that of the mRNA for MCP-1 after induction of the disease process, i.e. it increased after 30 min, decreased after 24 hrs and was increased again at 3 weeks. Within 30 min of the antibody injection an increased infiltration of monocyte-macrophages was observed in the glomeruli, which was maintained up to 3 wks of induction of the glomerulonephritis.

We conclude that MCP-1 is increased early on in glomeruli of rats with immune-mediated mesangial proliferative glomerulonephritis and might play an important role in the recruitment of monocyte-macrophages into glomeruli following *in situ* immune complex formation.

ROLE OF MACROPHAGE INFLAMMATORY PROTEIN (MIP)-1α IN GRANULOMATOUS INFLAMMATION

N.W. Lukacs, S.L. Kunkel, R.M. Strieter, K. Warmington, and S.W. Chensue

Dept. of Pathology and Internal Medicine, University of Michigan Medical Center, Ann Arbor, MI 48109, U.S.A.

Macrophage inflammatory protein (MIP)-1 is a low MW, LPS-inducible monocyte and neutrophil chemotactic cytokine which may be important in inflammation and pulmonary granuloma formation. The present study examined the kinetic generation, and *in vivo* relevance of murine MIP-1α utilizing synchronous pulmonary *Schistosoma mansoni* egg granulomas. Antigenic MIP-1α was measured in 24 hr supernatants from whole granulomas cultured (700/ml) with or without soluble egg antigen (SEA) utilizing an ELISA developed in our laboratory. Intact primary granulomas isolated at various times of development from normal mice showed low backgroud levels of MIP-1α production (<1 ng/ml), however, when challenged with antigen demonstrated significant production of MIP-1α beginning at day 8 (2.6 ng/ml) and peaking at day 16 (16.8 ng/ml). Intact pulmonary granulomas isolated from acutely infected mice demonstrated high backgroud levels of MIP-1α production (peaking at day 2; 7.8 ng/ml). Likewise, background levels from chronic infection granulomas (day 2; 5.6 ng/ml) were similar to acute granulomas. Antigen stimulation increased expression of MIP-1α at all time points with granulomas from acutely infected (peaking at day 2; 16 ± 4.6 ng/ml) but not chronically infected (peaking at day 8; 3.45 ± 1.3 ng/ml) mice. Treatment of mice with polyclonal rabbit anti-mouse MIP-1α (6,125 ± 310 um^2) abrogated 8 day primary pulmonary granuloma formation when compared to normal serum control group (12,704 ± 1154 um^2). Anti-MIP-1α sera also decreased granuloma formation in lungs of acutely (4-day; 27,897 ± 2400 um^2), but not chronically (4-day; 15257 + 1058 um^2) infected mice compared to normal serum treated groups (36,010 ± 2507 and 17319 ± 1016 um^2, respectively). Immunohistochemical staining of granuloma sections from acutely infected mice verified localization of MIP-1α production within the cells throughout the granuloma. Our findings demonstrate MIP-1α production is an important mediator in the cascade of events which lead to the formation of inflammatory granulomas.

NAP-1/IL-8 IS A WEAK INDUCER OF T CELL MIGRATION INTO SKIN

I.G. Colditz and D.L. Watson

CSIRO Division of Animal Health, Armidale, NSW 2350, Australia

The effect on lymphocyte migration into skin of intradermal injection of recombinant human IL-8 was compared to other chemotactic agonists and to cytokines known to induce leucocyte-endothelial adhesion molecules. Lymphocytes, collected from efferent prefemoral lymph of sheep and labelled with [111]In *in vitro*, migrated in large numbers to skin sites stimulated with PPD (delayed type hypersensitivity), TNFα or IFNγ. In contrast, IL-1α caused little accumulation of ^{111}In-lymphocytes at concentrations that induced intense accumulation of ^{111}In-neutrophils. The chemotactic agonists IL-8 and zymosan activated plasma (ZAP) recruited large numbers of neutrophils but only small numbers of lymphocytes, whereas PAF and LTB$_4$ failed to recruit lymphocytes into skin. IFNγ was the only mediator to recruit lymphocytes in preference to neutrophils into skin. The results suggest that lymphocyte chemotactic agonists (IL-8, C5a) lacking the ability to induce adhesion molecules on endothelium play a minor role in directing the migration of T cells into skin. Immunohistology confirmed these findings. There was a tendency for CD4[+] cells to outnumber CD8[+] cells in infiltrates induced by all mediators. In contrast to PPD none of the mediators induced migration of T19[+] subset of γδ TCR[+] cells into skin.

IL-8-STIMULATED LYMPHOCYTE LOCOMOTION IN ANTIGEN- OR
MITOGEN-ACTIVATED MONONUCLEAR CELL CULTURES

P.C. Wilkinson, and I. Newman

Department of Immunology, University of Glasgow (Western Infirmary), Glasgow G11 6NT, U.K.

When human mononuclear cells are cultured with antiCD3 or PPD for up to 72 h., the proportion of locomotor lymphocytes increases from 10% to 30-60%, reaching a maximum by 48 h. These activated lymphocytes respond to pure IL-8 by polarization, orientation, and invasion of collagen gels. IL-8 is found in the supernatant media of such cultures in increasing quantities up to 48h. and can also be detected at this time in the Golgi region of the cultured monocytes, but not in lymphocytes. Activated lymphocytes show locomotor responses to their own untreated culture supernatants but not to supernatants which have been preincubated with anti-IL-8. These results suggest that monocyte-derived IL-8 is released into the supernatant and is an important chemoattractant for immunologically-activated lymphocytes. Its release is stimulated by monocyte-lymphocyte clustering. Following clustering, lymphocytes enter the G$_1$ phase of cell cycle, increase in size, and become capable of responding to IL-8 in locomotion assays. These results suggest a role for IL-8 in immune activation and in recruitment of activated lymphocytes into sites of immune inflammation.

IN VIVO STUDIES OF RANTES AND THE RECRUITMENT OF LYMPHOCYTES TO INFLAMMATORY SITES

C.A. Michie, K.S. Soo, T. Schall[1], L.G. Bobrow, P.C.L. Beverley

Imperial Cancer Research Fund, London, U.K., [1]Genentech Inc., San Francisco, U.S.A.

The inflamed human appendix has been demonstrated to specifically recruit CD45RO+ve T lymphocytes from the peripheral circulation; lymphocytes with the CD45RA+ve phenotype are drained from the appendix in efferent lymphatics. This recruitment has been demonstrated to change the functional repertoire of the lymphocyte population in the appendix when tested in proliferation assays, with responses to recall antigen demonstrable in lymphocytes from inflamed, but not normal appendices. Increased expression of adhesion molecules in the tissue, or soluble molecules in the circulation has not been measurable in all inflamed appendices, and may therefore not completely explain this selectivity. We observed increased expression of Rantes peptide in the areas of inflamed appendix, and around the efferent lymphatics. There was minimal staining in normal appendices, restricted to the germinal centres. Facs analysis of appendiceal and peripheral lymphocytes showed Rantes positivity to be predominantly on the CD3+ve, CD4+ve CD45RO+ve population. Using a single cell secretion assay to test lymphocytes activated in vitro, we observed Rantes secretion by T lymphocytes activated by mitogen, CD3 and CD2. The frequency of secreting lymphocytes was increased by activating with costimuli such as CD28. We suggest the thesis that Rantes secretion by T lymphocytes is a significant component in the active recruitment of a CD45RO+ve lymphocyte population to the inflamed human appendix. This mechanism may be a general feature of inflammatory recruitment in vivo.

IL-1 AND TNF INDUCED UP-REGULATION OF IL-8 RECEPTORS ON HUMAN EPIDERMAL CELLS

L. Kemény, A. Sz. Kenderessy, G. Michel[1], R.U. Peter[1], A. Beetz[1], A. Dobozy, and T. Ruzicka[1]

Department of Dermatology, Albert Szent-Györgyi Medical University, Szeged, Hungary and [1]Department of Dermatology, University of Munich, Germany

Interleukin-8 (IL-8) is a novel cytokine with potent proinflammatory and mitogenic properties. Recently, we have found that IL-8 is chemotactic for epidermal cells, and human keratinocytes possess specific receptors for IL-8 (IL-8-R). Since IL-1α and tumor necrosis factor-α (TNF-α) are capable of modulating various cell surface receptors of keratinocytes, we have studied the effects of these cytokines on the IL-8-R expression of human epidermal cells. A transformed epidermal cell line (SCL-II cells), freshly separated and cultured normal human keratinocytes were used for the studies. The IL-8-R has been detected by flow cytometry using phycoerythrin-labeled IL-8 (IL-8-PE). Pretreatment of cells with IL-1α and/or TNF-α induced a dose-dependent increase in specific IL-8-PE binding to the cells. The observed increase in specific IL-8-PE binding was also time-dependent, reaching its maximum after a 24-hour pretreatment time. The up-regulation of IL-8-R on human epidermal cells by IL-1α and TNF-α indicates a very complex interaction between these inflammatory cytokines.

PROMISCUITY OF LIGAND BINDING IN THE CHEMOKINE BETA RECEPTOR FAMILY

D.J. Kelvin, Ji-Ming Wang, D. McVicar, and J.J. Oppenheim

LMI, BRMP, DCT, NCI-FCRDC, Frederick, MD 21702, U.S.A.

Ligand receptor interactions by chemokine beta cytokines stimulate a variety of cellular and subcellular events including monocyte and T-cell chemotaxis, adhesion, calcium mobilization and histamine release. To investigate the molecular nature of these events we have identified and analyzed the ligand binding characteristics of receptors for MIP-1α, MIP-1β, RANTES, and MCAF. Scatchard analysis of binding studies on THP-1 cells using ^{125}I labeled ligands revealed specific receptors for MIP-1α, MIP-1β, RANTES, and MCAF with affinities of 500pM, 600pM, 400pM, and 5nM respectively. Competitive inhibition studies showed MIP-1α and MIP-1β to inhibit 100% of MIP-1α, MIP-1β, and RANTES binding to the THP-1 cell line. MCAF could only partially inhibit MIP-1α and MIP-1β binding, however, MCAF like MIP-1α,β, could compete for 100% RANTES binding on THP-1 cells. Ligand desensitization studies for Ca^{2+} mobilization and chemotaxis using the above ligands yielded similar results in that MIP-1α and MIP-1β desensitized MIP-1α, MIP-1β and RANTES stimulation, and MCAF partially desensitized MIP-1α and MIP-1β stimulation, but completely desensitized RANTES induction. We hypothesize from these results that at least three distinct receptors are expressed on THP-1 cells, the MIP-1α,β, receptor that binds MIP-1α and β, the MCAF receptor that binds MCAF, and the RANTES receptor which binds RANTES, MIP-1α, MIP-1β, and MCAF. Therefore, a single chemokine β ligand can bind to different receptors and a given receptor can bind different chemokine β ligands.

DIFFERENTIATION-SPECIFIC EXPRESSION OF A NOVEL G PROTEIN-COUPLED CYTOKINE RECEPTOR FROM BURKITT'S LYMPHOMA

I. Wolf, T. Emrich, E. Kaiser, R. Förster, T. Dobner, and M. Lipp

Institut für Biochemie der Universität München, Am Klopferspitz 18a, D-8033 Martinsried, Germany

Deregulation of the proto-oncogene MYC by specific chromosomal translocations has been shown to be essential but not sufficient for the development of Burkitt's lymphoma

(BL). In order to identify other genes which either mark important steps in tumorigenesis or which reflect the cellular differentiation state of BL cells we have compared tumour cells to immortalized lymphoblastoid B cells by subtractive hybridization.

We have identified a complementary DNA clone which encodes a novel member of the superfamily of GTP-binding (G) protein-coupled receptors, designated BLR1. The corresponding mRNA is expressed in Burkitt's lymphoma and lymphatic tissues but not in other cell lines either of the B cell lineage or of other hematopoietic or non-hematopoietic origin. This exclusive expression of BLR1 and the oncogenic potential of this receptor class supports the hypothesis that BLR1 exerts a regulatory function in BL lymphomagenesis and/or B cell differentiation.

This assumption was further confirmed using the highly conserved mouse homologue of BLR1 to study the expression of the receptor in tissues of BALB/c versus immundeficient SCID mice and to analyse its regulation by cytokines during B cell development.

Moreover, the protein sequence is highly related to that of receptors for the cytokine interleukin-8 (IL-8) and other neutrophil chemoattractants. We conclude that BLR1 may represent a potential candidate involved in the process of physiologic trafficking, cell-cell interactions, and activation of mature B lymphocytes in lymphatic tissues.

SIGNAL TRANSDUCTION BY MONOCYTE CHEMOTACTIC PROTEIN-1 (MCP-1) AND CROSS-DESENSITIZATION WITH OTHER MEMBERS OF THE CHEMOKINE-β SUBFAMILY IN HUMAN MONOCYTES

S. Sozzani, M. Locati, M. Molino, D. Zhou, W. Luini, and A. Mantovani

Istituto di Ricerche Farmacologiche 'Mario Negri', Milan, and Consorzio Mario Negri Sud, Santa Maria Imbaro, Italy

Human monocytes show a rapid and transient increase of free cytosolic Ca^{2+} ($[Ca^{2+}]i$) after exposure to recombinant MCP-1 (rMCP-1; Sozzani et al. 1991. J. Immunol. 147: 2215). Influx across plasma membrane receptor operated Ca^{2+} channels (ROCs) appear to be the unique mechanism responsible for this effect. This conclusion is supported by several observations: i) the rise in $[Ca^{2+}]i$ is dependent on the presence of extracellular Ca^{2+}; ii) the rise in $[Ca^{2+}]i$ and Ca^{2+} influx, measured by Mn2+ Fura-2 fluorescence quenching, are completely blocked by Ni^{2+} and by SC38249, two Ca^{2+} channel inhibitors; iii) rMCP-1 and MG-63-derived MCP-1 do not induce IP_3 formation nor inositol lipids turnover. Chemotactic response to rMCP-1, Rantes and MIP-1α could be inhibited by C-I, a serine/threonine kinase inhibitor, and by Erbstatin and Genistein, two tyrosine kinase inhibitors. However, the three proteins showed a different degree of sensitivity with MIP-1α>Rantes>rMCP-1. On the contrary, neutrophil migration to IL-8 (a member of the Chemokine-α subfamily) was inhibited by tyrosine kinase inhibitors but increased by C-I. Finally, we observed that rMCP-1, Rantes and MIP-1α induced both homologous and heterologous desensitization assessed by Fura-2 fluorescence. rMCP-1 desensitized for Rantes and MIP-1α, and Rantes and MIP-1α desensitized each one for the other. However, both Rantes and MIP-1α were ineffective with respect to rMCP-1. These results strongly suggest that members of the Chemokine-β subfamily share a common or a closely related receptor.

MECHANISMS OF ZYMOSAN INDUCED IL-8 GENERATION IN VIVO

P.D. Collins, B.T. Au, P.J. Jose, and T.J. Williams

Dept. Applied Pharmacology, National Heart and Lung Institute, London SW3, U.K.

Acute peritonitis in the rabbit induced by intraperitoneal injection of zymosan is characterised by the sequential generation of at least three neutrophil chemoattractants (1). An initial phase of C5a generation (0-2 hours) is followed (2 hours onwards) by the appearance of two structurally related chemoattractant cytokines, rabbit IL-8 and rabbit MGSA (2). The time course of cytokine generation is similar to that of neutrophil accumulation, and phagocytosing neutrophils release IL-8 *in vitro*. We have therefore investigated whether the infiltrating neutrophils are the source of the IL-8 observed in the cavity at 6 hours. Depletion of circulating neutrophils (>98%) did not inhibit the generation of inflammatory mediators in the peritoneal cavity measured by both *in vivo* bioassay in skin and using a rabbit IL-8 specific RIA. In addition, HPLC separation of the mediators in exudate from depleted animals revealed that the inflammatory activity comprised both IL-8 and MGSA. Thus, it appears that infiltrating neutrophils are not necessary for production of these cytokines in this model, implicating other cells such as macrophages or mesothelial cells as the source.

1. Collins, P.D., P.J. Jose, and T.J. Williams. 1991. The sequential generation of neutrophil chemoattractant proteins in acute inflammation in the rabbit *in vivo:* relationship between C5a and a protein with the characteristics of IL-8. *J. Immunol.* 146: 677-684.
2. Jose, P.J., P.D. Collins, J.A. Perkins, B.C. Beaubien, N.F. Totty, M.D. Waterfield, J. Hsuan, and T.J. Williams. 1991. Identification of a second neutrophil chemoattractant cytokine generated during an inflammatory reaction in the rabbit peritoneal cavity *in vivo*: purification, partial amino acid sequence and structural relationship to melanoma growth stimulatory activity. *Biochem. J.* 278: 493-497.

The authors would like to thank the Wellcome Trust, Fisons, and the NAC for financial support.

THE EFFECTS OF INTRANASAL CHALLENGE WITH
INTERLEUKIN-8 (IL- 8)

J. Douglass, C. Gurr, J. Shute, M.K. Church, and S.T. Holgate

Immunopharmacology Group, Southampton General Hospital, Southampton, SO9 4XY, U.K.

Interleukin-8 is a cytokine which is a powerful neutrophil chemoattractant *in vitro*. We have investigated the effects of intranasal instillation of human recombinant IL-8 in 8 atopic (4 male) and 8 non-atopic (1 male) donors in a double-blind placebo controlled study. Changes in nasal resistance were monitored by posterior rhinomanometry for 4 hours after challenge. Symptom scores were recorded on a visual analogue scale. Nasal scrapings were taken and stained with Haema-Gurr for differential cell counts. Nasal resistance data in the whole group revealed an increase in nasal resistance 10 min after challenge compared with placebo (p<0.01 t-test). There was no significant difference between atopic and non-atopic subjects. Symptoms scores were significantly increased in all 16 subjects with IL-8 compared to placebo. Nasal scrapings from both groups revealed a significant increase in the proportion of neutrophils compared to the placebo control (p<0.001). There was no significant difference between atopic and non-atopic groups in the extent of neutrophil infiltration. There was no significant difference observed in eosinophil infiltration in the non-atopic group, however a marked eosinophil infiltrate was observed in 3 atopic subjects. Interleukin-8 therefore causes a significant increase in nasal obstruction, symptoms and cellular infiltrate in the nasal challenge model of respiratory disease.

This study was supported by the Sandoz Research Institute, Vienna.

INTERLEUKIN-8 DOWN-REGULATES THE OXIDATIVE BURST INDUCED BY TUMOR NECROSIS FACTOR ALPHA IN NEUTROPHILS ADHERENT TO FIBRONECTIN

L. Ottonello, I.J.D. Lindley[1], G. Pastorino, P. Dapino, F. Dallegri

Department of Internal Medicine, First Medical Clinic, University of Genova, Italy and [1]Sandoz Research Institute, Vienna, Austria

Various cytokines and chemotactic factors have been shown to induce a prolonged oxidative burst in neutrophils (PMN) adherent to biological substrates. PMN (5×10^4) incubated on fibronectin-precoated wells (1 µg/well) (flat-bottomed microtiter plates), released $2.83 \pm .25$ nmoles of superoxide (O_2^-) ($x \pm 1$ SEM, n = 15) in response to 100 ng/ml Tumor Necrosis Factor Alpha (TNF) (superoxide dismutase-inhibitable ferricytochrome c reduction method). On the contrary, the O_2^- production induced by Interleukin-8 (IL-8) (doses ranging from 10^{-10} M to 10^{-6} M) was comparable to that of "resting" cells (< .6 nmoles/5×10^4 PMN, n ≥ 5). IL-8 (10^{-7} M) did not affect the TNF-dependent O_2^- production when added with TNF (100 ng/ml) at the beginning of the assay, but reduced it by 70-80% when added with TNF after 1 h-preincubation of PMN on fibronectin. As compared with IL-8, N-formyl-methionyl-leucyl-phenylalanine (FMLP, 10^{-7} M) induced an efficient oxidative burst in adherent PMN. Moreover, the total O_2^- generation was additive for adherent PMN treated with both TNF and FMLP vs either alone, also when the cells were incubated (1 h) on fibronectin prior to the addition of the agonists. The data prove that: a) IL-8 does not trigger significant O_2^- production in PMN plated on fibronectin, whereas FMLP and TNF are effective; b) the interaction of PMN with fibronectin for 1 h uncovers the capacity of IL-8 to limit the cell response to TNF, without affecting the cell response to the combination of FMLP and TNF. Thus, although the chemotactic factors IL-8 and FMLP share the capacity of triggering the oxidative burst of PMN incubated in suspension, only IL-8 has the potential to down-regulate the responsiveness of fibronectin-adherent cells to TNF.

IDENTIFICATION OF A NOVEL TUMOR-DERIVED MONOCYTE CHEMOTACTIC FACTOR: RELATIONSHIP WITH TISSUE INHIBITOR OF METALLOPROTEINASE

R. Bertini[1], W. Luini[2], B. Bottazzi[2], A.R. Mackay[3], D. Boraschi[1],
J. Van Damme[4], and A. Mantovani[2]

[1]Dompe S.p.A., L'Aquila, Italy, [2]Mario Negri Institute, Milano, Italy, [3]Consorzio Biolaq., L'Aquila, Italy and [4]Rega Institute, Leuven, Belgium

Macrophages infiltrate human ovarian carcinomas and have been shown to release monocyte chemotactic activity (ovarian cancer-derived chemotactic factor, OCDCF). The aim of the present study was to characterize OCDCF. Ovarian SW626 cells released chemotactic activity *in vitro* and showed a considerable macrophage infiltrate when injected into nude mice. The cells did not express MCP-1, M-CSF and other known monocyte chemoattractants at the mRNA level. A chemotactic factor was purified to homogeneity by silicic acid adsorption, heparin-sepharose column, FPLC cation exchange and HPLC chromatography, and showed an apparent molecular weight of 25 kDa. N-terminal sequence analysis releaved identity over 17 amino acids with tissue inhibitor of metalloproteinase 1 (TIMP-1). The fractions containing purified OCDCF indeed exhibited metalloproteinase inhibitory activity consistent with that of TIMP-1. Purified OCDCF was active on monocytes but not on neutrophils, and checkerboard experiments indicated a true chemotactic effect. These results reveal an unexpected relationship between regulation of matrix degradation and attraction of monocytes.

α-MELANOCYTE STIMULATING HORMONE INHIBITS HUMAN EPIDERMAL CELL AND POLYMORPHONUCLEAR LEUKOCYTE CHEMOTAXIS

M. Csato, A. Sz. Kenderessy, and J.J. Nordlund[1]

Department of Dermatology, Albert Szent-Györgyi Medical University, Szeged, Hungary and [1]Department of Dermatology, University of Cincinnati Cincinnati, Ohio, U.S.A.

Evidence is accumulating that the neuropeptide α-melanocyte stimulating hormone (α-MSH) is capable of antagonizing immune/inflammatory reactions, particularly those orchestrated by interleukin-1 (IL-1) and also by other proinflammatory principles. In order to get further insights into this activity of α-MSH, we investigated the effect of the neuropeptide on the chemotactic/chemokinetic responsiveness of freshly separated human epidermal cells (EC) and polymorphonuclear leukocytes (PMN's) in the *in vitro* Boyden chamber assay. IL-1α, LTB$_4$, PAF, FINAP (fibroblast-derived neutrophil activating factor) and activated complement components were used as chemotactic/chemokinetic stimuli. Besides the already documented effect of IL-1α (*Immunol. Letters*, 22; 123), LTB$_4$, PAF and FINAP also exhibited potent epidermal cell chemotactic activity. α-MSH brought about a significant inhibition of the enhanced directed or random motility of both human

EC and PMN's, irrespective of the stimulus used. The optimum concentration of α-MSH for chemotaxis/chemokinesis inhibition was in the range of 10^{-7} M. The neuropeptide did not compete with the chemoattractants. A competitive inhibition/chemotactic deactivation is, therefore, not a likely mechanism underlying the action of the neuropeptide. To our knowledge, this is the first documentation of this novel aspect of a broad spectrum antiinflammatory activity of the neuropeptide on freshly obtained human cells, i.e. EC and PMN's.

MIP-1 ALPHA: A POTENT STEM CELL INHIBITOR

G.J. Graham, E.K. Parkinson, and I.B. Pragnell

The Beatson Institute for Cancer Research, Switchback Rd., Bearsden, Glasgow, G61 1BD, U.K.

We have recently characterised a potent inhibitor of haemopoietic stem cell proliferation and have shown it to be identical to a previously described cytokine, Macrophage Inflammatory Protein-1 alpha (MIP-1α) (1). This heat stable factor is reversible in its activity and is functional at picomolar concentrations. We have also identified the human homologue, which is identical to LD78 and have shown it to have an identical repertoire of activities.

We have also recently demonstrated the ability of MIP-1α to inhibit the growth of clonogenic human epidermal keratinocytes in *vitro*, and have demonstrated that the likely source of MIP-1α within the epidermis is the epidermal Langerhans cell. These results will be discussed in more detail.

Preliminary characterisation of the cell surface receptors for MIP-1α have revealed the presence of a single class of abundant receptor on primitive haemopoietic cells. Data relating to the properties of these receptors and their relevance to the mechanisms of stem cell inhibition will be discussed.

1. Graham, G.J. et al. 1990. Identification and characterisation of an inhibitor of haemopoietic stem cell proliferation. *Nature*. 344: 442-444.

INDEX

Nuclear factors 90
Nuclear run-off assays 89, 94
Peripheral blood mononuclear cells 87, 94, 139, 193, 201
Pertussis toxin 2, 5, 6, 35, 48, 104, 107, 185
PF-4 99, 101-106, 132, 185
Phagocytosis 9
Phorbol ester 88-90, 94, 95
Platelet activating factor 39, 119
Platelet activation 127, 128
Platelet basic protein 1, 99, 120, 128
Platelets 1, 20, 58, 59, 107, 121, 123, 126, 129
PMA 52, 53, 190, 192, 197
Postcapillary venules 83
Poxvirus 187
Promiscuity 148, 158, 218
Protein kinase A 90
Protein kinase C 2, 48, 49, 90
Psoriasis 84, 134, 171, 200, 208
RANTES 2, 5, 20, 30-36, 49-51, 99-102, 104, 115, 116, 124, 148, 149, 150-153, 139-145, 184-186, 217, 218, 219
RANTES receptor 152, 185, 218
Red blood cells 34, 36, 106
Red cell chemokine receptor 34, 36
Respiratory burst 2, 3, 5, 6, 135, 198, 210
Rheumatoid arthritis 22-25, 27, 77, 99, 134, 201, 204, 208
Sarcoidosis 211, 212
Sarcoma 96, 116
Septic shock 205
Serine protease 187

Shape change 2, 3, 139, 183
Smooth muscle cells 13, 70, 115, 200
Sodium nitroprusside 71, 72
Sputum 21, 27, 191
Stem cell 108, 223
Submandibulary glands 80
Superoxide anion 92, 93
Sweat 77-80, 84, 185
Synovial cells 88
Synovial fibroblasts 22-24
T lymphocytes 37, 86-88, 96, 140-145, 184, 186, 217
TCA-3 5
Thrombin 107, 121, 123, 126, 193, 197
Thrombosis 198
TNF 6, 20-25, 27, 29, 59-61, 64-66, 68, 71-73, 77, 88-92, 94, 104, 116, 130, 143, 185, 188, 189, 190, 195, 197, 198, 199, 201, 204, 205, 210, 211, 213, 214, 216-218, 221
Trafficking 39, 43, 199, 219
Transcription factors 72, 89-92, 95
Tuberculin 55-57, 63
Tuberculosis 159
Tumor-associated macrophages (TAM) 47, 193
U937 cells 30
Urinary tract infection 194
Urokinase-type plasminogen activator 51
Venules 3, 19, 83
Viruses 88, 187
Wortmannin 3
X-ray crystallography 171, 155
Zymosan 88, 196, 197, 216, 220